Instant Notes

ORGANIC CHEMISTRY

The INSTANT NOTES series

Series editor
B.D. Hames
School of Biochemistry and Molecular Biology, University of Leeds, Leeds, UK

Biochemistry
Animal Biology
Molecular Biology
Ecology
Genetics
Microbiology
Chemistry for Biologists
Immunology

Forthcoming titles
Biochemistry 2nd edition
Neuroscience
Plant Biology
Developmental Biology
Molecular Biology 2nd edition
Psychology

The INSTANT NOTES Chemistry Series
Consulting editor: Howard Stanbury
Organic Chemistry

Forthcoming titles
Inorganic Chemistry
Physical Chemistry

Instant Notes

ORGANIC CHEMISTRY

G. L. Patrick
Department of Chemistry and Chemical Engineering,
Paisley University, Paisley, Scotland

First published 2000

A CIP catalogue record for this book is available from the British Library.

ISBN 1 85996 158 4

BIOS Scientific Publishers Ltd
9 Newtec Place, Magdalen Road, Oxford OX4 1RE, UK.
Tel. +44 (0) 1865 726286. Fax +44 (0) 1865 246823
World Wide Web home page: http://www.bios.co.uk/

Published in the United States of America, its dependent territories and Canada by Springer-Verlag New York Inc., 175 Fifth Avenue, New York, NY 10010-7858, in association with BIOS Scientific Publishers Ltd

Published in Hong Kong, Taiwan, Singapore, Thailand, Cambodia, Korea, The Philippines, Indonesia, The People's Republic of China, Brunei, Laos, Malaysia, Macau and Vietnam by Springer-Verlag Singapore Pte. Ltd, 1 Tannery Road, Singapore 347719, in association with BIOS Scientific Publishers Ltd.

Consulting Editor: Howard Stanbury

Production Editor: Fran Kingston
Typeset by J&L Composition Ltd, Filey, UK
Printed by Biddles Ltd, Guildford, UK

CONTENTS

Preface

This textbook aims to provide a comprehensive set of basic notes in organic chemistry, which will be suitable for undergraduate students taking chemistry, chemistry-related courses, or courses which involve organic chemistry as an ancillary subject. The book concentrates on core topics which are most likely to be common to those organic chemistry courses which follow on from a foundation or introductory general chemistry course.

Organic chemistry is a subject which can lead some students to the heights of ecstasy, yet drive others up the wall. Some students 'switch on' to it immediately, while others can make neither head nor tail of it, no matter how hard they try. Certainly, one of the major problems in studying the subject is the vast amount of material which often has to be covered. Many students blanche at the prospect of having to learn a seemingly endless number of reactions, and when it comes to drawing mechanisms and curly arrows, they see only a confusing maze of squiggly lines going everywhere yet nowhere. The concepts or organic reaction mechanisms are often the most difficult to master. These difficulties are often compounded by the fact that current textbooks in organic chemistry are typically over 1200 pages long and can be quite expensive to buy.

This book attempts to condense the essentials of organic chemistry into a manageable text of 310 pages which is student friendly and which does not cost an arm and a leg. It does this by concentrating purely on the basics of the subject without going into exhaustive detail or repetitive examples. Furthermore, key notes at the start of each topic summarize the essential facts covered and help focus the mind on the essentials.

Organic chemistry is a peculiar subject in that it becomes easier as you go along! This might seem an outrageous statement to make, especially to a first-year student who is struggling to come to terms with the rules of nomenclature, trying to memorize a couple of dozen reactions and making sense of mechanisms at the same time. However, these topics are the basics of the subject and once they have been grasped, the overall picture becomes clear.

Understanding the mechanism of how a reaction takes place is particularly crucial in this. It brings a logic to the reactions of the different functional groups. This in turn transforms a list of apparently unrelated fats into a sensible theme which makes remembering the reactions a 'piece of cake' (well, nearly).

Once this happy state of affairs has been reached, the relevance of organic chemistry to other subjects such as genetics and biochemistry suddenly leaps off the page. Understanding organic chemistry leads to a better understanding of life chemistry and how the body works at the molecular level. It also helps in the understanding of the molecular mechanisms involved in disease and bodily malfunction, leading in turn to an understanding of how drugs can be designed to cure these disease states – the science of medicinal chemistry.

And that's not all. An understanding of organic chemistry will help the industrial chemist or chemical engineer faced with unexpected side-reactions in a chemical process, and the agro-scientist trying to understand the molecular processes taking place within plants and crops; and it will assist in the design and synthesis of new herbicides and fungicides which will be eco-friendly. It will aid the forensic scientist wishing to analyze a nondescript white powder – is it heroin or flour?

The list of scientific subject areas involving organic chemistry is endless – designing spacesuits, developing new photographic dyes, inventing new molecular technology in microelectronics – one could go on and on. Organic chemistry is an exciting subject since it leads to an essential understanding of molecules and their properties.

The order in which the early topics of this book are presented is important. The first two sections cover structure and bonding, which are crucial to later sections. Just why does carbon form four bonds? What is hybridization?

The third section on functional groups is equally crucial if students are to be capable of categorizing the apparent maze of reactions which organic compounds can undergo. It is followed by stereochemistry, sections E and F, in which the basic theory of reactions and mechanisms is covered. What are nucleophiles and electrophiles? What does a mechanism represent? What does a curly arrow mean?

The remaining sections can be used in any order and look at the reactions and mechanisms of the common functional groups which are important in chemistry and biochemistry.

It is hoped that students will find this textbook useful in their studies and that once they have grasped what organic chemistry is all about they will read more widely and enter a truly exciting world of molecular science.

A1 ATOMIC STRUCTURE OF CARBON

Key Notes

Atomic orbitals

The atomic orbitals available for the six electrons of carbon are the *s* orbital in the first shell, the *s* orbital in the second shell and the three *p* orbitals in the second shell. The 1*s* and 2*s* orbitals are spherical in shape. The 2*p* orbitals are dumbbell in shape and can be assigned $2p_x$, $2p_y$ or $2p_z$ depending on the axis along which they are aligned.

Energy levels

The 1*s* orbital has a lower energy than the 2*s* orbital which has a lower energy than the 2*p* orbitals. The 2*p* orbitals have equal energy (i.e. degenerate).

Electronic configuration

Carbon is in the second row of the periodic table and has six electrons which will fill up lower energy atomic orbitals before entering higher energy orbitals (aufbau principle). Each orbital is allowed a maximum of two electrons of opposite spin (Pauli exclusion principle). When orbitals of equal energy are available, electrons will occupy separate orbitals before pairing up (Hund's rule). Thus, the electronic configuration of a carbon atom is $1s^2\ 2s^2\ 2p_x^{\ 1}\ 2p_y^{\ 1}$.

Related topic

Covalent bonding and hybridization (A2)

Atomic orbitals

Carbon has six electrons and is in row 2 of the periodic table. This means that there are two shells of atomic orbitals available for these electrons. The first shell closest to the nucleus has a single *s* orbital – the 1*s* orbital. The second shell has a single *s* orbital (the 2*s* orbital) and three *p* orbitals (3 × 2*p*). Therefore, there are a total of five atomic orbitals into which these six electrons can fit. The *s* orbitals are spherical in shape with the 2*s* orbital being much larger then the 1*s* orbital. The *p* orbitals are dumbbell-shaped and are aligned along the *x*, *y* and *z* axes. Therefore, they are assigned the $2p_x$, $2p_y$ and $2p_z$ atomic orbitals (*Fig. 1*).

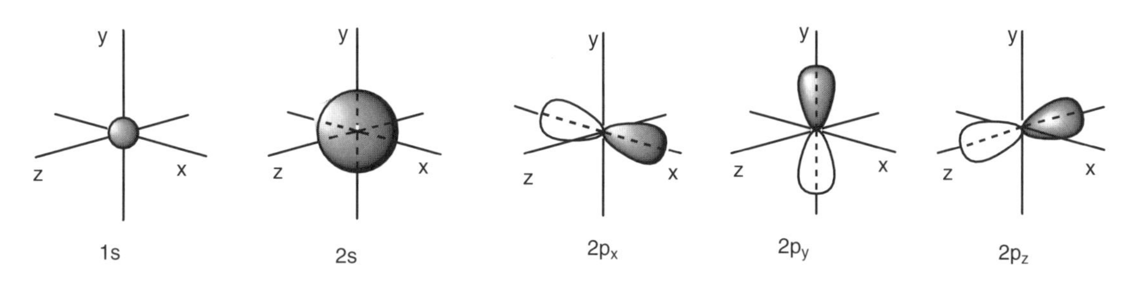

Fig. 1. Atomic orbitals.

Energy levels

The atomic orbitals described above are not of equal energy (*Fig. 2*). The 1*s* orbital has the lowest energy. The 2*s* orbital is next in energy and the 2*p* orbitals have the highest energies. The three 2*p* orbitals have the same energy, meaning that they are **degenerate**.

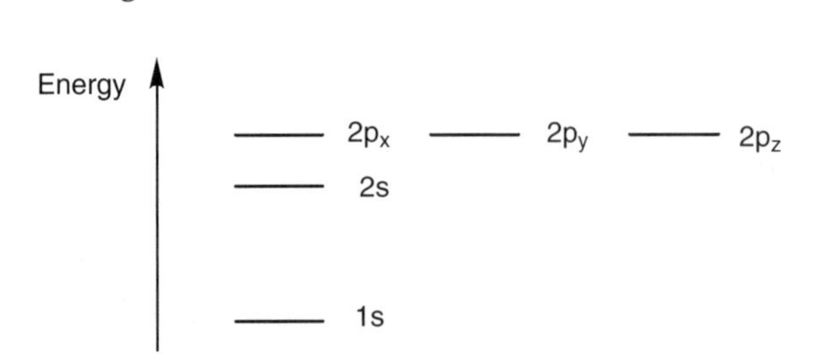

Fig. 2. Energy levels of atomic orbitals.

Electronic configuration

Carbon is in the second row of the periodic table and has six electrons which will fill up the lower energy atomic orbitals first. This is known as the **aufbau principle**. The 1*s* orbital is filled up before the 2*s* orbital, which is filled up before the 2*p* orbitals. The **Pauli exclusion principle** states that each orbital is allowed a maximum of two electrons and that these electrons must have opposite spins. Therefore, the first four electrons fill up the 1*s* and 2*s* orbitals. The electrons in each orbital have opposite spins and this is represented in *Fig. 3* by drawing the arrows pointing up or down. There are two electrons left to fit into the remaining 2*p* orbitals. These go into separate orbitals such that there are two half-filled orbitals and one empty orbital. Whenever there are orbitals of equal energy, electrons will only start to pair up once all the degenerate orbitals are half filled. This is known as **Hund's rule**.

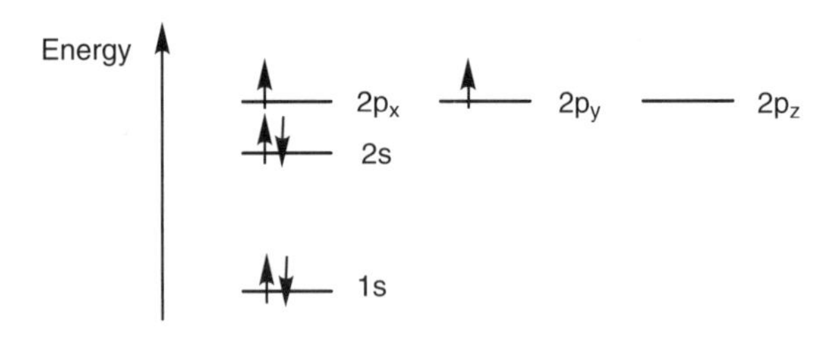

Fig. 3. Electronic configuration for carbon.

The electronic configuration for carbon is $1s^2\ 2s^2\ 2p_x^{\ 1}\ 2p_y^{\ 1}$. The numbers in superscript refer to the numbers of electrons in each orbital. The letters refer to the types of atomic orbital involved and the numbers in front refer to which shell the orbital belongs.

A2 Covalent bonding and hybridization

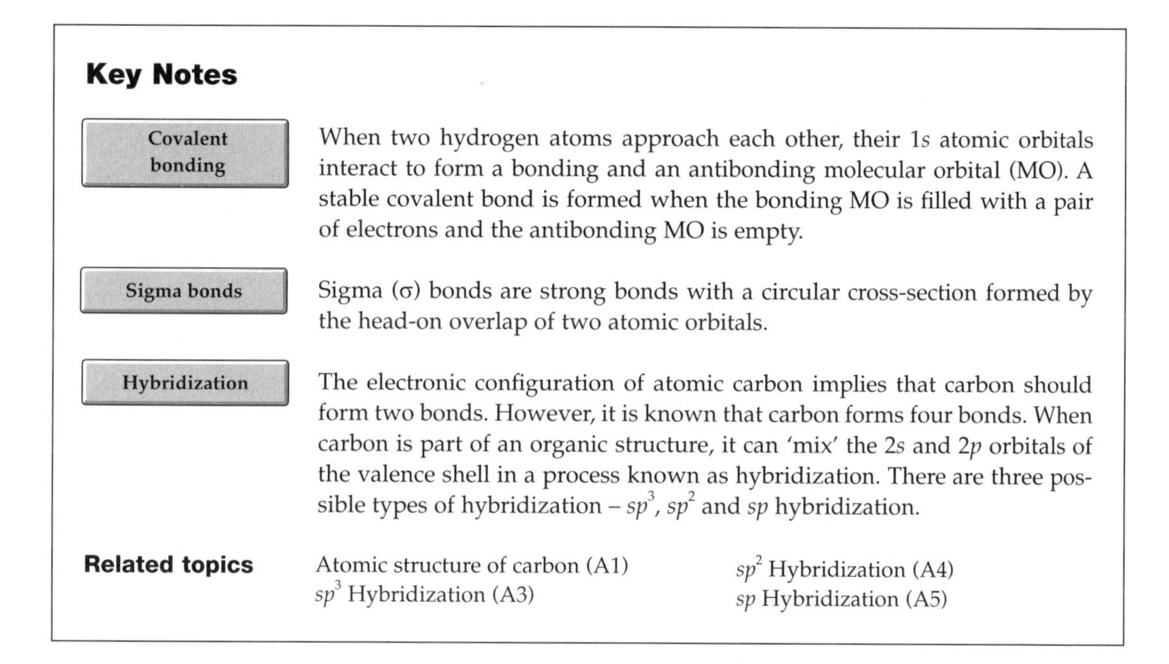

Key Notes

Covalent bonding	When two hydrogen atoms approach each other, their 1*s* atomic orbitals interact to form a bonding and an antibonding molecular orbital (MO). A stable covalent bond is formed when the bonding MO is filled with a pair of electrons and the antibonding MO is empty.
Sigma bonds	Sigma (σ) bonds are strong bonds with a circular cross-section formed by the head-on overlap of two atomic orbitals.
Hybridization	The electronic configuration of atomic carbon implies that carbon should form two bonds. However, it is known that carbon forms four bonds. When carbon is part of an organic structure, it can 'mix' the 2*s* and 2*p* orbitals of the valence shell in a process known as hybridization. There are three possible types of hybridization – sp^3, sp^2 and *sp* hybridization.
Related topics	Atomic structure of carbon (A1); sp^3 Hybridization (A3); sp^2 Hybridization (A4); *sp* Hybridization (A5)

Covalent bonding

A covalent bond binds two atoms together in a molecular structure and is formed when atomic orbitals overlap to produce a **molecular orbital** – so called because the orbital belongs to the molecule as a whole rather than to one specific atom. A simple example is the formation of a hydrogen molecule (H_2) from two hydrogen atoms. Each hydrogen atom has a half-filled 1*s* atomic orbital and when the atoms approach each other, the atomic orbitals interact to produce two MOs (the number of resulting MOs must equal the number of original atomic orbitals, *Fig. 1).*

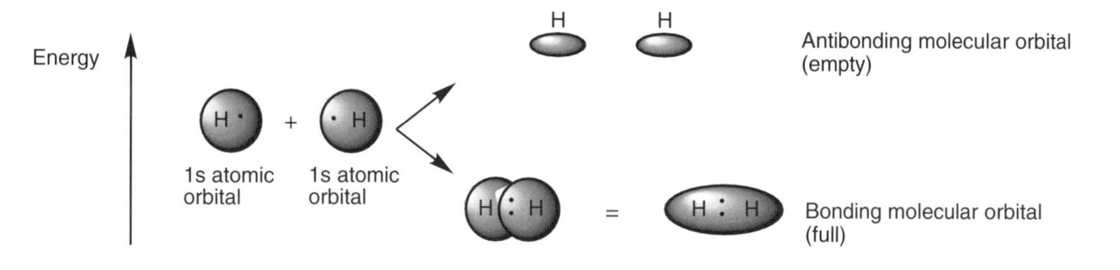

Fig. 1. Molecular orbitals for hydrogen (H_2).

The MOs are of different energies. One is more stable than the original atomic orbitals and is called the **bonding** MO. The other is less stable and is called the **antibonding** MO. The bonding MO is shaped like a rugby ball and results from the combination of the 1*s* atomic orbitals. Since this is the more stable MO, the valence electrons (one from each hydrogen) enter this orbital and pair up. The antibonding MO is of higher energy and consists of two deformed spheres. This remains empty. Since the electrons end up in a bonding MO which is more stable than the original atomic orbitals, energy is released and bond formation is favoured. In the subsequent discussions, we shall concentrate solely on the bonding MOs to describe bonding and molecular shape, but it is important to realize that antibonding molecular orbitals also exist.

Sigma bonds

The bonding molecular orbital of hydrogen is an example of a sigma (σ) bond: σ bonds have a circular cross-section and are formed by the head-on overlap of two atomic orbitals. This is a strong interaction and so sigma bonds are strong bonds. In future discussions, we shall see other examples of σ bonds formed by the interaction of atomic orbitals other than the 1*s* orbital.

Hybridization

Atoms can form bonds with each other by sharing unpaired electrons such that each bond contains two electrons. In Topic A1, we identified that a carbon atom has two unpaired electrons and so we would expect carbon to form two bonds. However, carbon forms four bonds! How does a carbon atom form four bonds with only two unpaired electrons?

So far, we have described the electronic configuration of an isolated carbon atom. However, when a carbon atom forms bonds and is part of a molecular structure, it can 'mix' the *s* and *p* orbitals of its second shell (the valence shell). This is known as **hybridization** and it allows carbon to form the four bonds which we observe in reality.

There are three ways in which this mixing process can take place.

- the 2*s* orbital is mixed with all three 2*p* orbitals. This is known as sp^3 hybridization;
- the 2*s* orbital is mixed with two of the 2*p* orbitals. This is known as sp^2 hybridization;
- the 2*s* orbital is mixed with one of the 2*p* orbitals. This is known as *sp* hybridization.

A3 SP^3 HYBRIDIZATION

Key Notes

Definition

In sp^3 hybridization, the *s* and the *p* orbitals of the second shell are 'mixed' to form four hybridized sp^3 orbitals of equal energy.

Electronic configuration

Each hybridized orbital contains a single unpaired electron and so four bonds are possible.

Geometry

Each sp^3 orbital is shaped like a deformed dumbbell with one lobe much larger than the other. The hybridized orbitals arrange themselves as far apart from each other as possible such that the major lobes point to the corners of a tetrahedron. sp^3 Hybridization explains the tetrahedral carbon in saturated hydrocarbon structures.

Sigma bonds

Sigma (σ) bonds are strong bonds formed between two sp^3 hybridized carbons or between an sp^3 hybridized carbon and a hydrogen atom. A σ bond formed between two sp^3 hybridized carbon atoms involves the overlap of half filled sp^3 hybridized orbitals from each carbon atom. A σ bond formed between an sp^3 hybridized carbon and a hydrogen atom involves a half-filled sp^3 orbital from carbon and a half-filled 1*s* orbital from hydrogen.

Nitrogen, oxygen and chlorine

Nitrogen, oxygen, and chlorine atoms are also sp^3 hybridized in organic molecules. This means that nitrogen has three half-filled sp^3 orbitals and can form three bonds which are pyramidal in shape. Oxygen has two half-filled sp^3 orbitals and can form two bonds which are angled with respect to each other. Chlorine has a single half-filled sp^3 orbital and can only form a single bond. All the bonds which are formed are σ bonds.

Related topics

Covalent bonding and hybridization (A2)

Bonds and hybridized centers (A6)

Definition

In sp^3 hybridization, the 2*s* orbital is mixed with all three of the 2*p* orbitals to give a set of four sp^3 hybrid orbitals. (The number of hybrid orbitals must equal the number of original atomic orbitals used for mixing.) The hybrid orbitals will each have the same energy but will be different in energy from the original atomic orbitals. That energy difference will reflect the mixing of the respective atomic orbitals. The energy of each hybrid orbital is greater than the original *s* orbital but less than the original *p* orbitals (*Fig. 1*).

Electronic configuration

The valence electrons for carbon can now be fitted into the sp^3 hybridized orbitals (*Fig. 1*). There was a total of four electrons in the original 2*s* and 2*p* orbitals. The *s* orbital was filled and two of the *p* orbitals were half filled. After hybridization, there is a total of four hybridized sp^3 orbitals all of equal energy. By Hund's rule, they are all half filled with electrons which means that there are four unpaired electrons. Four bonds are now possible.

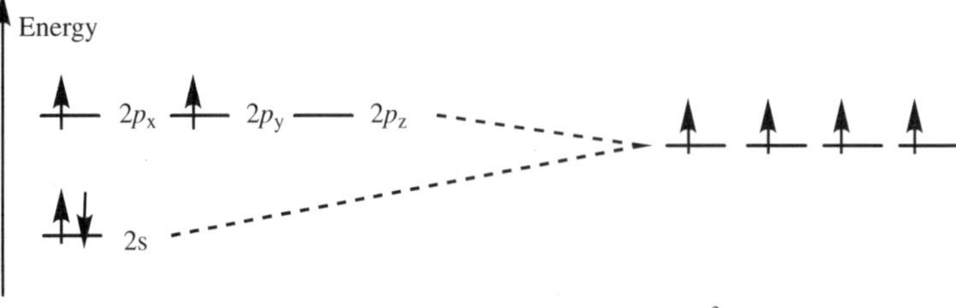

Fig. 1. sp^3 *Hybridization.*

Geometry

Each of the sp^3 hybridized orbitals has the same shape – a rather deformed looking dumbbell (*Fig. 2*). This deformed dumbbell looks more like a *p* orbital than an *s* orbital since more *p* orbitals were involved in the mixing process.

Fig. 2. sp^3 *Hybridized orbital.*

Each sp^3 orbital will occupy a space as far apart from each other as possible by pointing to the corners of a tetrahedron (*Fig. 3*). Here, only the major lobe of each hybridized orbital has been shown and the angle between each of these lobes is 109.5°. This is what is meant by the expression **tetrahedral carbon**. The three-dimensional shape of the tetrahedral carbon can be represented by drawing a nor-

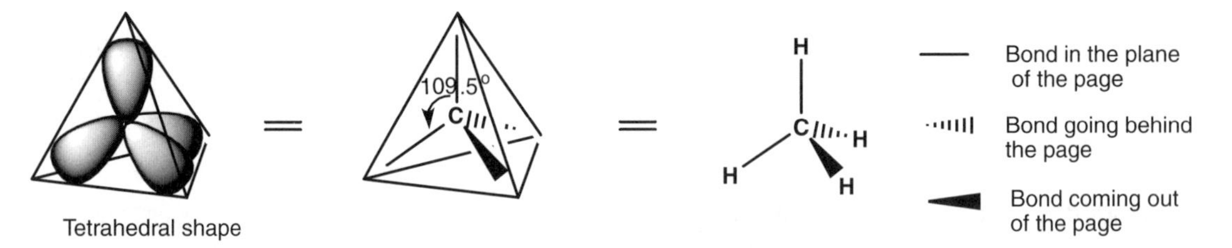

Fig. 3. *Tetrahedral shape of an* sp^3 *hybridized carbon*

mal line for bonds in the plane of the page. Bonds going behind the page are represented by a hatched wedge, and bonds coming out the page are represented by a solid wedge.

Sigma bonds

A half-filled sp^3 hybridized orbital from one carbon atom can be used to form a bond with a half-filled sp^3 hybridized orbital from another carbon atom. In *Fig. 4a*, the major lobes of the two sp^3 orbitals overlap directly leading to a strong σ bond. It is the ability of hybridized orbitals to form strong σ bonds that explains why hybridization takes place in the first place. The deformed dumbbell shapes allow a much better orbital overlap than would be obtained from a pure *s* orbital or a pure *p* orbital. A σ bond between an sp^3 hybridized carbon atom and a hydrogen atom involves the carbon atom using one of its half-filled sp^3 orbitals and the hydrogen atom using its half-filled 1*s* orbital (*Fig. 4b*).

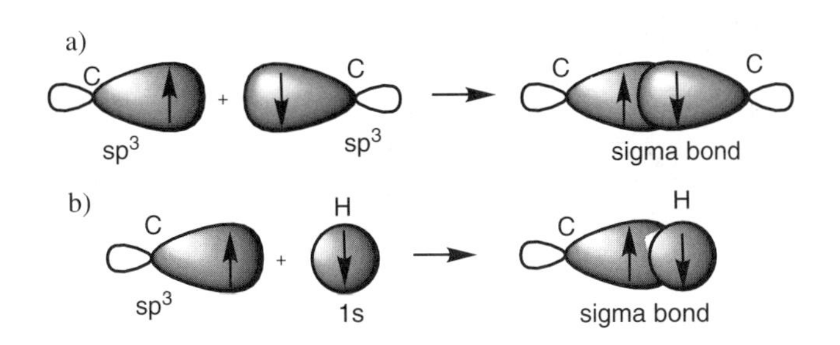

Fig. 4. (a) σ Bond between two sp^3 *hybridized carbons; (b) σ bond between an* sp^3 *hybridized carbon and hydrogen*

Nitrogen, oxygen, and chlorine

Nitrogen, oxygen and chlorine atoms can also be sp^3 hybridized in organic structures. Nitrogen has five valence electrons in its second shell. After hybridization, it will have three half-filled sp^3 orbitals and can form three bonds. Oxygen has six valence electrons. After hybridization, it will have two half-filled sp^3 orbitals and will form two bonds. Chlorine has seven valence electrons. After hybridization, it will have one half-filled sp^3 orbital and will form one bond.

The four sp^3 orbitals for these three atoms form a tetrahedral arrangement with one or more of the orbitals occupied by a lone pair of electrons. Considering the atoms alone, nitrogen forms a pyramidal shape where the bond angles are slightly less than 109.5° (c. 107°) (*Fig. 5a*). This compression of the bond angles is due to the orbital containing the lone pair of electrons, which demands a slightly greater amount of space than a bond. Oxygen forms an angled or bent shape where two lone pairs of electrons compress the bond angle from 109.5° to c. 104° (*Fig. 5b*).

Alcohols, amines, alkyl halides, and ethers all contain sigma bonds involving nitrogen, oxygen, or chlorine. Bonds between these atoms and carbon are formed by the overlap of half-filled sp^3 hybridized orbitals from each atom. Bonds involving hydrogen atoms (e.g. O–H and N–H) are formed by the overlap of the half-filled 1*s* orbital from hydrogen and a half-filled sp^3 orbital from oxygen or nitrogen.

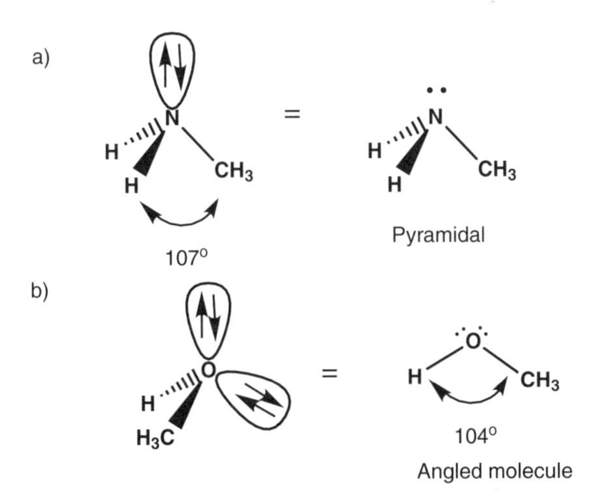

Fig. 5. (a) Geometry of sp^3 *hybridized nitrogen; (b) geometry of* sp^3 *hybridized oxygen.*

A4 sp^2 HYBRIDIZATION

Key Notes

Definition

In sp^2 hybridization, a 2*s* orbital is 'mixed' with two of the 2*p* orbitals to form three hybridized sp^2 orbitals of equal energy. A single 2*p* orbital is left over which has a slightly higher energy than the hybridized orbitals.

Electronic configuration

For carbon, each sp^2 hybridized orbital contains a single unpaired electron. There is also a half-filled 2*p* orbital. Therefore, four bonds are possible.

Geometry

Each sp^2 orbital is shaped like a deformed dumbbell with one lobe much larger than the other. The remaining 2*p* orbital is a symmetrical dumbbell. The major lobes of the three sp^2 hybridized orbitals point to the corners of a triangle, with the 2*p* orbital perpendicular to the plane.

Alkenes

Each sp^2 hybridized carbon forms three σ bonds using three sp^2 hybridized orbitals. The remaining 2*p* orbital overlaps 'side on' with a neighboring 2*p* orbital to form a pi (π) bond. The π bond is weaker than the σ bond, but is strong enough to prevent rotation of the C=C bond. Therefore, alkenes are planar, with each carbon being trigonal planar.

Carbonyl groups

The oxygen and carbon atoms are both sp^2 hybridized. The carbon has three sp^2 hybridized orbitals and can form three σ bonds, one of which is to the oxygen. The oxygen has one sp^2 orbital which is used in the σ bond with carbon. The *p* orbitals on carbon and oxygen are used to form a π bond.

Aromatic rings

Aromatic rings are made up of six sp^2 hybridized carbons. Each carbon forms three σ bonds which results in a planar ring. The remaining 2*p* orbital on each carbon is perpendicular to the plane and can overlap with a neighboring 2*p* orbital on either side. This means that a molecular orbital is formed round the whole ring such that the six π electrons are delocalized around the ring. This results in increased stability such that aromatic rings are less reactive than alkenes.

Conjugated systems

Conjugated systems such as conjugated alkenes and α,β-unsaturated carbonyl compounds involve alternating single and double bonds. In such systems, the *p* orbitals of one π bond are able to overlap with the *p* orbitals of a neighboring π bond, and thus give a small level of double bond character to the connecting bond. This partial delocalization gives increased stability to the conjugated system.

Related topics

Properties of alkenes and alkynes (H2)
Conjugated dienes (H11)
Aromaticity (I1)
Properties (J2)
α,β-Unsaturated aldehydes and ketones (J11)
Structure and properties (K1)

Definition

In sp^2 hybridization, the s orbital is mixed with two of the 2*p* orbitals (e.g. $2p_x$ and $2p_z$) to give three sp^2 hybridized orbitals of equal energy. The remaining $2p_y$ orbital is unaffected. The energy of each hybridized orbital is greater than the original s orbital but less than the original *p* orbitals. The remaining 2*p* orbital (in this case the $2p_y$ orbital) remains at its original energy level (*Fig. 1*).

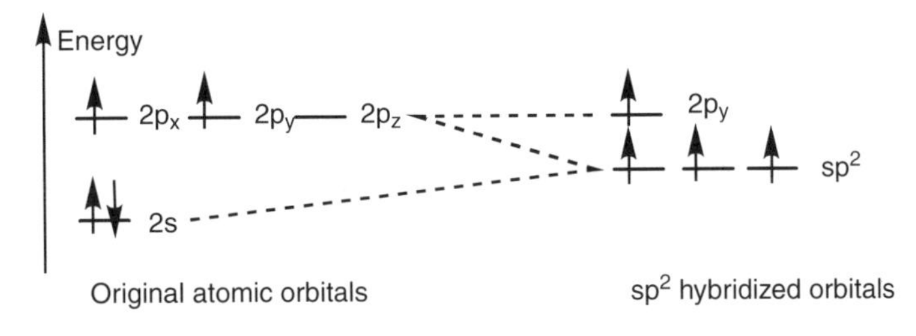

Fig. 1. sp^2 *Hybridization.*

Electronic configuration

For carbon, there are four valence electrons to fit into the three hybridized sp^2 orbitals and the remaining 2*p* orbital. The first three electrons are fitted into each of the hybridized orbitals according to Hund's rule such that they are all half-filled. This leaves one electron still to place. There is a choice between pairing it up in a half-filled sp^2 orbital or placing it into the vacant $2p_y$ orbital. The usual principle is to fill up orbitals of equal energy before moving to an orbital of higher energy. However, if the energy difference between orbitals is small (as here) it is easier for the electron to fit into the higher energy $2p_y$ orbital resulting in three half-filled sp^2 orbitals and one half-filled orbital (*Fig. 1*). Four bonds are possible.

Geometry

The $2p_y$ orbital has the usual dumbbell shape. Each of the sp^2 hybridized orbitals has a deformed dumbbell shape similar to an sp^3 hybridized orbital. However, the difference between the sizes of the major and minor lobes is larger for the sp^2 hybridized orbital.

The hybridized orbitals and the $2p_y$ orbital occupy spaces as far apart from each other as possible. The lobes of the $2p_y$ orbital occupy the space above and below the plane of the *x* and *z* axes (*Fig. 2a*). The three sp^2 orbitals (major lobes shown only) will then occupy the remaining space such that they are as far apart from the $2p_y$ orbital and from each other as possible. As a result, they are all placed in the *x–z* plane pointing toward the corner of a triangle (trigonal planar shape; *Fig. 2b*). The angle between each of these lobes is 120°. We are now ready to look at the bonding of an sp^2 hybridized carbon.

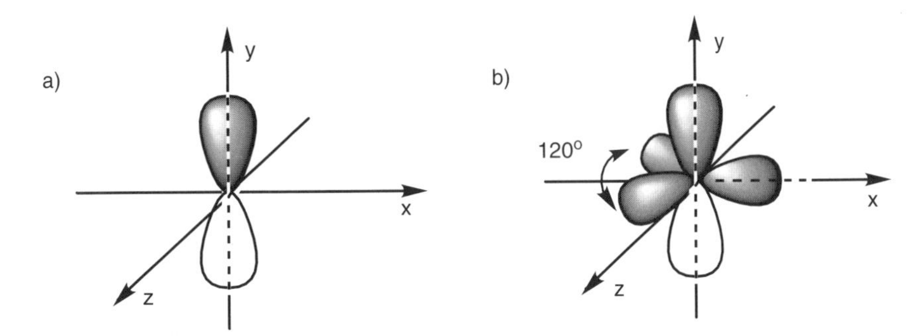

Fig. 2. (a) Geometry of the $2p_y$ *orbital; (b) geometry of the* $2p_y$ *orbital and the* sp^2 *hybridized orbitals.*

Alkenes

sp^2 Hybridization results in three half-filled sp^2 hybridized orbitals which form a trigonal planar shape. The use of these three orbitals in bonding explains the shape of an alkene, for example ethene ($H_2C{=}CH_2$). As far as the C–H bonds are concerned, the hydrogen atom uses a half-filled 1*s* orbital to form a strong σ bond with a half filled sp^2 orbital from carbon (*Fig. 3a*). A strong σ bond is also possible between the two carbon atoms of ethene due to the overlap of sp^2 hybridized orbitals from each carbon (*Fig. 3b*).

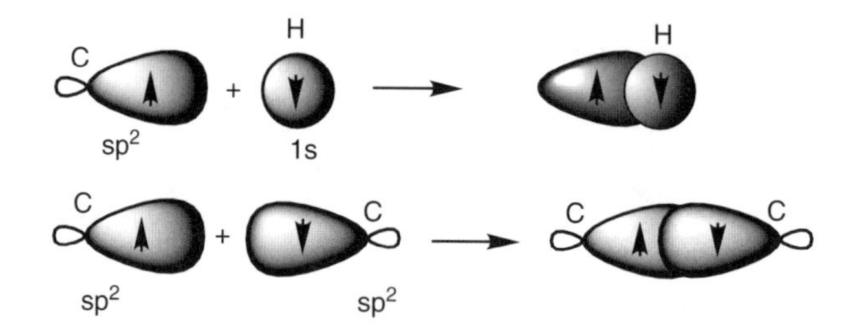

Fig. 3. (a) Formation of a C–H σ bond; (b) formation of a C–C σ bond.

The full σ bonding diagram for ethene is shown in *Fig. 4a* and can be simplified as shown in *Fig. 4b*. Ethene is a flat, rigid molecule where each carbon is trigonal planar. We have seen how sp^2 hybridization explains the trigonal planar carbons but we have not explained why the molecule is rigid and planar. If the σ bonds were the only bonds present in ethene, the molecule would not remain planar since rotation could occur round the C–C σ bond (*Fig. 5*). Therefore, there must be further bonding which 'locks' the alkene into this planar shape. This bond involves

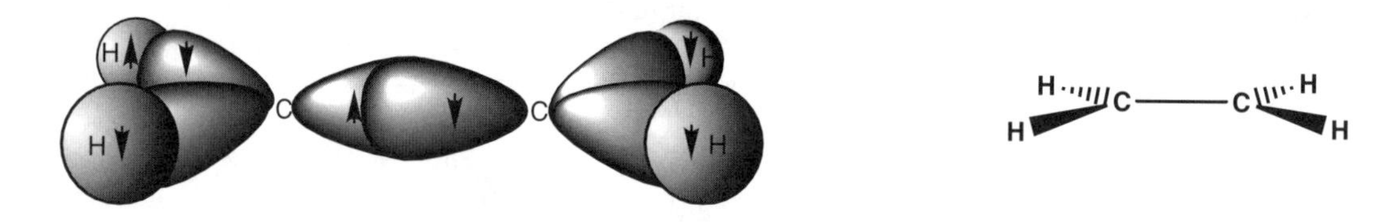

Fig. 4. (a) σ Bonding diagram for ethene; (b) simple representation of σ bonds for ethene.

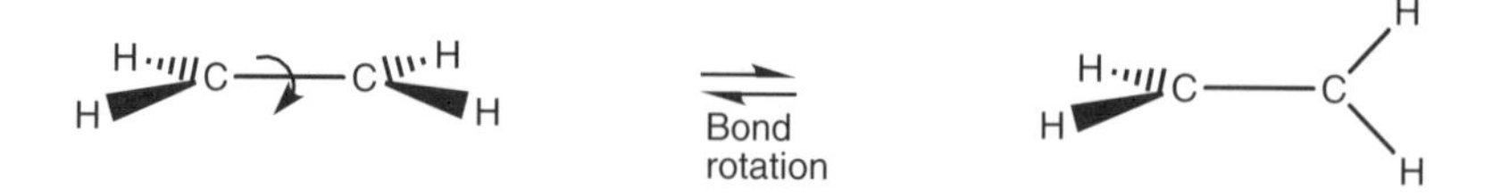

Fig. 5. Bond rotation around a σ bond.

the remaining half-filled $2p_y$ orbitals on each carbon which overlap side-on to produce a **pi (π) bond**), with one lobe above and one lobe below the plane of the molecule (*Fig. 6*). This π bond prevents rotation round the C–C bond since the π bond would have to be broken to allow rotation. A π bond is weaker than a σ bond since the $2p_y$ orbitals overlap side-on, resulting in a weaker overlap. The presence of a π bond also explains why alkenes are more reactive than alkanes, since a π bond is more easily broken and is more likely to take part in reactions.

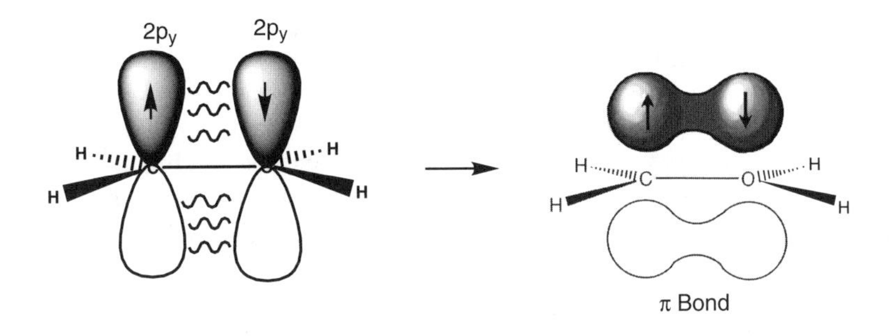

Fig. 6. *Formation of a π bond.*

Carbonyl groups

The same theory explains the bonding within a carbonyl group (C=O) where both the carbon and oxygen atoms are sp^2 hybridized. The following energy level diagram (*Fig. 7*) shows how the valence electrons of oxygen are arranged after sp^2 hybridization. Two of the sp^2 hybridized orbitals are filled with lone pairs of electrons, which leaves two half-filled orbitals available for bonding. The sp^2 orbital can be used to form a strong σ bond, while the $2p_y$ orbital can be used for the weaker π bond. *Figure 8* shows how the σ and π bonds are formed in the carbonyl group and explains why carbonyl groups are planar with the carbon atom having a trigonal planar shape. It also explains the reactivity of carbonyl groups since the π bond is weaker than the σ bond and is more likely to be involved in reactions.

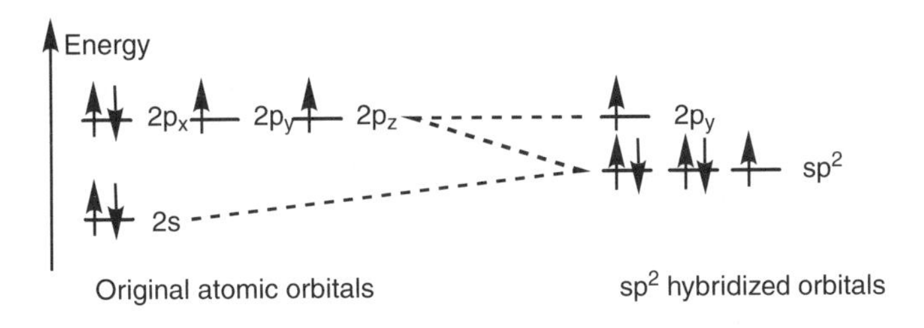

Fig. 7. *Energy level diagram for* sp^2 *hybridized oxygen.*

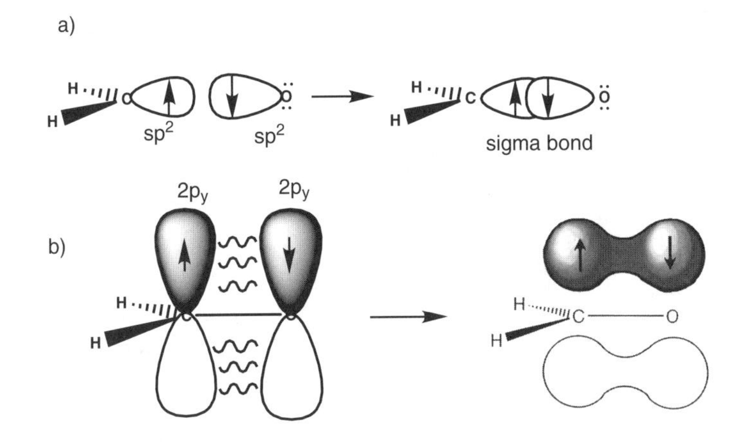

Fig. 8. *(a) Formation of the carbonyl σ bond; (b) formation of the carbonyl π bond.*

Aromatic rings

All the carbons in an aromatic ring are sp^2 hybridized which means that each carbon can form three σ bonds and one π bond. In *Fig. 9a*, all the single bonds are σ while each double bond consists of one σ bond and one π bond. However, this is an oversimplification of the aromatic ring. For example, double bonds are shorter than single bonds and if benzene had this exact structure, the ring would be deformed with longer single bonds than double bonds (*Fig. 9b*).

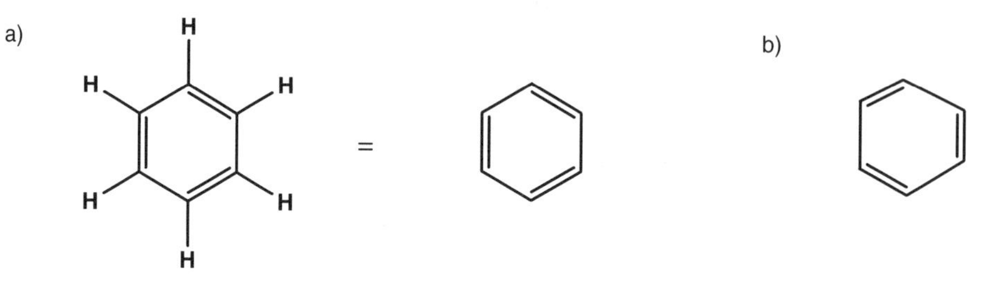

Fig. 9. (a) Representation of the aromatic ring; (b) 'deformed' structure resulting from fixed bonds.

In fact, the C–C bonds in benzene are all the same length. In order to understand this, we need to look more closely at the bonding which takes place. *Figure 10a* shows benzene with all its σ bonds and is drawn such that we are looking into the plane of the benzene ring. Since all the carbons are sp^2 hybridized, there is a $2p_y$ orbital left over on each carbon which can overlap with a $2p_y$ orbital on either side of it (*Fig. 10b*). From this, it is clear that each $2p_y$ orbital can overlap with its neighbors right round the ring. This leads to a molecular orbital which involves all the $2p_y$ orbitals where the upper and lower lobes merge to give two doughnut-like lobes above and below the plane of the ring (*Fig. 11a*). The molecular orbital is symmetrical and the six π electrons are said to be delocalized around the aromatic ring since they are not localized between any two particular carbon atoms. The aromatic ring is often represented as shown in *Fig. 11b* to represent this delocalization of the π

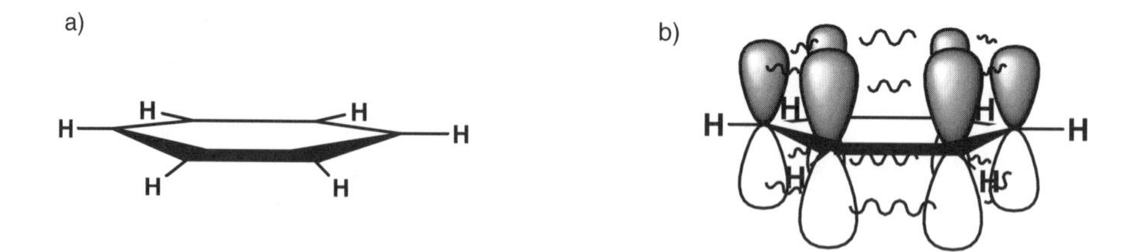

Fig. 10. (a) σ Bonding diagram for benzene, (b) π Bonding diagram for benzene.

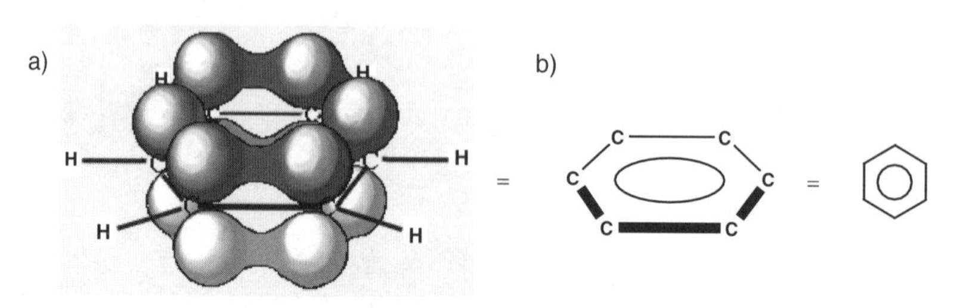

Fig. 11. Bonding molecular orbital for benzene; (b) representation of benzene to illustrate delocalization.

electrons. Delocalization increases the stability of aromatic rings such that they are less reactive than alkenes (i.e. it requires more energy to disrupt the delocalized π system of an aromatic ring than it does to break the isolated π bond of an alkene).

Conjugated systems

Aromatic rings are not the only structures where delocalization of π electrons can take place. Delocalization occurs in conjugated systems where there are alternating single and double bonds (e.g. 1,3-butadiene). All four carbons in 1,3-butadiene are sp^2 hybridized and so each of these carbons has a half-filled *p* orbital which can interact to give two π bonds (*Fig. 12a*). However, a certain amount of overlap is also possible between the *p* orbitals of the middle two carbon atoms and so the bond connecting the two alkenes has some double bond character (*Fig. 12b*) – borne out by the observation that this bond is shorter in length than a typical single bond. This delocalization also results in increased stability. However, it is important to realize that the conjugation in a conjugated alkene is not as great as in the aromatic system. In the latter system, the π electrons are completely delocalized round the ring and all the bonds are equal in length. In 1,3-butadiene, the π electrons are not fully delocalized and are more likely to be found in the terminal C–C bonds. Although there is a certain amount of π character in the middle bond, the latter is more like a single bond than a double bond.

Other examples of conjugated systems include α,β-unsaturated ketones and α,β-unsaturated esters (*Fig. 13*). These too have increased stability due to conjugation.

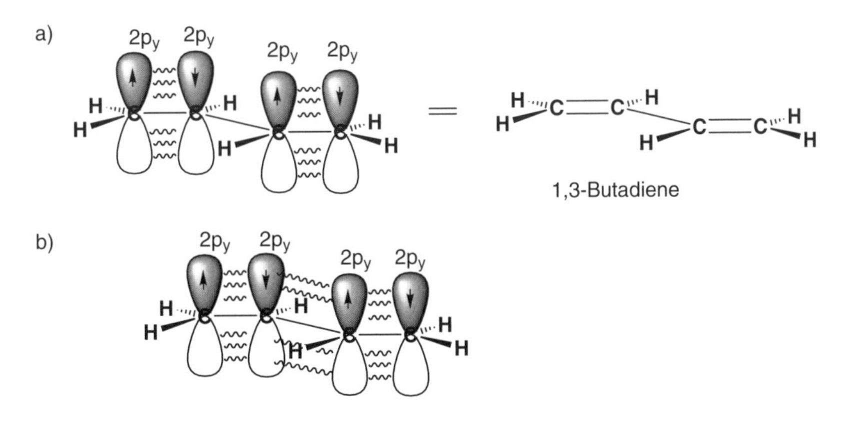

Fig. 12. (a) π Bonding in 1,3-butadiene; (b) delocalization in 1,3-butadiene.

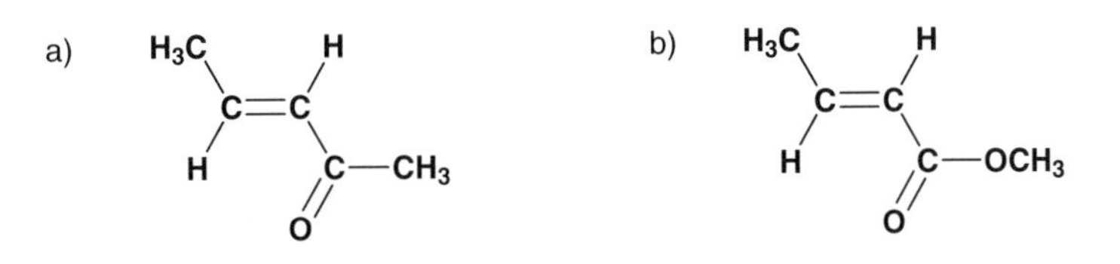

Fig. 13. (a) α,β-Unsaturated ketone; (b) α,β-unsaturated ester.

A5 SP HYBRIDIZATION

Key Notes

Definition

In *sp* hybridization, the 2s orbital and one of the three 2*p* orbitals are 'mixed' to form two hybridized *sp* orbitals of equal energy. Two 2*p* orbitals are left over and have slightly higher energy than the unhybridized orbitals.

Electronic configuration

For carbon, each *sp* hybridized orbital contains a single unpaired electron. There are also two half-filled 2*p* orbitals. Therefore, four bonds are possible.

Geometry

Each *sp* orbital is shaped like a deformed dumbbell with one lobe much larger than the other. The remaining 2*p* orbitals are symmetrical dumbbells. If we define the 2*p* orbitals as being aligned along the y and the z axes, the two *sp* hybridized orbitals point in opposite directions along the x axis.

Alkynes

Each *sp* hybridized carbon of an alkyne can form two σ bonds using *sp* hybridized orbitals. The remaining 2*p* orbitals can overlap 'side-on' to form two π bonds. Alkynes are linear molecules and are reactive due to the π bonds.

Nitrile groups

The nitrogen and carbon atoms of a nitrile group (C≡N) are both *sp* hybridized. The carbon has two *sp* hybridized orbitals and can form two σ bonds, one of which is to nitrogen. The nitrogen has one *sp* orbital which is used in the σ bond with carbon. Both the carbon and the nitrogen have two 2*p* orbitals which can be used to form two π bonds.

Related topics

Properties of alkenes and alkynes (H2)

Chemistry of nitriles (O4)

Definition

In *sp* hybridization, the 2*s* orbital is mixed with one of the 2*p* orbitals (e.g. $2p_x$) to give two *sp* hybrid orbitals of equal energy. This leaves two 2*p* orbitals unaffected ($2p_y$ and $2p_z$) with slightly higher energy than the hybridized orbitals (*Fig. 1*).

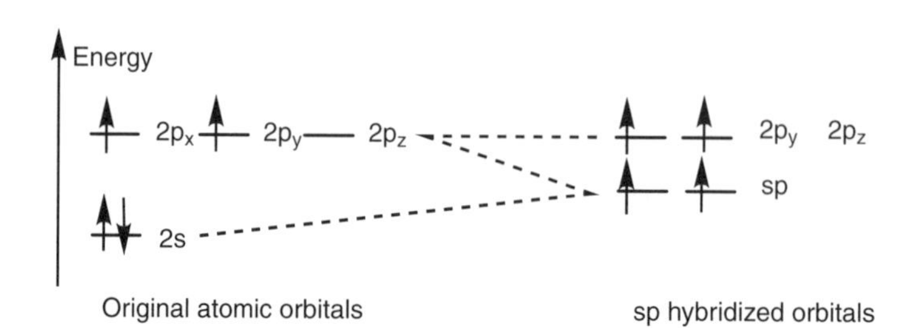

Fig. 1. sp *Hybridization of carbon.*

Electronic configuration

For carbon, the first two electrons fit into each *sp* orbital according to Hund's rule such that each orbital has a single unpaired electron. This leaves two electrons which can be paired up in the half-filled *sp* orbitals or placed in the vacant $2p_y$ and $2p_z$ orbitals. The energy difference between the orbitals is small and so it is easier for the electrons to fit into the higher energy orbitals than to pair up. This leads to two half-filled *sp* orbitals and two half-filled $2p$ orbitals (*Fig. 1*), and so four bonds are possible.

Geometry

The $2p$ orbitals are dumbbell in shape while the *sp* hybridized orbitals are deformed dumbbells with one lobe much larger than the other. The $2p_y$ and $2p_z$ orbitals are at right angles to each other (*Fig. 2a*). The *sp* hybridized orbitals occupy the space left over and are in the x axis pointing in opposite directions (only the major lobe of the *sp* orbitals are shown in black; *Fig. 2b*).

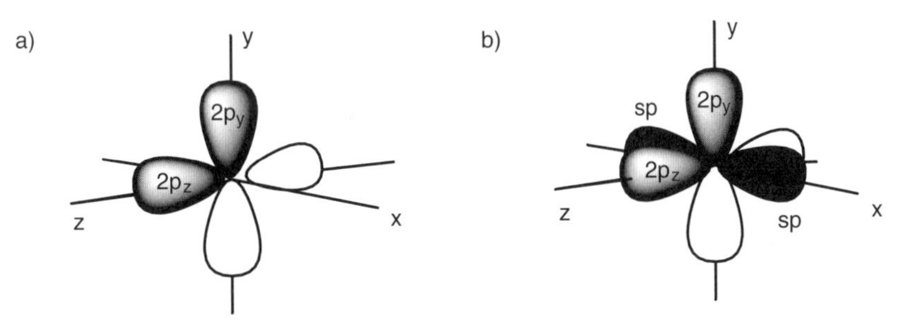

Fig. 2. (a) $2p_y$ and $2p_z$ orbitals of an sp *hybridized carbon; (b) $2p_y$, $2p_z$ and* sp *hybridized orbitals of an* sp *hybridized carbon.*

A molecule using the two *sp* orbitals for bonding will be linear in shape. There are two common functional groups where such bonding takes place – alkynes and nitriles.

Alkynes

Let us consider the bonding in ethyne (*Fig. 3*) where each carbon is *sp* hybridized. The C–H bonds are strong σ bonds where each hydrogen atom uses its half-filled

H—C≡C—H

Fig. 3. Ethyne.

1*s* orbital to bond with a half-filled *sp* orbital on carbon. The remaining *sp* orbital on each carbon is used to form a strong σ carbon–carbon bond. The full σ bonding diagram for ethyne is linear (*Fig. 4a*) and can be simplified as shown (*Fig. 4b*).

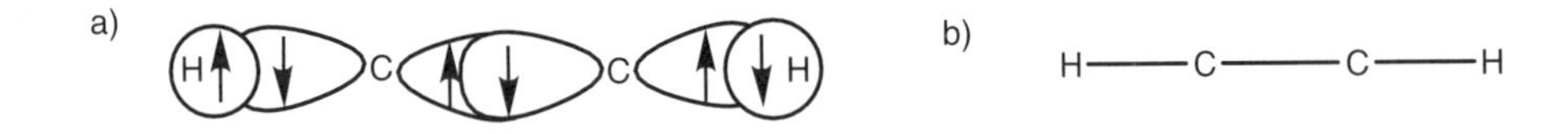

Fig. 4. (a) σ Bonding for ethyne; (b) representation of σ bonding.

Further bonding is possible since each carbon has half-filled *p* orbitals. Thus, the $2p_y$ and $2p_z$ orbitals of each carbon atom can overlap side-on to form two π bonds (*Fig. 5*). The π bond formed by the overlap of the $2p_y$ orbitals is represented in dark gray. The π bond resulting from the overlap of the $2p_z$ orbitals is represented in light gray. Alkynes are linear molecules and are reactive due to the relatively weak π bonds.

Nitrile groups

Exactly the same theory can be used to explain the bonding within a nitrile group (C≡N) where both the carbon and the nitrogen are *sp* hybridized. The energy level diagram in *Fig. 6* shows how the valence electrons of nitrogen are arranged after *sp* hybridization. A lone pair of electrons occupies one of the *sp* orbitals, but the other *sp* orbital can be used for a strong σ bond. The $2p_y$ and $2p_z$ orbitals can be used for two π bonds. *Figure 7* represents the σ bonds of HCN as lines and how the remaining 2*p* orbitals are used to form two π bonds.

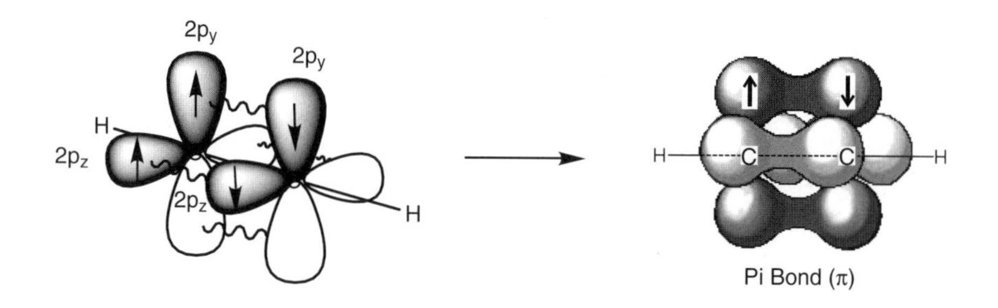

Fig. 5. π-Bonding in ethyne.

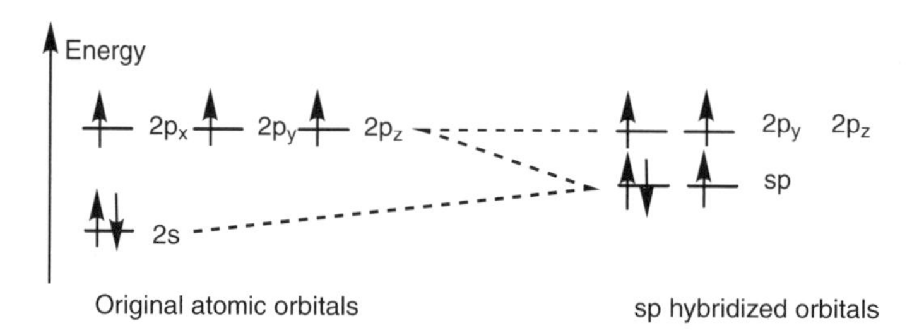

Fig. 6. sp *Hybridization of nitrogen.*

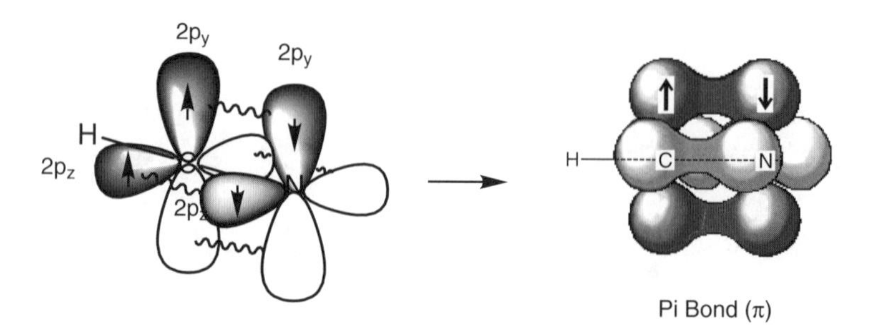

Fig. 7. π-Bonding in HCN.

A6 Bonds and hybridized centers

Key Notes

σ and π bonds — Every bond in an organic structure is a σ bond or a π bond. Every atom in a structure is linked to another by a single σ bond. If there is more than one bond between any two atoms, the remaining bonds are π bonds.

Hybridized centers — All atoms in an organic structure (except hydrogen) are either sp, sp^2 or sp^3 hybridized. Atoms linked by single bonds are sp^3 hybridized, atoms linked by double bonds are sp^2 hybridized* and atoms linked by triple bonds are *sp* hybridized.*

Shape — sp^3 Hybridized centers are tetrahedral, sp^2 hybridized centers are trigonal planar and *sp* centers are linear. This determines the shape of functional groups. Functional group containing sp^2 hybridized centers are planar while functional groups containing *sp* hybridized centers are linear.

Reactivity — Functional groups containing π bonds tend to be reactive since the π bond is weaker than a σ bond and is more easily broken.

Related topics

sp^3 Hybridization (A3)
sp^2 Hybridization (A4)
sp Hybridization (A5)

(* with the exception of allenes $R_2C{=}C{=}CR_2$)

σ and π bonds

Identifying σ and π bonds in a molecule (*Fig. 1*) is quite easy as long as you remember the following rules:

- all bonds in organic structures are either sigma (σ) or pi (π) bonds;
- all single bonds are σ bonds;
- all double bonds are made up of one σ bond and one π bond;
- all triple bonds are made up of one σ bond and two π bonds.

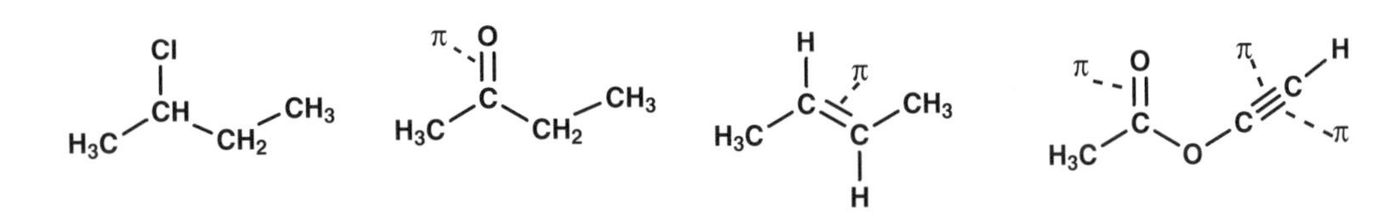

Fig. 1. Examples – all the bonds shown are σ bonds except those labelled as π.

Hybridized centers

All the atoms in an organic structure (except hydrogen) are either *sp*, sp^2 or sp^3 hybridized (*Fig. 2*).

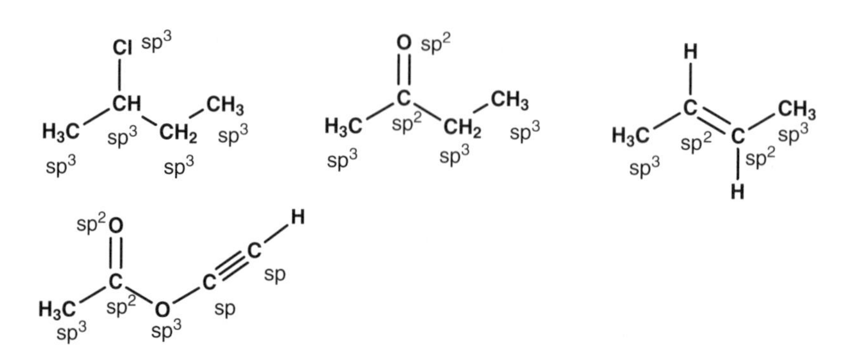

Fig. 2. Examples of sp, sp^2 *and* sp^3 *hybridized centers.*

The identification of *sp*, sp^2 and sp^3 centers is simple if you remember the following rules:

- all atoms linked by a single bond are sp^3 hybridized (except hydrogen).
- both carbon atoms involved in the double bond of an alkene (C=C) must be sp^2 hybridized.*
- both the carbon and the oxygen of a carbonyl group (C=O) must be sp^2 hybridized.
- all aromatic carbons must be sp^2 hybridized.
- both atoms involved in a triple bond must be *sp* hybridized.
- hydrogen uses a 1*s* orbital for bonding and is not hybridized.

Hydrogen atoms cannot be hybridized. They can only bond by using an *s* orbital since there are no *p* orbitals in the first electron shell. It is therefore impossible for a hydrogen to take part in π bonding. Oxygen, nitrogen and halogens on the other hand can form hybridized orbitals which are either involved in bonding or in holding lone pairs of electrons.

Shape

The shape of organic molecules and the functional groups within them is determined by the hybridization of the atoms present. For example, functional groups containing trigonal planar sp^2 centers are planar while functional groups containing *sp* centers are linear:

- planar functional groups – aldehyde, ketone, alkene, carboxylic acid, acid chloride, acid anhydride, ester, amide, aromatic.
- linear functional groups – alkyne, nitrile.
- functional groups with tetrahedral carbons – alcohol, ether, alkyl halide.

Reactivity

Functional groups which contain π bonds are reactive since the π bond is weaker than a σ bond and can be broken more easily. Common functional groups which contain π bonds are aromatic rings, alkenes, alkynes, aldehydes, ketones, carboxylic acids, esters, amides, acid chlorides, acid anhydrides, and nitriles.

* Functional groups known as allenes ($R_2C{=}C{=}CR_2$) have an *sp* hybridized carbon located at the center of two double bonds, but these functional groups are beyond the scope of this text.

B1 DEFINITION

Key Notes

Alkanes

Alkanes are organic molecules consisting solely of carbon and hydrogen atoms linked by single σ bonds. All the carbon atoms are tetrahedral and sp^3 hybridized. Alkanes are stable molecules and unreactive to most chemical reagents. They have the general formula C_nH_{2n+2}

Cycloalkanes

Cycloalkanes are cyclic alkane structures. They have the general formula C_nH_{2n}. Most cycloalkanes are unreactive to chemical reagents. However, three- and four-membered rings are reactive due to ring strain and behave like alkenes.

Related topics sp^3 Hybridization (A3) Conformational isomers (D4)

Alkanes

Alkanes are organic molecules with the general formula C_nH_{2n+2}, which consist of carbon and hydrogen atoms linked together by C–C and C–H single bonds. They are often referred to as **saturated hydrocarbons** – saturated because all the bonds are single bonds, hydrocarbons because the only atoms present are carbon and hydrogen. All the carbon atoms in an alkane are sp^3 hybridized and tetrahedral in shape. The C–C and C–H bonds are strong σ bonds, and so alkanes are unreactive to most chemical reagents.

Alkanes are sometimes referred to as **straight chain** or **acyclic** alkanes to distinguish them from **cycloalkanes** or **alicyclic** compounds.

Cycloalkanes

Cycloalkanes are cyclic alkanes (**alicyclic** compounds) having the general formula C_nH_{2n} where the carbon atoms have been linked together to form a ring. All sizes of ring are possible. However, the most commonly encountered cycloalkane in organic chemistry is the six-membered ring (cyclohexane). Most cycloalkanes are unreactive to chemical reagents. However, small three- and four-membered rings are reactive and behave like alkenes. Such cyclic structures are highly strained since it is impossible for the carbon atoms to adopt their preferred tetrahedral shape.

B2 DRAWING STRUCTURES

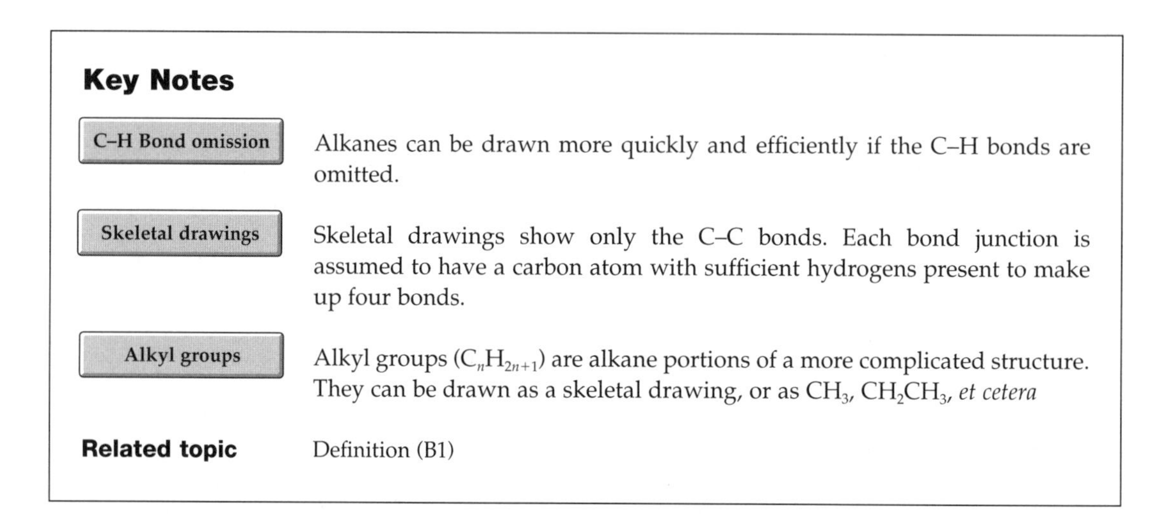

Key Notes

C–H Bond omission	Alkanes can be drawn more quickly and efficiently if the C–H bonds are omitted.
Skeletal drawings	Skeletal drawings show only the C–C bonds. Each bond junction is assumed to have a carbon atom with sufficient hydrogens present to make up four bonds.
Alkyl groups	Alkyl groups (C_nH_{2n+1}) are alkane portions of a more complicated structure. They can be drawn as a skeletal drawing, or as CH_3, CH_2CH_3, *et cetera*
Related topic	Definition (B1)

C–H Bond omission

There are several ways of drawing organic molecules. A molecule such as ethane can be drawn showing every C–C and C–H bond. However, this becomes tedious, especially with more complex molecules, and it is much easier to miss out the C–H bonds (*Fig. 1*).

Skeletal drawings

A further simplification is often used where only the carbon–carbon bonds are shown. This is a skeletal drawing of the molecule (*Fig. 2*). With such drawings, it is understood that a carbon atom is present at every bond junction and that every carbon has sufficient hydrogens attached to make up four bonds.

Straight chain alkanes can also be represented by drawing the C–C bonds in a zigzag fashion (*Fig. 3*).

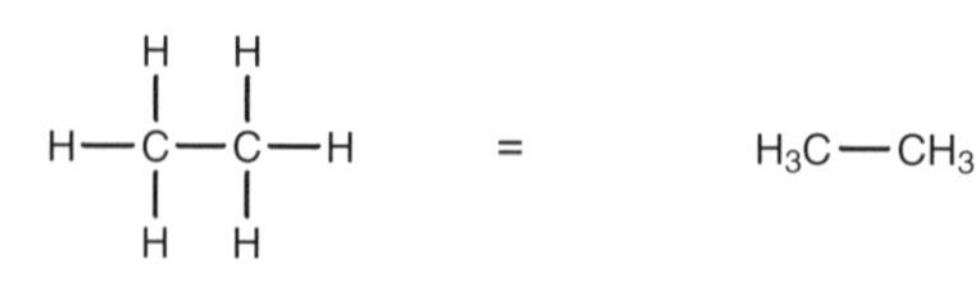

Fig. 1. Ethane.

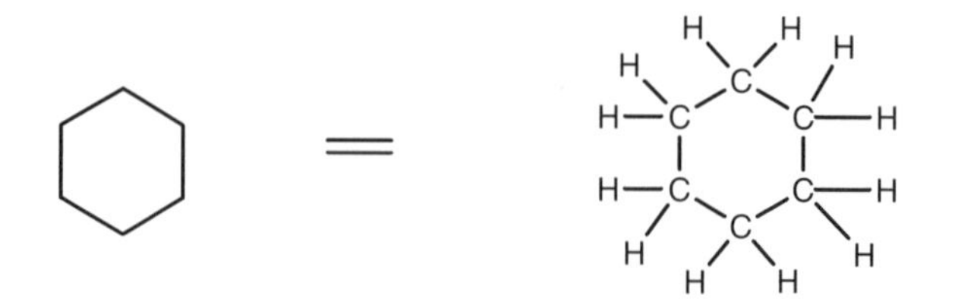

Fig. 2. Skeletal drawing of cyclohexane.

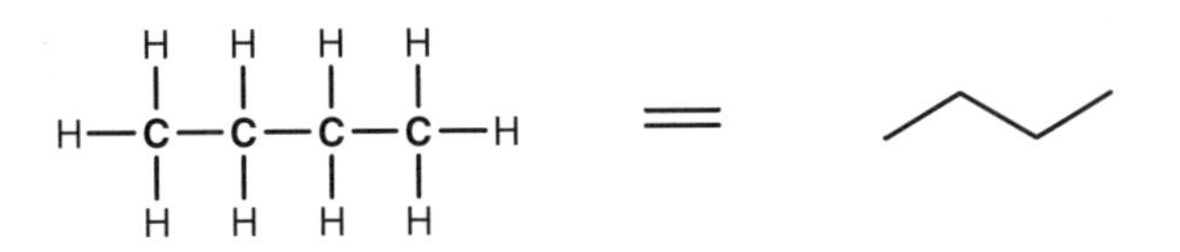

Fig. 3. Skeletal drawing of butane.

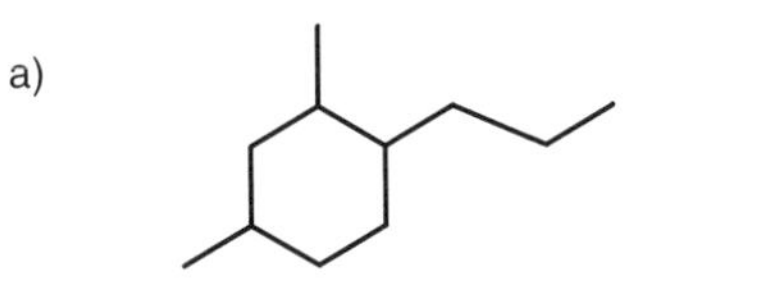

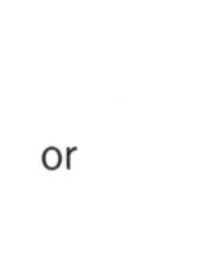

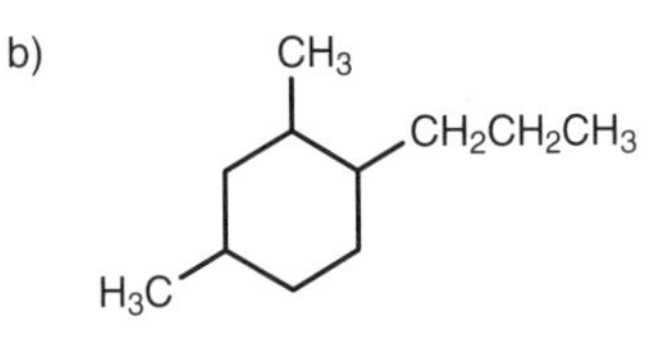

Fig. 4. Drawings of an alkyl substituted cyclohexane.

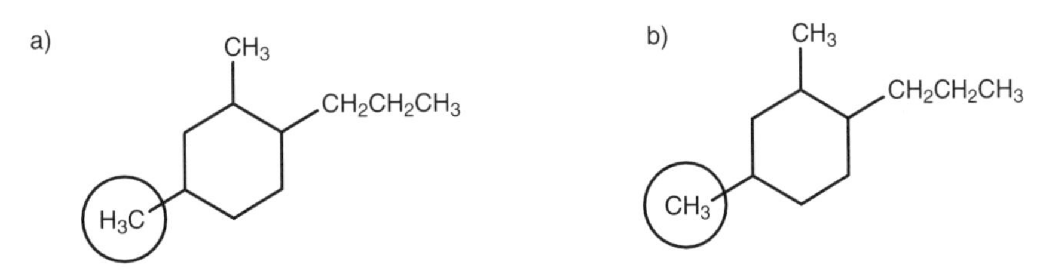

Fig. 5. (a) Correct depiction of methyl group; (b) wrong depiction of methyl group.

Alkyl groups

Alkyl groups (C_nH_{2n+1}) are alkane substituents of a complex molecule. Simple alkyl groups can be indicated in skeletal form (*Fig. 4a*), or as CH_3, CH_2CH_3, $CH_2CH_2CH_3$, *et cetera*. (*Fig. 4b*).

Notice how the CH_3 groups have been written in *Fig. 5*. The structure in *Fig. 5a* is more correct than the structure in *Fig. 5b* since the bond shown is between the carbons.

B3 NOMENCLATURE

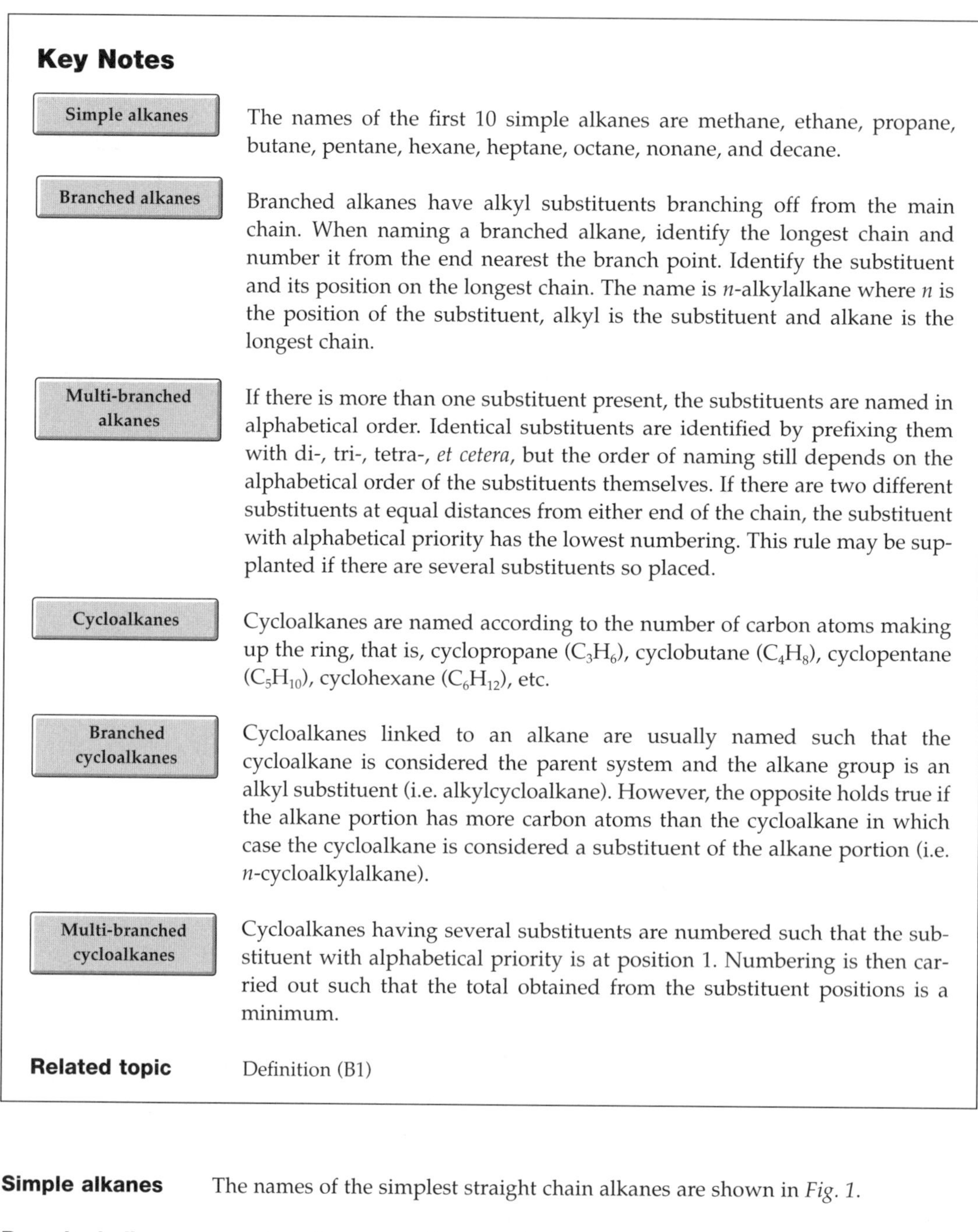

Key Notes

Simple alkanes

The names of the first 10 simple alkanes are methane, ethane, propane, butane, pentane, hexane, heptane, octane, nonane, and decane.

Branched alkanes

Branched alkanes have alkyl substituents branching off from the main chain. When naming a branched alkane, identify the longest chain and number it from the end nearest the branch point. Identify the substituent and its position on the longest chain. The name is *n*-alkylalkane where *n* is the position of the substituent, alkyl is the substituent and alkane is the longest chain.

Multi-branched alkanes

If there is more than one substituent present, the substituents are named in alphabetical order. Identical substituents are identified by prefixing them with di-, tri-, tetra-, *et cetera*, but the order of naming still depends on the alphabetical order of the substituents themselves. If there are two different substituents at equal distances from either end of the chain, the substituent with alphabetical priority has the lowest numbering. This rule may be supplanted if there are several substituents so placed.

Cycloalkanes

Cycloalkanes are named according to the number of carbon atoms making up the ring, that is, cyclopropane (C_3H_6), cyclobutane (C_4H_8), cyclopentane (C_5H_{10}), cyclohexane (C_6H_{12}), etc.

Branched cycloalkanes

Cycloalkanes linked to an alkane are usually named such that the cycloalkane is considered the parent system and the alkane group is an alkyl substituent (i.e. alkylcycloalkane). However, the opposite holds true if the alkane portion has more carbon atoms than the cycloalkane in which case the cycloalkane is considered a substituent of the alkane portion (i.e. *n*-cycloalkylalkane).

Multi-branched cycloalkanes

Cycloalkanes having several substituents are numbered such that the substituent with alphabetical priority is at position 1. Numbering is then carried out such that the total obtained from the substituent positions is a minimum.

Related topic Definition (B1)

Simple alkanes

The names of the simplest straight chain alkanes are shown in *Fig. 1*.

Branched alkanes

Branched alkanes are alkanes with alkyl substituents branching off from the main chain. They are named by the following procedure:

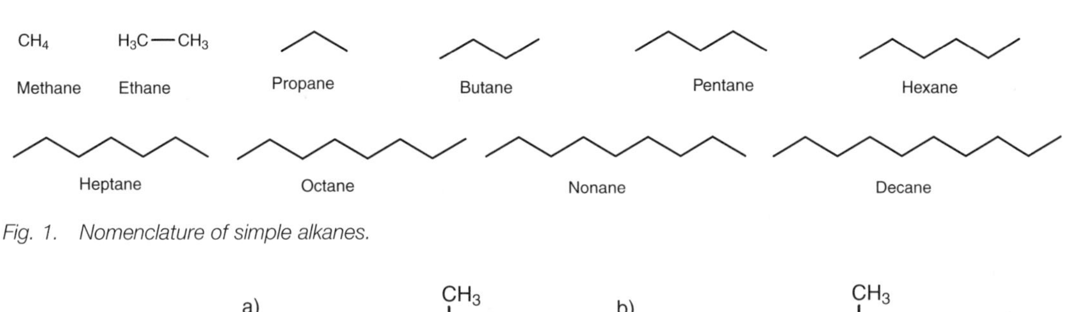

Fig. 1. Nomenclature of simple alkanes.

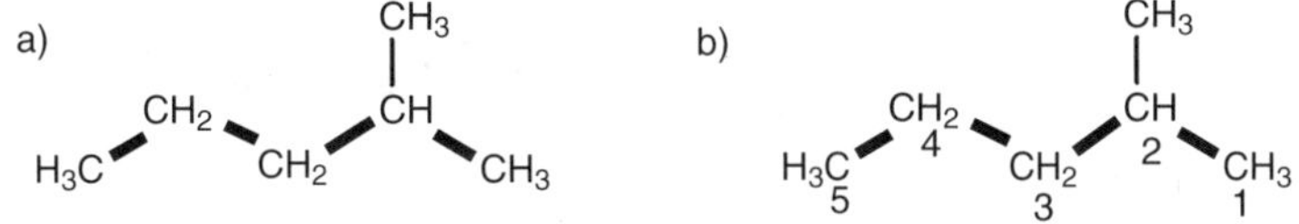

Fig. 2. (a) identify the longest chain; (b) number the longest chain.

- identify the longest chain of carbon atoms. In the example shown (*Fig. 2a*), the longest chain consists of five carbon atoms and a pentane chain;
- number the longest chain of carbons, starting from the end nearest the branch point (*Fig. 2b*);
- identify the carbon with the branching group (number 2 in *Fig. 2b*);
- identify and name the branching group. (In this example it is CH_3. Branching groups (or **substituents**) are referred to as **alkyl groups** (C_nH_{2n+1}) rather than alkanes (C_nH_{2n+2}). Therefore, CH_3 is called meth**yl** and not meth**ane**.)
- name the structure by first identifying the substituent and its position in the chain, then naming the longest chain. The structure in *Fig. 1* is called 2-methylpentane. Notice that the substituent and the main chain is one complete word, that is, 2-methylpentane rather than 2-methyl pentane.

Multi-branched alkanes

If there is more than one alkyl substituent present in the structure then the substituents are named in alphabetical order, numbering again from the end of the chain nearest the substituents. The structure in *Fig. 3* is 4-ethyl-3-methyloctane and not 3-methyl-4-ethyloctane.

If a structure has identical substituents, then the prefixes di-, tri-, tetra-, *et cetera* are used to represent the number of substituents. For example, the structure in *Fig. 4* is called 2,2-dimethylpentane and not 2-methyl-2-methylpentane.

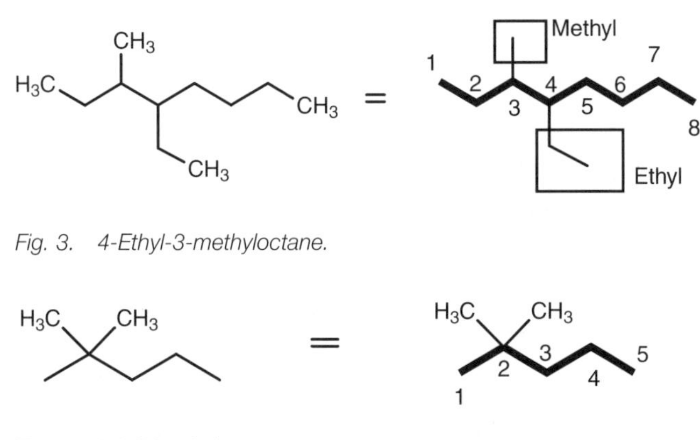

Fig. 3. 4-Ethyl-3-methyloctane.

Fig. 4. 2,2-Dimethylpentane.

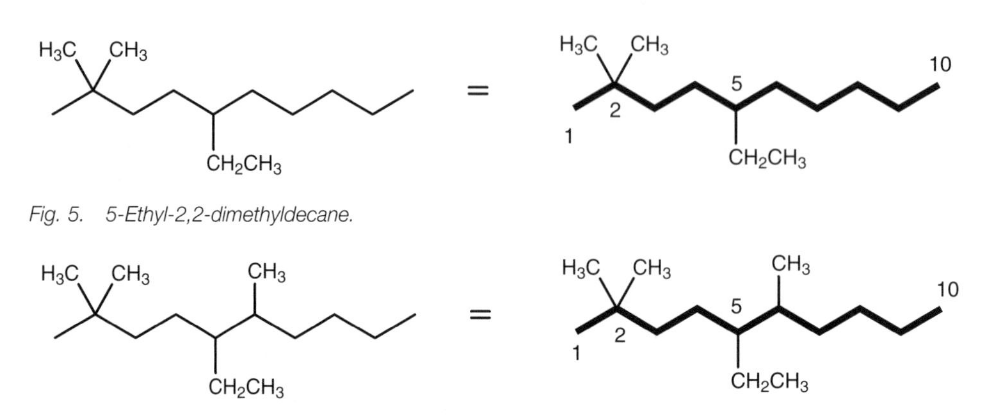

Fig. 5. 5-Ethyl-2,2-dimethyldecane.

Fig. 6. 5-Ethyl-2,2,6-trimethyldecane.

The prefixes di-, tri-, tetra- etc. are used for identical substituents, but the order in which they are written is still dependent on the alphabetical order of the substituents themselves (i.e. ignore the di-, tri-, tetra-, *et cetera*). For example, the structure in *Fig. 5* is called 5-ethyl-2,2-dimethyldecane and not 2,2-dimethyl-5-ethyldecane.

Identical substituents can be in different positions on the chain, but the same rules apply. For example, the structure in *Fig. 6* is called 5-ethyl-2,2,6-trimethyldecane.

In some structures, it is difficult to decide which end of the chain to number from. For example, two different substituents might be placed at equal distances from either end of the chain. If that is the case, the group with alphabetical priority should be given the lowest numbering. For example, the structure in *Fig. 7a* is 3-ethyl-5-methylheptane and not 5-ethyl-3-methylheptane.

However, there is another rule which might take precedence over the above rule. The structure (*Fig. 7c*) has ethyl and methyl groups equally placed from each end of the chain, but there are two methyl groups to one ethyl group. Numbering should be chosen such that the smallest total is obtained. In this example, the structure is called 5-ethyl-3,3-dimethylheptane (*Fig. 7c*) rather than 3-ethyl-5,5-dimethylheptane (*Fig. 7b*) since 5+3+3 = 11 is less than 3+5+5 = 13.

Cycloalkanes

Cycloalkanes are simply named by identifying the number of carbons in the ring and prefixing the alkane name with cyclo (*Fig. 8*).

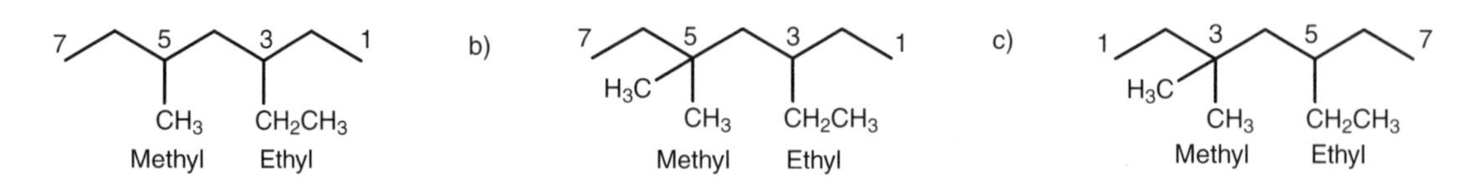

Fig. 7. (a) 3-Ethyl-5-methylheptane; (b) incorrect numbering; (c) 5-ethyl-3,3-dimethylheptane.

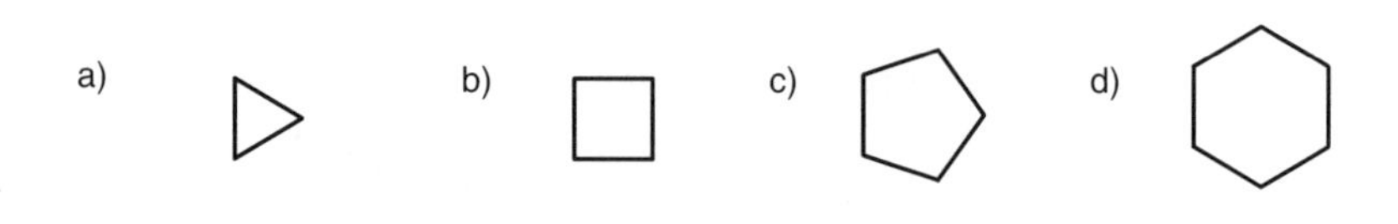

Fig. 8. (a) Cyclopropane; (b) cyclobutane; (c) cyclopentane; (d) cyclohexane.

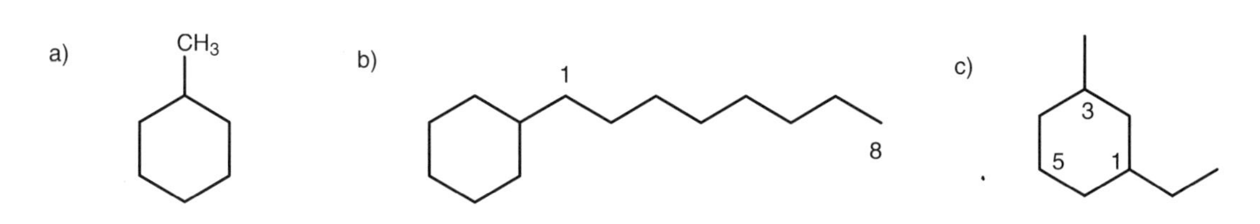

Fig. 9. (a) Methylcyclohexane; (b) 1-cyclohexyloctane; (c) 1-ethyl-3-methylcyclohexane.

Branched cyclohexanes

Cycloalkanes consisting of a cycloalkane moiety linked to an alkane moiety are usually named such that the cycloalkane is the parent system and the alkane moiety is considered to be an alkyl substituent. Therefore, the structure in *Fig. 9a* is methylcyclohexane and not cyclohexylmethane. Note that there is no need to number the cycloalkane ring when only one substituent is present.

If the alkane moiety contains more carbon atoms than the ring, the alkane moiety becomes the parent system and the cycloalkane group becomes the substituent. For example, the structure in *Fig. 9b* is called 1-cyclohexyloctane and not octylcyclohexane. In this case, numbering is necessary to identify the position of the cycloalkane on the alkane chain.

Multi-branched cycloalkanes

Branched cycloalkanes having different substituents are numbered such that the alkyl substituent having alphabetical priority is at position 1. The numbering of the rest of the ring is then carried out such that the substituent positions add up to a minimum. For example, the structure in *Fig. 9c* is called 1-ethyl-3-methylcyclohexane rather than 1-methyl-3-ethylcyclohexane or 1-ethyl-5-methylcyclohexane. The last name is incorrect since the total obtained by adding the substituent positions together is $5+1 = 6$ which is higher than the total obtained from the correct name (i.e. $1+3=4$).

C1 RECOGNITION OF FUNCTIONAL GROUPS

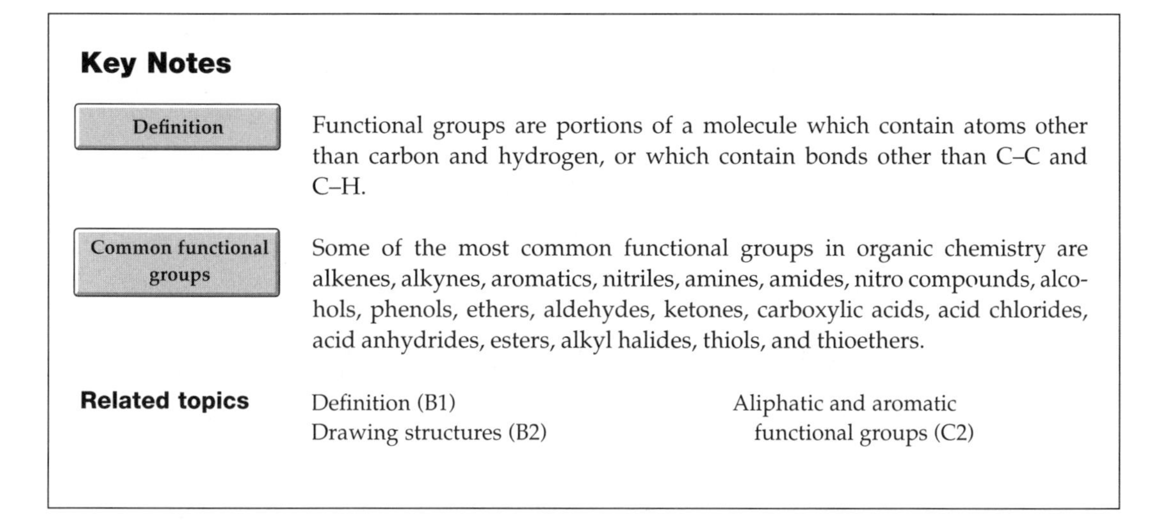

Key Notes

Definition — Functional groups are portions of a molecule which contain atoms other than carbon and hydrogen, or which contain bonds other than C–C and C–H.

Common functional groups — Some of the most common functional groups in organic chemistry are alkenes, alkynes, aromatics, nitriles, amines, amides, nitro compounds, alcohols, phenols, ethers, aldehydes, ketones, carboxylic acids, acid chlorides, acid anhydrides, esters, alkyl halides, thiols, and thioethers.

Related topics — Definition (B1); Drawing structures (B2); Aliphatic and aromatic functional groups (C2)

Definition

A **functional group** is a portion of an organic molecule which consists of atoms other than carbon and hydrogen, or which contains bonds other than C–C and C–H bonds. For example, ethane (*Fig. 1a*) is an alkane and has no functional group. All the atoms are carbon and hydrogen and all the bonds are C–C and C–H. Ethanoic acid on the other hand (*Fig. 1b*), has a portion of the molecule (boxed portion) which contains atoms other than carbon and hydrogen, and bonds other than C—H and C—C. This portion of the molecule is called a functional group – in this case a carboxylic acid.

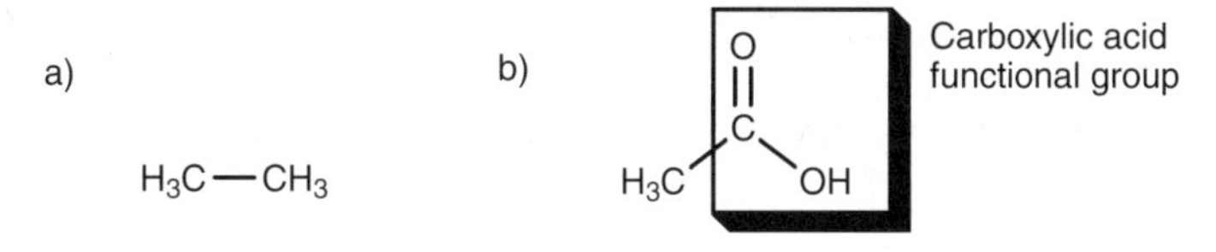

Fig. 1. (a) Ethane; (b) ethanoic acid.

Common functional groups

The following are some of the more common functional groups in organic chemistry.

- functional groups which contain carbon and hydrogen only (*Fig. 2*);

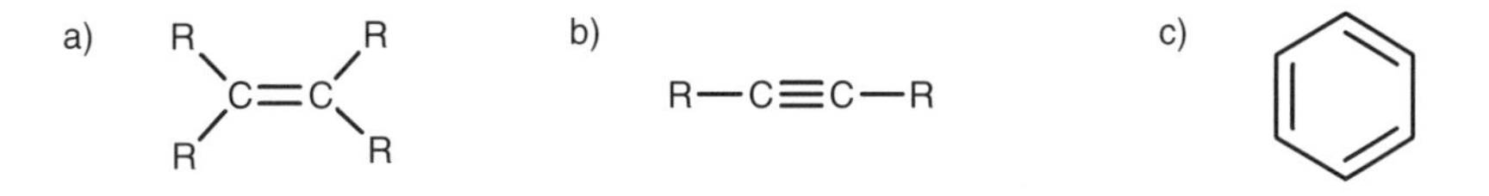

Fig. 2. (a) Alkene; (b) alkyne; (c) aromatic.

- functional groups which contain nitrogen (*Fig. 3*);

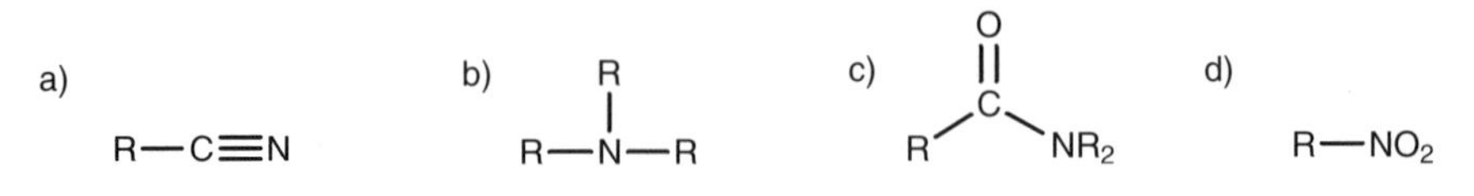

Fig. 3. (a) Nitrile; (b) amine; (c) amide; (d) nitro.

- functional groups involving single bonds and which contain oxygen (*Fig. 4*);

a) R—OH b) R—O—R

Fig. 4. (a) Alcohol or alkanol; (b) ether.

- functional groups involving double bonds and which contain oxygen (*Fig. 5*);

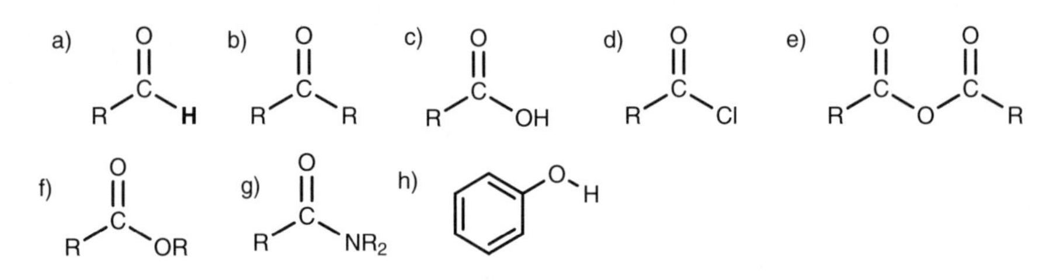

Fig. 5. (a) Aldehyde or alkanal; (b) ketone or alkanone; (c) carboxylic acid; (d) carboxylic acid chloride; (e) carboxylic acid anhydride; (f) ester; (g) amide; (h) phenol.

- functional groups which contain a halogen atom (*Fig. 6*);

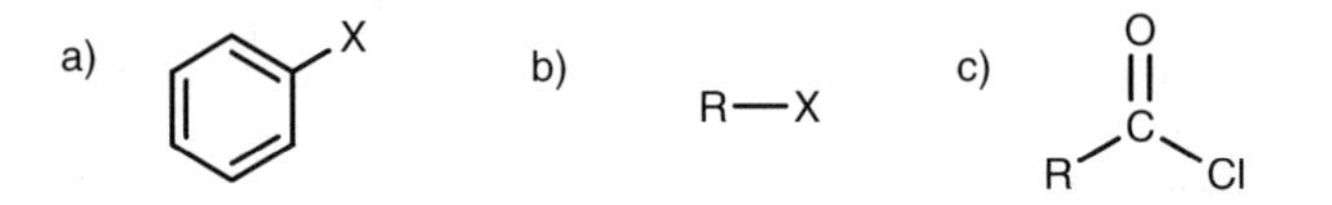

Fig. 6. (a) Aryl halide (X = F, Cl, Br, I); (b) alkyl halide or halogenoalkane (X = F, Cl, Br, I); (c) acid chloride.

- functional groups which contain sulfur (*Fig. 7*).

a) R—SH b) R—S—R

Fig. 7. (a) Thiol; (b) thioether.

C2 Aliphatic and aromatic functional groups

Key Notes

Aliphatic functional groups Functional groups are defined as aliphatic if there is no aromatic ring directly attached to them. It is possible to have an aromatic molecule containing an aliphatic functional group if the aromatic ring is not directly attached to the functional group.

Aromatic functional groups Functional groups are defined as aromatic if they have an aromatic ring directly attached to them. In the case of esters and amides, the aromatic ring must be attached to the carbonyl side of the functional group. If the aromatic ring is attached to the heteroatom, the functional groups are defined as aliphatic.

Related topic Recognition of functional groups (C1)

Aliphatic functional groups

Functional groups can be classed as aliphatic or aromatic. An aliphatic functional group is one where there is no aromatic ring directly attached to the functional group (*Fig. 1a* and *b*).

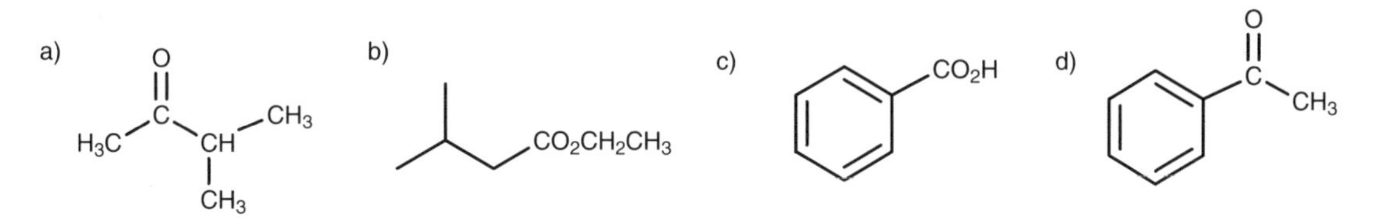

Fig. 1. (a) Aliphatic ketone; (b) aliphatic ester; (c) aromatic carboxylic acid; (d) aromatic ketone.

Aromatic functional groups

An aromatic functional group is one where an aromatic ring is directly attached to the functional group (*Fig. 1c* and *d*).

There is one complication involving esters and amides. These functional groups are defined as aromatic or aliphatic depending on whether the aryl group is directly attached to the **carbonyl** end of the functional group, that is, Ar–CO–X. If the aromatic ring is attached to the heteroatom instead, then the ester or amide is classed as an aliphatic amide (*Fig. 2*).

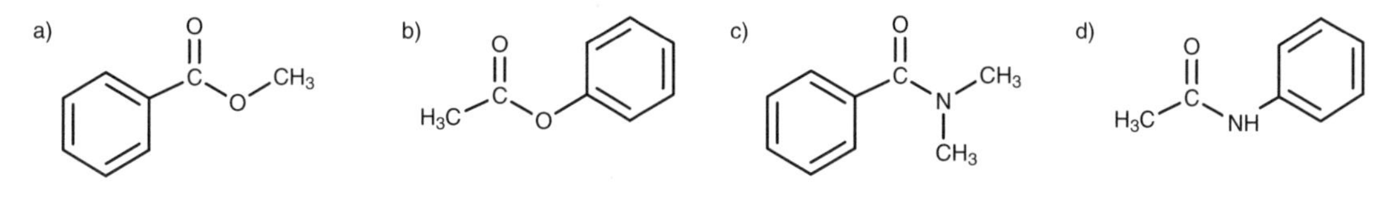

Fig. 2. (a) Aromatic ester; (b) aliphatic ester; (c) aromatic amide; (d) aliphatic amide.

C3 Intermolecular bonding

Key Notes

Definition

Intermolecular bonding takes place between different molecules. This can take the form of ionic bonding, hydrogen bonding, dipole–dipole interactions and van der Waals interactions. The type of bonding involved depends on the functional groups present.

Ionic bonding

Ionic bonds are possible between ionized functional groups such as carboxylic acids and amines

Hydrogen bonding

Intermolecular hydrogen bonding is possible for alcohols, carboxylic acids, amides, amines, and phenols. These functional groups contain a hydrogen atom bonded to nitrogen or oxygen. Hydrogen bonding involves the interaction of the partially positive hydrogen on one molecule and the partially negative heteroatom on another molecule. Hydrogen bonding is also possible with elements other than nitrogen or oxygen.

Dipole–dipole interactions

Dipole–dipole interactions are possible between molecules having polarizable bonds, in particular the carbonyl group (C=O). Such bonds have a dipole moment and molecules can align themselves such that their dipole moments are parallel and in opposite directions. Ketones and aldehydes are capable of interacting through dipole–dipole interactions.

van der Waals interactions

van der Waals interactions are weak intermolecular bonds between regions of different molecules bearing transient positive and negative charges. These transient charges are caused by the random fluctuation of electrons. Alkanes, alkenes, alkynes and aromatic rings interact through van der Waals interactions.

Related topic

Recognition of functional groups (C1)

Definition

Intermolecular bonding is the bonding interaction which takes place between different molecules. This can take the form of **ionic bonding, hydrogen bonding, dipole–dipole interactions** or **van der Waals interactions**. These bonding forces are weaker than the covalent bonds, but they have an important influence on the physical and biological properties of a compound.

Ionic bonding

Ionic bonding takes place between molecules having opposite charges and involves an **electrostatic** interaction between the two opposite charges. The functional groups which most easily ionize are amines and carboxylic acids (*Fig. 1*).

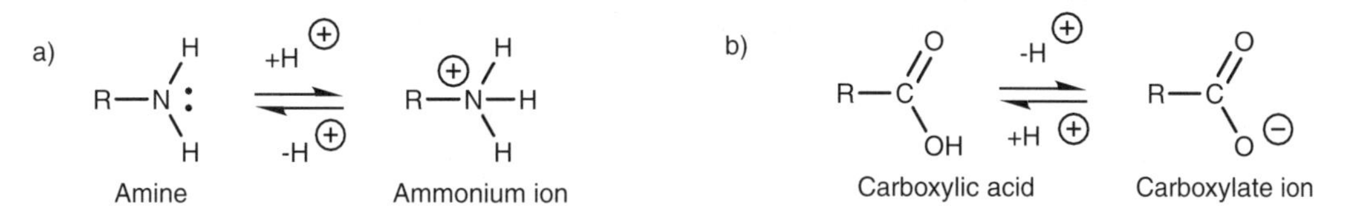

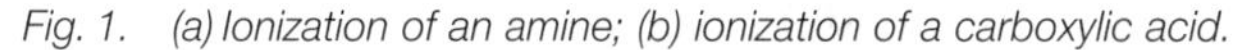

Fig. 1. (a) Ionization of an amine; (b) ionization of a carboxylic acid.

Ionic bonding is possible between a molecule containing an ammonium ion and a molecule containing a carboxylate ion. Some important naturally occurring molecules contain both groups – the amino acids. Both these functional groups are ionized to form a structure known as a **zwitterion** (a neutral molecule bearing both a positive and a negative charge) and intermolecular ionic bonding can take place (*Fig. 2*).

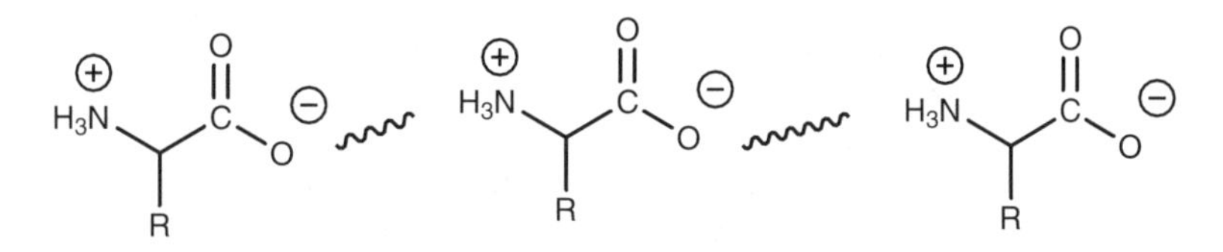

Fig. 2. Intermolecular ionic bonding of amino acids.

Hydrogen bonding

Hydrogen bonding can take place when molecules have a hydrogen atom attached to a heteroatom such as nitrogen or oxygen. The common functional groups which can participate in hydrogen bonding are alcohols, phenols, carboxylic acids, amides, and amines. Hydrogen bonding is possible due to the polar nature of the N–H or O–H bond. Nitrogen and oxygen are more electronegative than hydrogen. As a result, the heteroatom gains a slightly negative charge and the hydrogen gains a slightly positive charge. Hydrogen bonding involves the partially charged hydrogen of one molecule (the H bond donor) interacting with the partially charged heteroatom of another molecule (the H bond acceptor) (*Fig. 3*).

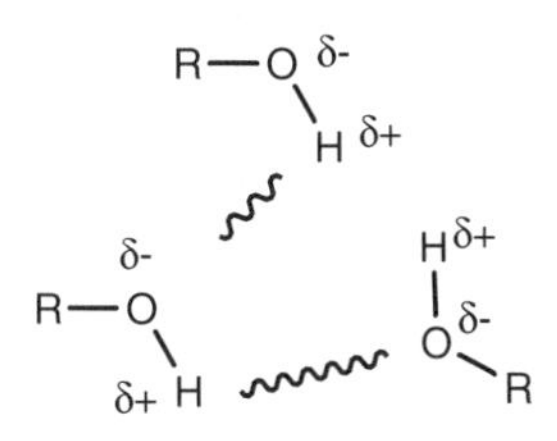

Fig. 3. Intermolecular hydrogen bonding between alcohols.

Dipole–dipole interactions

Dipole–dipole interactions are possible between polarized bonds other than N–H or O–H bonds. The most likely functional groups which can interact in this way are those containing a carbonyl group (C=O). The electrons in the carbonyl bond are polarized towards the more electronegative oxygen such that the oxygen gains a slight negative charge and the carbon gains a slight positive charge. This results in a dipole moment which can be represented by the arrow shown in *Fig. 4*. The arrow points to the negative end of the dipole moment. Molecules containing dipole moments can align themselves with each other such that the dipole moments are pointing in opposite directions (*Fig. 4b*).

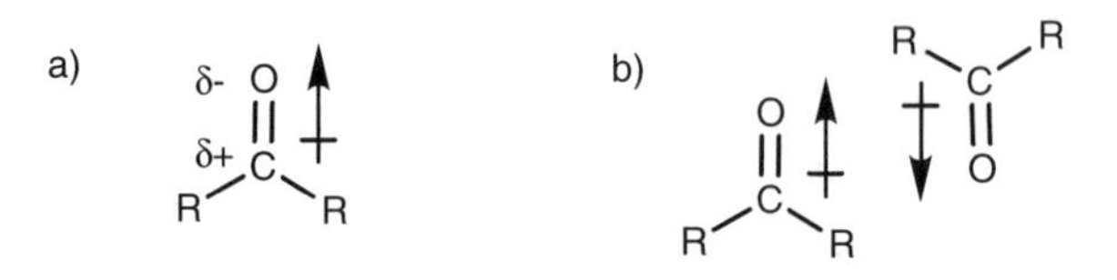

Fig. 4. (a) Dipole moment of a ketone; (b) intermolecular dipole–dipole interaction between ketones.

van der Waals interactions

van der Waals interactions are the weakest of the intermolecular bonding forces and involve the transient existence of partial charges in a molecule. Electrons are continually moving in an unpredictable fashion around any molecule. At any moment of time, there is a slight excess of electrons in one part of the molecule and a slight deficit in another part. Although the charges are very weak and fluctuate around the molecule, they are sufficiently strong to allow a weak interaction between molecules, where regions of opposite charge in different molecules attract each other.

Alkane molecules can interact in this way and the strength of the interaction increases with the size of the alkane molecule. van der Waals interactions are also important for alkenes, alkynes and aromatic rings. The types of molecules involved in this form of intermolecular bonding are 'fatty' molecules which do not dissolve easily in water and such molecules are termed **hydrophobic** (water-hating). Hydrophobic molecules can dissolve in nonpolar, hydrophobic solvents due to van der Waals interactions and so this form of intermolecular bonding is sometimes referred to as a hydrophobic interaction.

C4 Properties and reactions

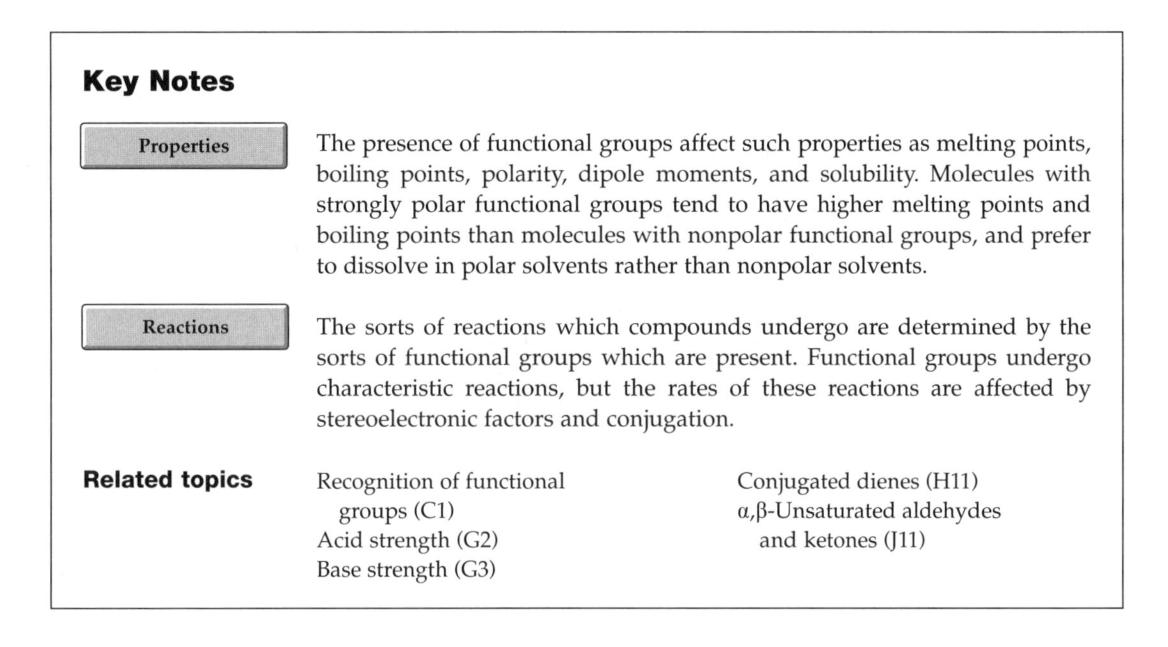

Key Notes

Properties

The presence of functional groups affect such properties as melting points, boiling points, polarity, dipole moments, and solubility. Molecules with strongly polar functional groups tend to have higher melting points and boiling points than molecules with nonpolar functional groups, and prefer to dissolve in polar solvents rather than nonpolar solvents.

Reactions

The sorts of reactions which compounds undergo are determined by the sorts of functional groups which are present. Functional groups undergo characteristic reactions, but the rates of these reactions are affected by stereoelectronic factors and conjugation.

Related topics

Recognition of functional groups (C1)
Acid strength (G2)
Base strength (G3)
Conjugated dienes (H11)
α,β-Unsaturated aldehydes and ketones (J11)

Properties

The chemical and physical properties of an organic compound are determined by the sort of intermolecular bonding forces present, which in turn depends on the functional group present. A molecule such as methane has a low boiling point and is a gas at room temperature because its molecules are bound together by weak van der Waals forces (*Fig. 1a*). In contrast, methanol is a liquid at room temperature since hydrogen bonding is possible between the alcoholic functional groups (*Fig. 1b*).

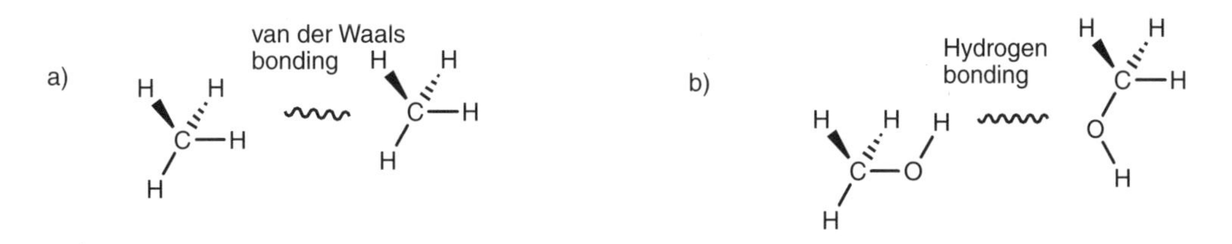

Fig. 1. (a) Intermolecular van der Waals (methane); (b) intermolecular hydrogen bonding (methanol).

The polarity of molecules depends on which functional groups are present. A molecule will be polar and have a dipole moment if it contains polar functional groups such as an alcohol, amine, or ketone. Polarity also determines solubility in different solvents. Polar molecules prefer to dissolve in polar solvents such as water or alcohols, whereas nonpolar molecules prefer to dissolve in nonpolar solvents such as ether and chloroform. Polar molecules which can dissolve in water are termed **hydrophilic** (water-loving) while nonpolar molecules are termed **hydrophobic** (water-hating).

In most cases, the presence of a polar functional group will determine the physical properties of the molecule. However, this is not always true. If a molecule has a polar group such as a carboxylic acid, but has a long hydrophobic alkane chain, then the molecule will tend to be hydrophobic.

Reactions

The vast majority of organic reactions take place at functional groups and are characteristic of that functional group. However, the reactivity of the functional group is affected by **stereoelectronic** effects. For example, a functional group may be surrounded by bulky groups which hinder the approach of a reagent and slow down the rate of reaction. This is referred to as **steric shielding**. **Electronic effects** can also influence the rate of a reaction. Neighboring groups can influence the reactivity of a functional group if they are electron-withdrawing or electron-donating and influence the electronic density within the functional group. Conjugation and aromaticity also has an important effect on the reactivity of functional groups. For example, an aromatic ketone reacts at a different rate from an aliphatic ketone. The aromatic ring is in conjugation with the carbonyl group and this increases the stability of the overall system, making it less reactive.

C5 NOMENCLATURE OF FUNCTIONAL GROUPS

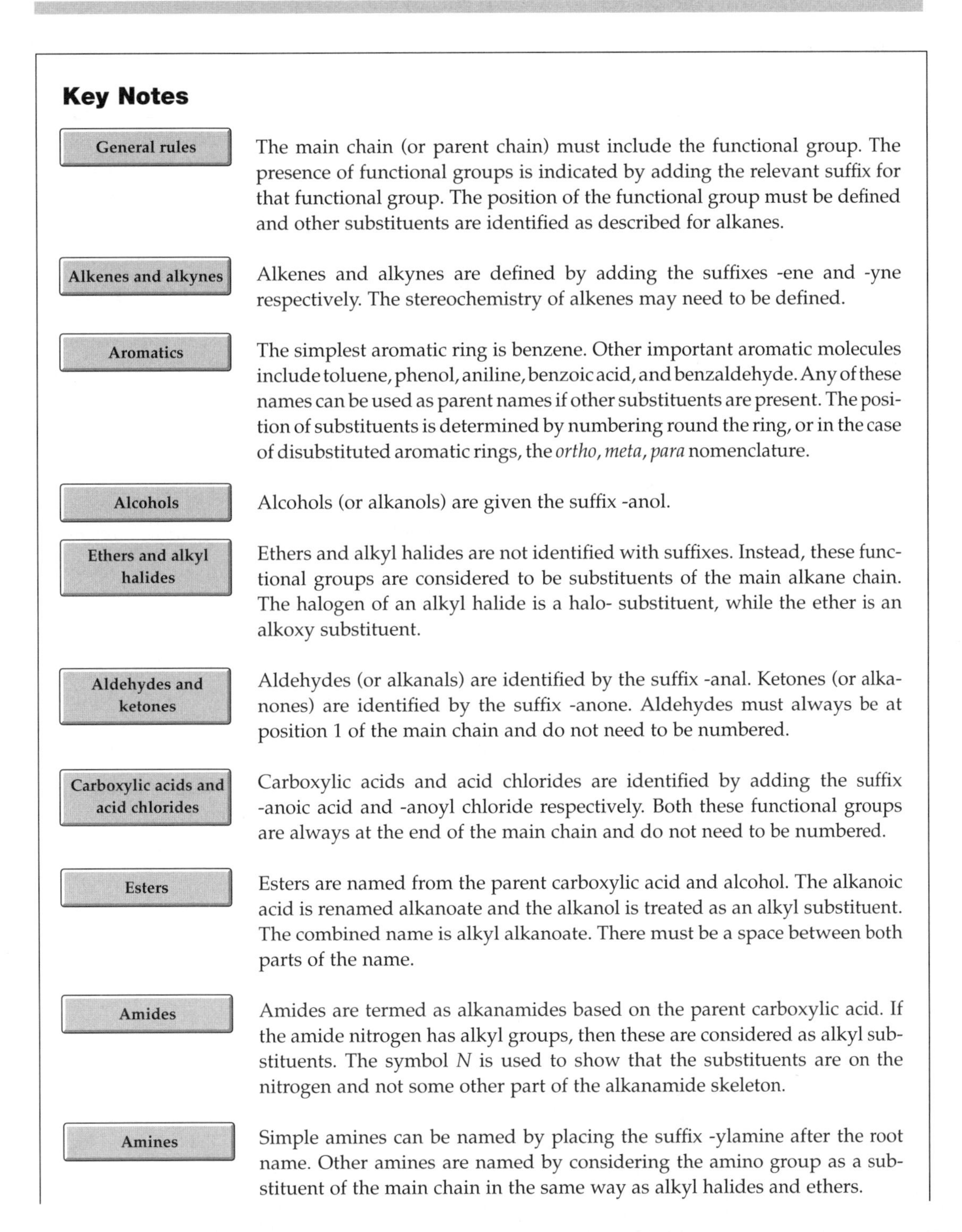

Key Notes

General rules

The main chain (or parent chain) must include the functional group. The presence of functional groups is indicated by adding the relevant suffix for that functional group. The position of the functional group must be defined and other substituents are identified as described for alkanes.

Alkenes and alkynes

Alkenes and alkynes are defined by adding the suffixes -ene and -yne respectively. The stereochemistry of alkenes may need to be defined.

Aromatics

The simplest aromatic ring is benzene. Other important aromatic molecules include toluene, phenol, aniline, benzoic acid, and benzaldehyde. Any of these names can be used as parent names if other substituents are present. The position of substituents is determined by numbering round the ring, or in the case of disubstituted aromatic rings, the *ortho*, *meta*, *para* nomenclature.

Alcohols

Alcohols (or alkanols) are given the suffix -anol.

Ethers and alkyl halides

Ethers and alkyl halides are not identified with suffixes. Instead, these functional groups are considered to be substituents of the main alkane chain. The halogen of an alkyl halide is a halo- substituent, while the ether is an alkoxy substituent.

Aldehydes and ketones

Aldehydes (or alkanals) are identified by the suffix -anal. Ketones (or alkanones) are identified by the suffix -anone. Aldehydes must always be at position 1 of the main chain and do not need to be numbered.

Carboxylic acids and acid chlorides

Carboxylic acids and acid chlorides are identified by adding the suffix -anoic acid and -anoyl chloride respectively. Both these functional groups are always at the end of the main chain and do not need to be numbered.

Esters

Esters are named from the parent carboxylic acid and alcohol. The alkanoic acid is renamed alkanoate and the alkanol is treated as an alkyl substituent. The combined name is alkyl alkanoate. There must be a space between both parts of the name.

Amides

Amides are termed as alkanamides based on the parent carboxylic acid. If the amide nitrogen has alkyl groups, then these are considered as alkyl substituents. The symbol *N* is used to show that the substituents are on the nitrogen and not some other part of the alkanamide skeleton.

Amines

Simple amines can be named by placing the suffix -ylamine after the root name. Other amines are named by considering the amino group as a substituent of the main chain in the same way as alkyl halides and ethers.

Thiols and thioethers

Thiols are named by adding the suffix thiol to the name of the main alkane chain. Thioethers are named in the same way as ethers where the major alkyl substituent is considered to be the main chain with an alkylthio substituent. Simple thioethers can be identified as dialkylsulfides.

Related topics

Nomenclature (B3)

Configurational isomers – alkenes and cycloalkanes (D2)

General rules

Many of the nomenclature rules for alkanes (Topic B3) hold true for molecules containing a functional group, but extra rules are needed in order to define the type of functional group present and its position within the molecule. The main rules are as follows:

(i) The main (or parent) chain must include the functional group, and so may not necessarily be the longest chain (Fig. 1);

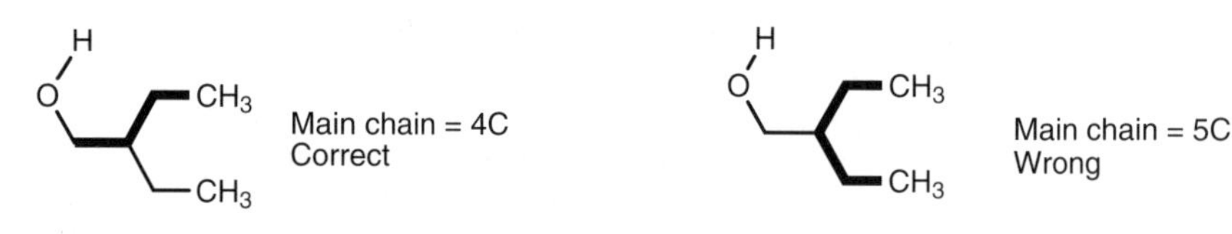

Fig. 1. Identification of the main chain.

(ii) The presence of some functional groups is indicated by replacing -ane for the parent alkane chain with the following suffixes:

functional group	suffix	functional group	suffix
alkene	-ene	alkyne	-yne
alcohol	-anol	aldehyde	-anal
ketone	-anone	carboxylic acid	-anoic acid
acid chloride	-anoyl chloride	amine	-ylamine.

The example in *Fig. 1* is a but**anol**.

(iii) Numbering must start from the end of the main chain nearest the functional group. Therefore, the numbering should place the alcohol (*Fig. 2*) at position 1 and not position 4.

Fig. 2. Numbering of the longest chain.

(iv) The position of the functional group must be defined in the name. Therefore, the alcohol (*Fig. 2*) is a 1-butanol.

(v) Other substituents are named and ordered in the same way as for alkanes. The alcohol (*Fig. 2*) has an ethyl group at position 3 and so the full name for the structure is 3-ethyl-1-butanol.

There are other rules designed for specific situations. For example, if the functional group is an equal distance from either end of the main chain, the numbering starts from the end of the chain nearest to any substituents. For example, the alcohol (*Fig. 3*) is 2-methyl-3-pentanol and not 4-methyl-3-pentanol.

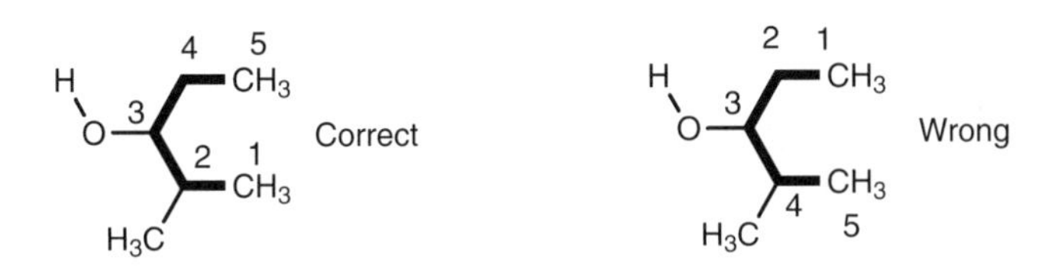

Fig. 3. 2-Methyl-3-pentanol.

Alkenes and alkynes

Alkenes and alkynes have the suffixes -ene and -yne respectively (*Fig. 4*). With some alkenes it is necessary to define the stereochemistry of the double bond (Topic D2).

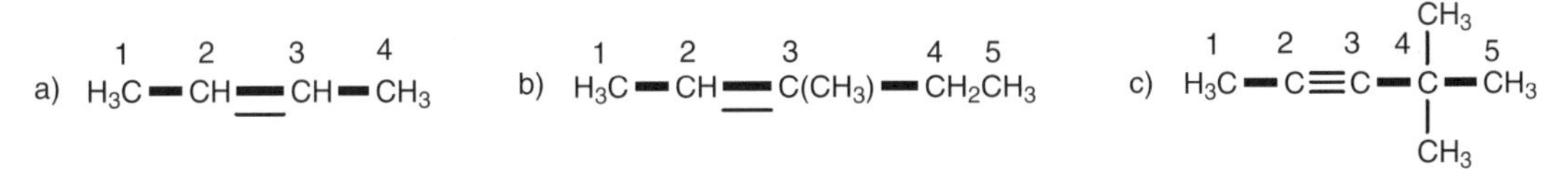

Fig. 4. (a) 2-Butene; (b) 3-methyl-2-pentene; (c) 4,4-dimethyl-2-pentyne.

Aromatics

The best known aromatic structure is benzene. If an alkane chain is linked to a benzene molecule, then the alkane chain is usually considered to be an alkyl substituent of the benzene ring. However, if the alkane chain contains more than six carbons, then the benzene molecule is considered to be a **phenyl** substituent of the alkane chain (*Fig. 5*).

Note that a **benzyl** group consists of an aromatic ring and a methylene group (*Fig. 6*).

Benzene is not the only parent name which can be used for aromatic compounds (*Fig. 7*).

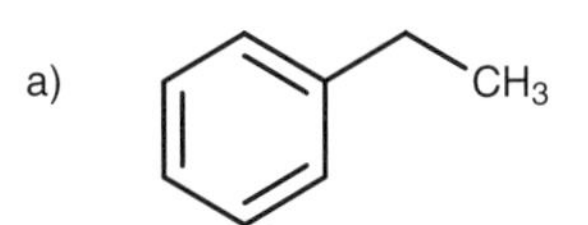

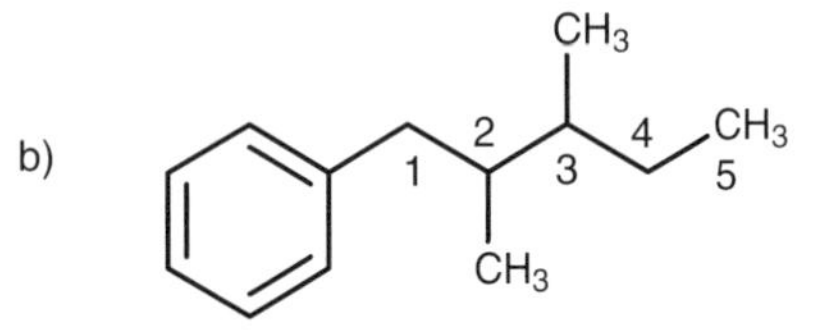

Fig. 5. (a) Ethylbenzene; (b) 1-phenyl-2,3-dimethylpentane.

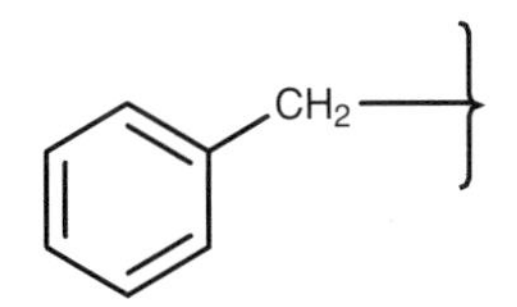

Fig. 6. Benzyl group.

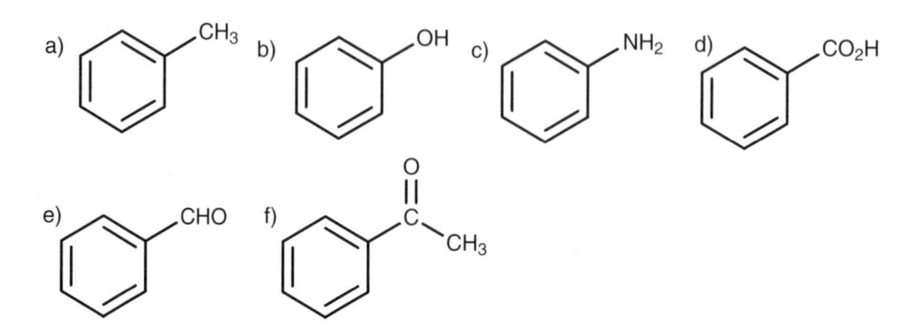

Fig 7. (a) Toluene; (b) phenol; (c) aniline; (d) benzoic acid; (e) benzaldehyde; (f) acetophenone.

With disubstituted aromatic rings, the position of substituents must be defined by numbering around the ring such that the substituents are positioned at the lowest numbers possible, for example, the structure (*Fig. 8*) is 1,3-dichlorobenzene and not 1,5-dichlorobenzene.

Fig. 8. 1,3-Dichlorobenzene

Alternatively, the terms *ortho*, *meta*, and *para* can be used. These terms define the relative position of one substituent to another (*Fig. 9*). Thus, 1,3-dichlorobenzene can also be called *meta*-dichlorobenzene. This can be shortened to *m*-dichlorobenzene. The examples in *Fig. 10* illustrate how different parent names may be used. Notice that the substituent which defines the parent name is defined as position 1. For example, if the parent name is toluene, the methyl group must be at position 1.

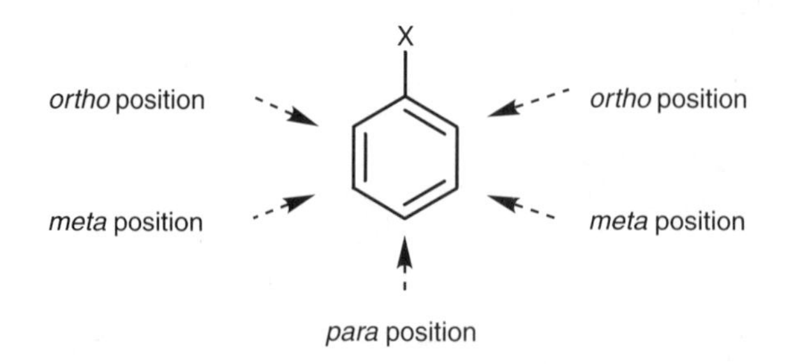

Fig. 9. ortho, meta *and* para *positions of an aromatic ring.*

When more than two substituents are present on the aromatic ring, the *ortho*, *meta*, *para* nomenclature is no longer valid and numbering has to be used (*Fig. 11*).

Once again, the relevant substituent has to be placed at position 1 if the

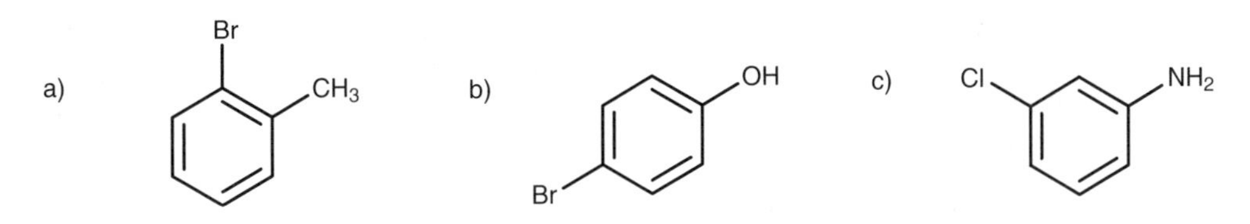

Fig. 10. *(a) 2-Bromotoluene or* o-*bromotoluene; (b) 4-bromophenol or* p-*bromophenol; (c) 3-chloroaniline or* m-*chloroaniline.*

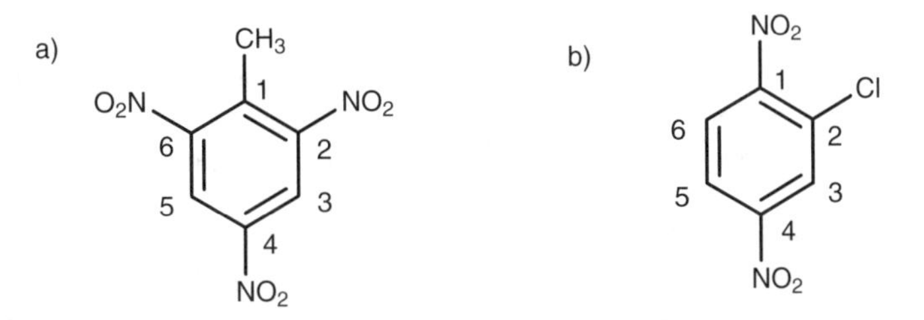

Fig. 11. *(a) 2,4,6-Trinitrotoluene; (b) 2-chloro-1,4-dinitrobenzene.*

parent name is toluene, aniline, *et cetera*. If the parent name is benzene, the numbering is chosen such that the lowest possible numbers are used. In the example shown, any other numbering would result in the substituents having higher numbers (*Fig. 12*).

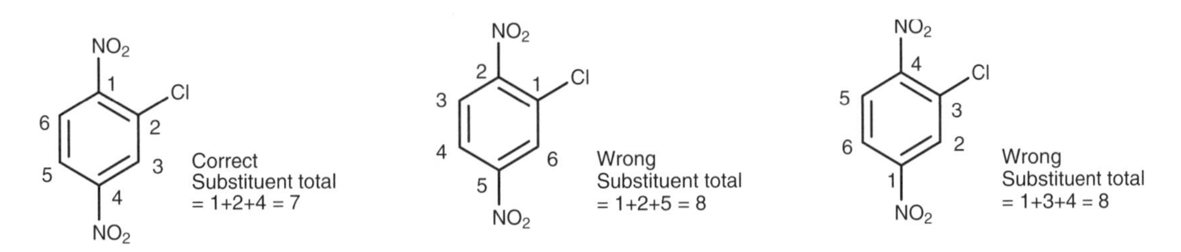

Fig. 12. *Possible numbering systems of tri-substituted aromatic ring.*

Alcohols

Alcohols or **alkanols** are identified by using the suffix -anol. The general rules described earlier can be used to name alcohols (*Fig. 13*).

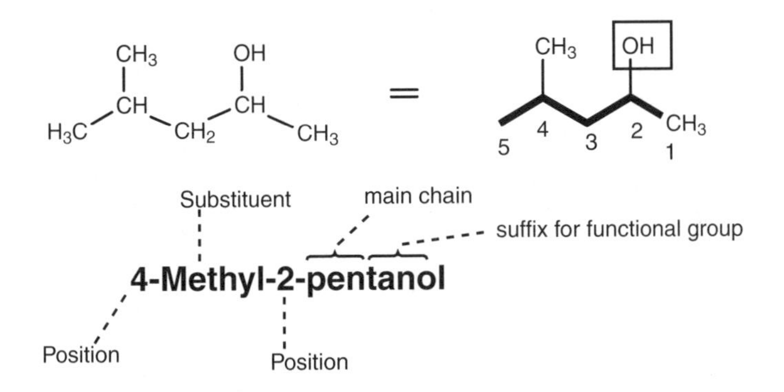

Fig. 13. *4-Methyl-2-pentanol.*

Ethers and alkyl halides

The nomenclature for these compounds is slightly different from previous examples in that the functional group is considered to be a substituent of the main alkane chain. The functional group is numbered and named as a substituent (*Fig. 14*).

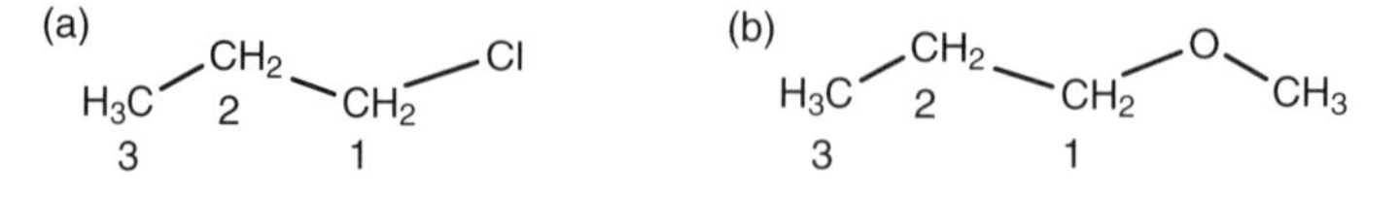

Fig. 14. (a) 1-Chloropropane; (b) 1-methoxypropane.

Note that ethers have two alkyl groups on either side of the oxygen. The larger alkyl group is the parent alkane. The smaller alkyl group along with the oxygen is the substituent and is known as an **alkoxy** group.

Aldehydes and ketones

The suffix for an aldehyde (or alkanal) is -anal, while the suffix for a ketone (or alkanone) is -anone. The main chain must include the functional group and the numbering is such that the functional group is at the lowest number possible. If the functional group is in the center of the main chain, the numbering is chosen to ensure that other substituents have the lowest numbers possible (e.g. 2,2-dimethyl-3-pentanone and not 4,4-dimethyl-3-pentanone; *Fig. 15*). 3-Methyl-

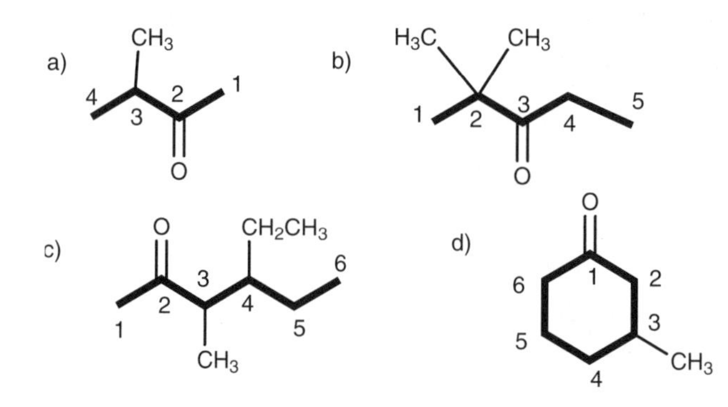

Fig. 15. (a) 3-Methyl-2-butanone; (b) 2,2-dimethyl-3-pentanone; (c) 4-ethyl-3-methyl-2-hexanone; (d) 3-methylcyclohexanone.

2-butanone can in fact be simplified to 3-methylbutanone. There is only one possible place for the ketone functional group in this molecule. If the carbonyl C=O group was at the end of the chain, it would be an aldehyde and not a ketone. Numbering is also not necessary in locating an aldehyde group since it can only be at the end of a chain (*Fig. 16*).

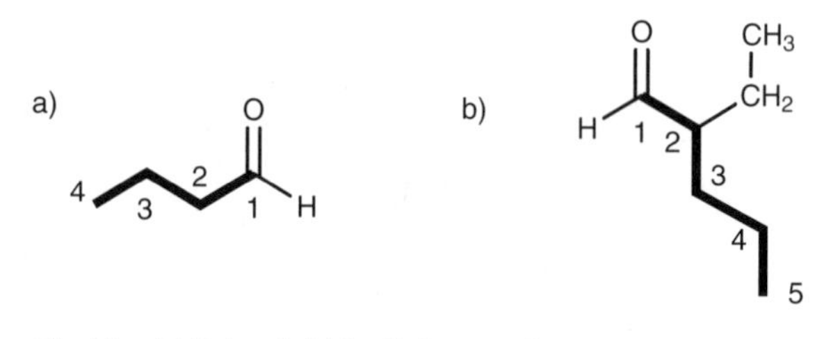

Fig. 16. (a) Butanal; (b) 2-ethylpentanal.

Carboxylic acids and acid chlorides

Carboxylic acids and acid chlorides are identified by adding the suffix -anoic acid and -anoyl chloride respectively. Both these functional groups are always at the end of the main chain and do not need to be numbered (*Fig. 17*).

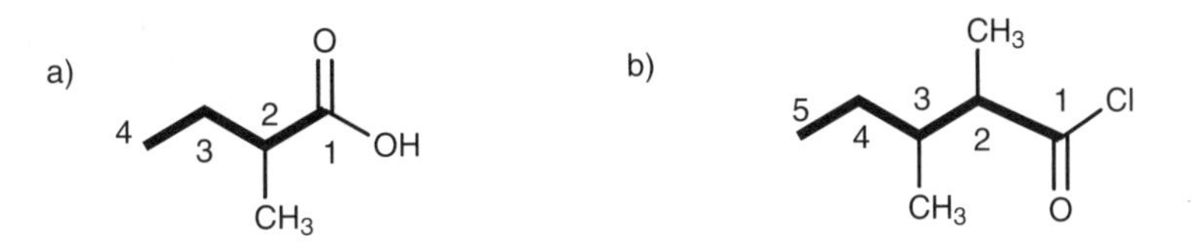

Fig. 17. (a) 2-Methylbutanoic acid; (b) 2,3-dimethylpentanoyl chloride.

Esters

To name an ester, the following procedure is carried out:

(i) identify the carboxylic acid (alkanoic acid) from which it was derived;
(ii) change the name to an alkanoate rather than an alkanoic acid;
(iii) identify the alcohol from which the ester was derived and consider this as an alkyl substituent;
(iv) the name becomes an alkyl alkanoate.

For example, the ester (*Fig. 18*) is derived from ethanoic acid and methanol (Topic K5). The ester would be an alkyl ethanoate since it is derived from ethanoic acid. The alkyl group comes from methanol and is a methyl group. Therefore, the full name is methyl ethanoate. (Note that there is a space between both parts of the name.)

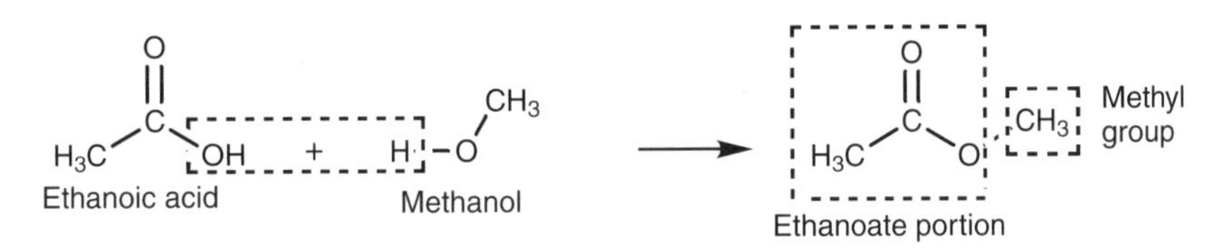

Fig. 18. Ester formation.

Amides

Amides are also derivatives of carboxylic acids. This time the carboxylic acid is linked with ammonia or an amine. As with esters, the parent carboxylic acid is identified. This is then termed an **alkanamide** and includes the nitrogen atom. For example, linking ethanoic acid with ammonia gives ethanamide (*Fig. 19*).

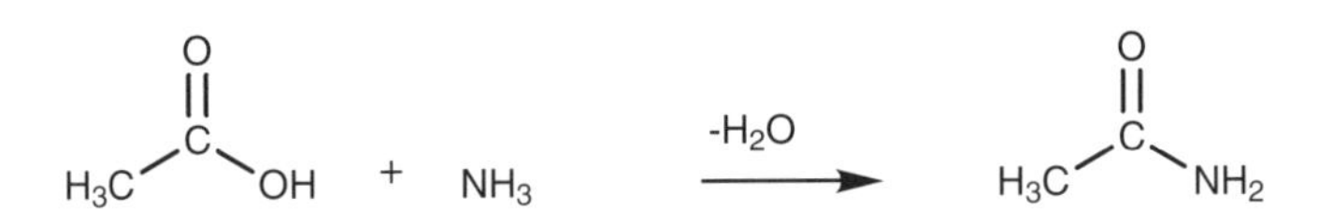

Fig. 19. Formation of ethanamide.

If the carboxylic acid is linked with an amine, then the amide will have alkyl groups on the nitrogen. These are considered as alkyl substituents and come at the beginning of the name. The symbol *N* is used to show that the substituents are on the nitrogen and not some other part of the alkanamide skeleton. For example, the structure in *Fig. 20* is called *N*-ethylethanamide.

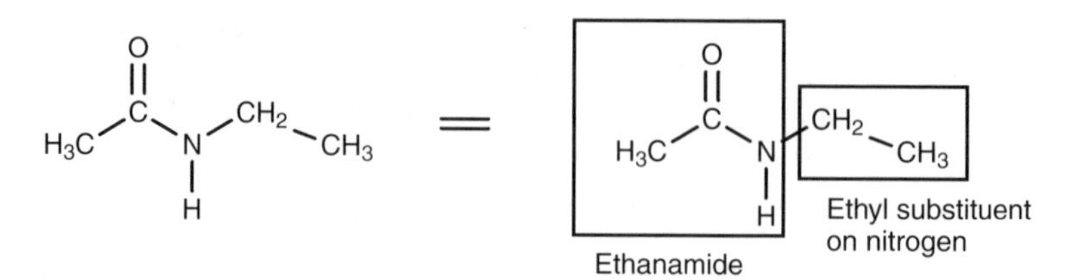

Fig. 20. N-Ethylethanamide.

Amines

The nomenclature for amines is similar to alkyl halides and ethers in that the main part (or root) of the name is an alkane and the amino group is considered to be a substituent (*Fig. 21*). Simple amines are sometimes named by placing the suffix -ylamine after the main part of the name (*Fig. 22*).

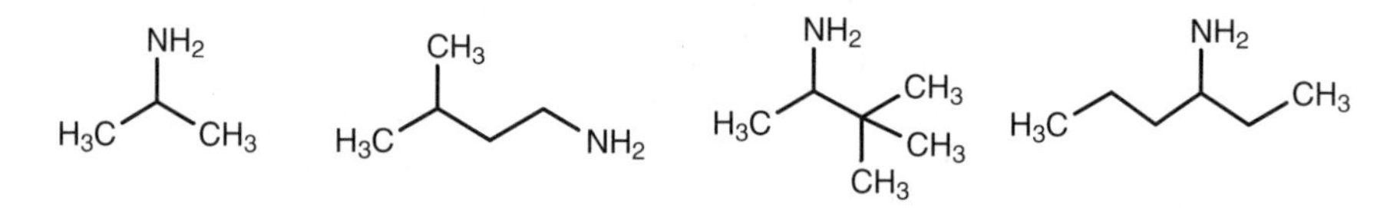

Fig. 21. (a) 2-Aminopropane; (b) 1-amino-3-methylbutane; (c) 2-amino-3,3-dimethylbutane; (d) 3-aminohexane.

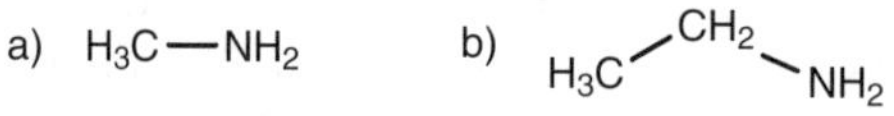

a) $H_3C—NH_2$ b) $H_3C—CH_2—NH_2$

Fig. 22. (a) Methylamine; (b) ethylamine.

Amines having more than one alkyl group attached are named by identifying the longest carbon chain attached to the nitrogen. In the example (*Fig. 23*), that is an ethane chain and so this molecule is an aminoethane (*N,N*-dimethylaminoethane).

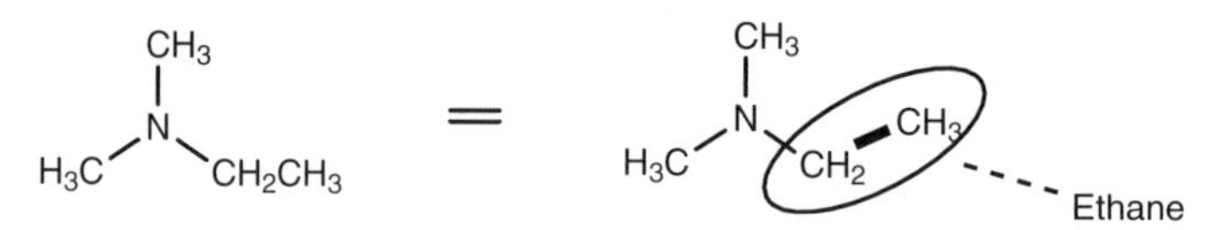

Fig. 23. N,N-Dimethylaminoethane.

Some simple secondary and tertiary amines have common names (*Fig. 24*).

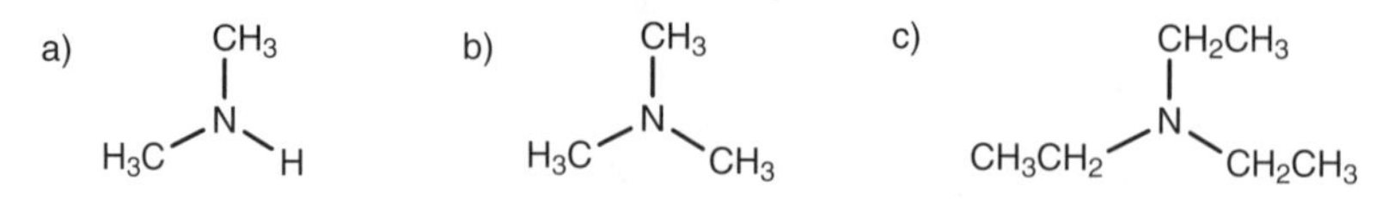

Fig. 24. (a) Dimethylamine; (b) trimethylamine; (c) triethylamine.

Thiols and thioethers

Thiols are named by adding the suffix **thiol** to the name of the parent alkane (*Fig. 25a*). Thioethers are named in the same way as ethers using the prefix **alkylthio,** for example, 1-(methylthio)propane (*Fig. 25c*). Simple thioethers can be named by identifying the thioether as a **sulfide** and prefixing this term with the alkyl substituents, for example, dimethyl sulfide (*Fig. 25b*).

a) $CH_3CH_2—SH$ b) $H_3C—S—CH_3$ c) $H_3C—S—CH_2CH_2CH_3$

Fig. 25. (a) Ethanethiol; (b) dimethylsulfide; (c) 1-(methylthio)propane.

C6 PRIMARY, SECONDARY, TERTIARY AND QUATERNARY NOMENCLATURE

Key Notes

Definition

Carbon centers, as well as some functional groups (alcohols, alkyl halides, amines and amides), can be defined as primary (1°), secondary (2°), tertiary (3°) or quaternary (4°).

Carbon centers

Carbon centers can be identified as primary, secondary, tertiary, or quaternary depending on the number of bonds leading to other carbon atoms. A methyl group contains a primary carbon center. A methylene group (CH_2) contains a secondary carbon center. The methine group (CH) contains a tertiary carbon center while a carbon atom having four substituents is a quaternary center.

Amines and amides

Amines and amides can be defined as being primary, secondary, tertiary, or quaternary depending on the number of bonds leading from nitrogen to carbon.

Alcohols and alkyl halides

Alcohols and alkyl halides are defined as primary, secondary, or tertiary depending on the carbon to which the alcohol or halide is attached. The assignment depends on the number of bonds from that carbon to other carbon atoms. It is not possible to get quaternary alcohols or quaternary alkyl halides.

Related topic

Recognition of functional groups (C1)

Definition

The primary (1°), secondary (2°), tertiary (3°) and quaternary (4°) nomenclature is used in a variety of situations: to define a carbon center, or to define functional groups such as alcohols, halides, amines and amides. Identifying functional groups in this way can be important since the properties and reactivities of these groups may vary depending on whether they are primary, secondary, tertiary, or quaternary.

Carbon centers

One of the easiest ways of determining whether a carbon center is 1°, 2°, 3° or 4° is to count the number of bonds leading from that carbon center to another carbon atom (*Fig. 1*). A **methyl** group (CH_3) is a primary carbon center, a **methylene** group (CH_2) is a secondary carbon center, a **methine** group (CH) is a tertiary carbon center, and a carbon center with four alkyl substituents (C) is a quaternary carbon center (*Fig. 2*).

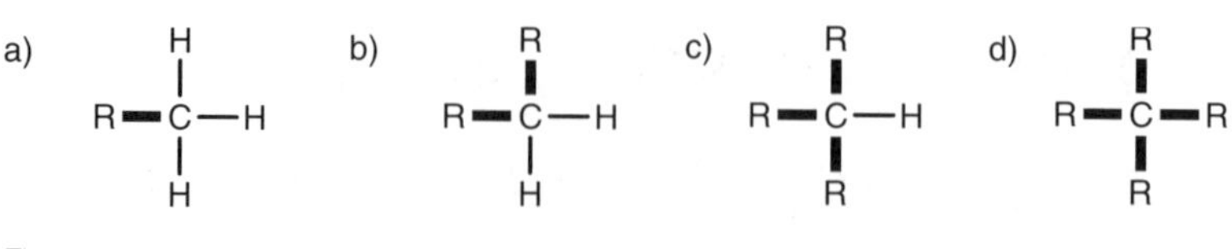

Fig. 1. Carbon centers; (a) primary; (b) secondary; (c) tertiary; (d) quaternary.

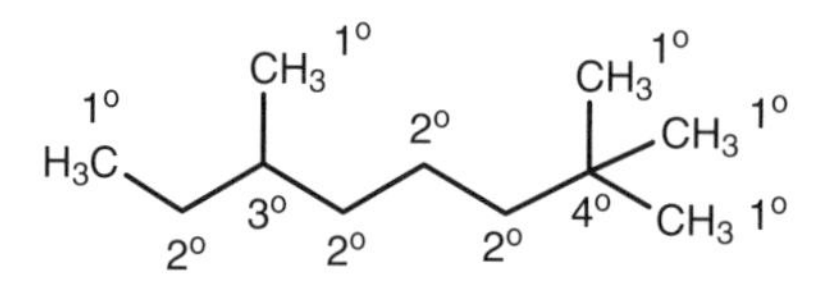

Fig. 2. Primary, secondary, tertiary, and quaternary carbon centers.

Amines and amides

Amines and amides can be defined as being primary, secondary, tertiary, or quaternary depending on the number of bonds from **nitrogen** to carbon (*Fig. 3*). Note that a quaternary amine is positively charged and is therefore called a quaternary ammonium ion. Note also that it is not possible to get a quaternary amide.

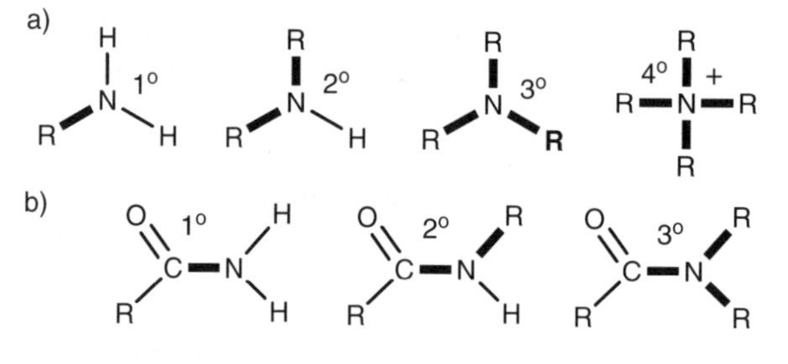

Fig. 3. (a) Amines; (b) amides.

Alcohols and alkyl halides

Alcohols and alkyl halides can also be defined as being primary, secondary, or tertiary (*Fig. 4*). However, the definition depends on the carbon to which the alcohol or halide is attached and it ignores the bond to the functional group. Thus, quaternary alcohols or alkyl halides are not possible.

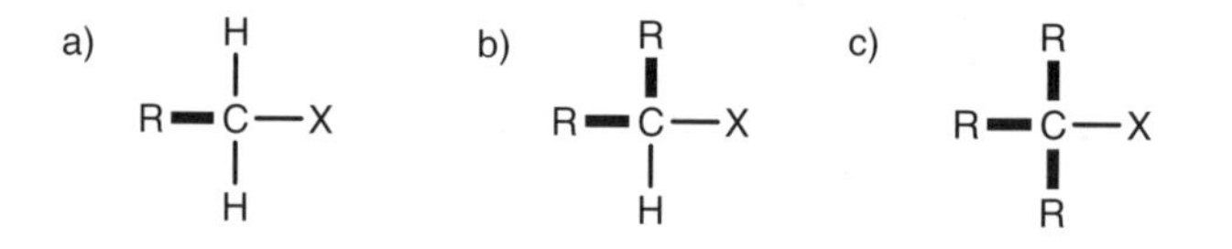

Fig. 4. Alcohols and alkyl halides; (a) primary; (b) secondary; (c) tertiary.

The following examples (*Fig. 5*) illustrate different types of alcohols and alkyl halides.

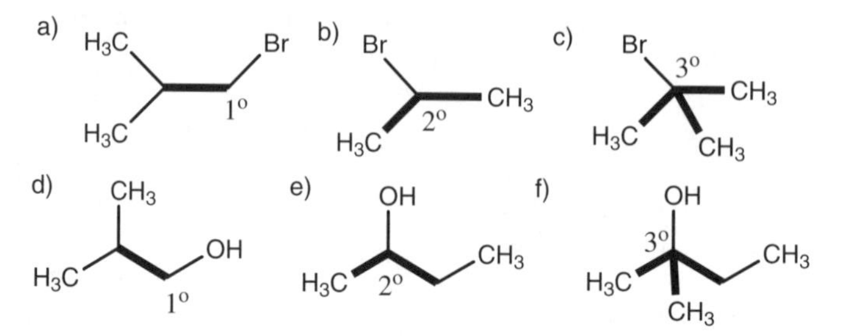

Fig. 5. (a) 1° alkyl bromide; (b) 2° alkyl bromide; (c) 3° alkyl bromide; (d) 1° alcohol; (e) 2° alcohol; (f) 3° alcohol.

D1 CONSTITUTIONAL ISOMERS

Key Notes

Introduction Isomers are compounds which have the same molecular formula, but differ in the way the atoms are arranged. There are three types of isomers – constitutional, configurational, and conformational.

Definition Constitutional isomers are compounds which have the same molecular formula but have the atoms joined together in a different way. Constitutional isomers have different physical and chemical properties.

Alkanes Alkanes of a particular molecular formula can have various constitutional isomers. The larger the alkane, the more isomers which are possible.

Related topics

Nomenclature (B3)
Nomenclature of functional groups (C5)
Configurational isomers – alkenes and cycloalkanes (D2)
Configurational isomers – optical isomers (D3)
Conformational isomers (D4)

Introduction

Isomers are compounds which have the same molecular formula (i.e. they have the same atoms), but differ in the way these atoms are arranged. There are three types of isomers – constitutional isomers, configurational isomers, and conformational isomers. Constitutional isomers are isomers where the atoms are linked together in a different skeletal framework and are different compounds. Configurational isomers are structures having the same atoms and bonds, but which have different geometrical shapes which cannot be interconverted without breaking covalent bonds. Configurational isomers can be separated and are different compounds with different properties. Conformational isomers are different shapes of the same molecule and cannot be separated.

Definition

Constitutional isomers are compounds which have the same molecular formula but have the atoms joined together in a different way. In other words, they have different carbon skeletons. Constitutional isomers have different physical and chemical properties.

Alkanes

Alkanes of a particular molecular formula can exist as different constitutional isomers. For example, the alkane having the molecular formula C_4H_{10} can exist as two constitutional isomers – the straight chain alkane (butane) or the branched alkane (2-methylpropane; *Fig. 1*). These are different compounds with different physical and chemical properties.

a) $H_3C-CH_2-CH_2-CH_3$

b) $(H_3C)_2CH-CH_3$

Fig. 1. (a) Butane (C_4H_{10}); (b) 2-methylpropane (C_4H_{10}).

D2 Configurational isomers – alkenes and cycloalkanes

Key Notes

Definition

Configurational isomers have the same molecular formula and the same bonds. However, some of the atoms are arranged differently in space with respect to each other, and the isomers cannot be interconverted without breaking a covalent bond. Substituted alkenes and cycloalkanes can exist as configurational isomers.

Alkenes – *cis* and *trans* isomerism

Alkenes having two different substituents at each end of the double bond can exist as two configurational isomers. Simple alkenes can be defined as *cis* or *trans* depending on whether substituents at different ends of the alkene are on the same side of the alkene (i.e. *cis*) or on opposite sides (i.e. *trans*).

Alkenes – (*Z*) and (*E*) nomenclature

Alkenes can be assigned as (*Z*) or (*E*) depending on the relative positions of priority groups. If the priority groups at each end of the alkene are on the same side of the double bond, the alkene is the (*Z*) isomer. If they are on opposite sides, the alkene is defined as the (*E*) isomer. Priority groups are determined by the atomic numbers of the atoms directly attached to the alkene. If there is no distinction between these atoms, the next atom of each substituent is compared.

Cycloalkanes

Substituted cycloalkanes can exist as configurational isomers where the substituents are *cis* or *trans* with respect to each other.

Related topics

Nomenclature (B3)
Nomenclature of functional groups (C5)
Properties of alkenes and alkynes (H2)

Definition

Configurational isomers are isomers which have the same molecular formula and the same molecular structure. In other words, they have the same atoms and the same bonds. However, the isomers are different because some of the atoms are arranged differently in space, and the isomers cannot be interconverted without breaking and remaking covalent bonds. As a result, configurational isomers are different compounds having different properties. Common examples of configurational isomers are substituted alkenes and substituted cycloalkanes where the substituents are arranged differently with respect to each other.

Alkenes – *cis* and *trans* isomerism

Alkenes having identical substituents at either end of the double bond can only exist as one molecule. However, alkenes having different substituents at each end of the double bond can exist as two possible isomers. For example, 1-butene (*Fig. 1a*) has two hydrogens at one end of the double bond and there is only one way of constructing it. On the other hand, 2-butene has different substituents at both ends of the double bond (H and CH_3) and can be constructed in two ways. The methyl groups can be on the same side of the double bond (the *cis* isomer; *Fig. 1b*), or on opposite sides (the *trans* isomer; *Fig. 1c*). The *cis* and *trans* isomers of an alkene are configurational isomers (also called **geometric** isomers) because they have different shapes and cannot interconvert since the double bond of an alkene cannot rotate. Therefore, the substituents are 'fixed' in space relative to each other. The structures are different compounds with different chemical and physical properties.

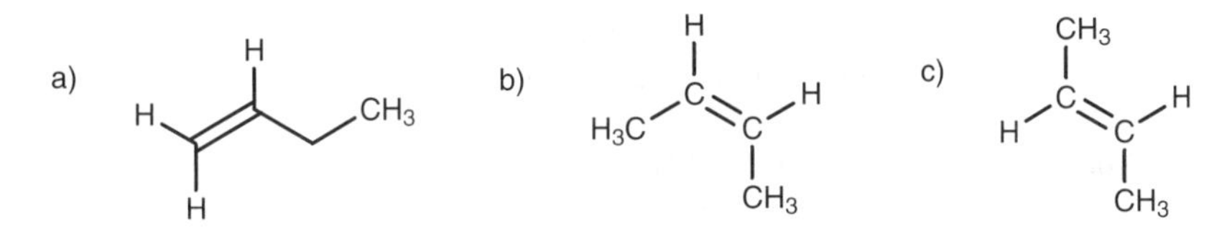

Fig. 1. (a) 1-Butene; (b) cis*-2-butene; (c)* trans*-2-butene.*

Alkenes – (*Z*) and (*E*) nomenclature

The *cis* and *trans* nomenclature for alkenes is an old method of classifying the configurational isomers of alkenes and is still commonly used. However, it is only suitable for simple 1,2-disubstituted alkenes where one can compare the relative position of the two substituents with respect to each other. When it comes to trisubstituted and tetrasubstituted alkenes, a different nomenclature is required.

The (Z)/(*E*) nomenclature allows a clear, unambiguous definition of the configuration of alkenes. The method by which alkenes are classified as (Z) or (*E*) is illustrated in *Fig. 2*. First of all, the atoms directly attached to the double bond are identified and given their atomic number (*Fig. 2b*). The next stage is to compare the two atoms at each end of the alkene. The one with the highest atomic number takes priority over the other (*Fig. 2c*). At the left hand side, oxygen has a higher atomic number than hydrogen and takes priority. At the right hand side, both atoms are the same (carbon) and we cannot choose between them.

Therefore, we now need to identify the atom of highest atomic number attached to each of these identical carbons. These correspond to a hydrogen for the methyl substituent and a carbon for the ethyl substituent. These are now different and so a priority can be made (*Fig. 3a*). Having identified which groups have priority, we can now see whether the priority groups are on the same side of the double bond or on opposite sides. If the two priority groups are on the same side of the double bond, the alkene is designated as (Z) (from the German word 'zusammen' meaning *together*). If the two priority groups are on opposite sides of the double bond,

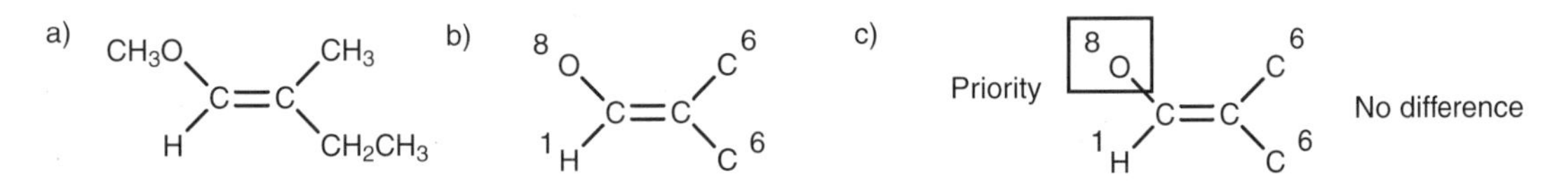

Fig. 2. (a) Alkene; (b) atomic numbers; (c) priority groups.

Fig. 3. (a) Choosing priority groups; (b) (E)-1-methoxy-2-methyl-1-butene.

Fig. 4. (a) cis-*1,2-Dimethylcyclopropane; (b)* trans-*1,2-dimethylcyclopropane.*

the alkene is designated as (*E*) (from the German word 'entgegen' meaning *across*). In this example, the alkene is (*E*) (*Fig. 3b*).

Cycloalkanes

Substituted cycloalkanes can also exist as configurational isomers. For example, there are two configurational isomers of 1,2-dimethylcyclopropane depending on whether the methyl groups are on the same side of the ring or on opposite sides (*Fig. 4*). The relative positions of the methyl groups can be defined by the bonds. A solid wedge indicates that the methyl group is coming out the page towards you, whereas a hatched wedge indicates that the methyl group is pointing into the page away from you. If the substituents are on the same side of the ring, the structure is defined as *cis*. If they are on opposite sides, the structure is defined as *trans*.

D3 CONFIGURATIONAL ISOMERS – OPTICAL ISOMERS

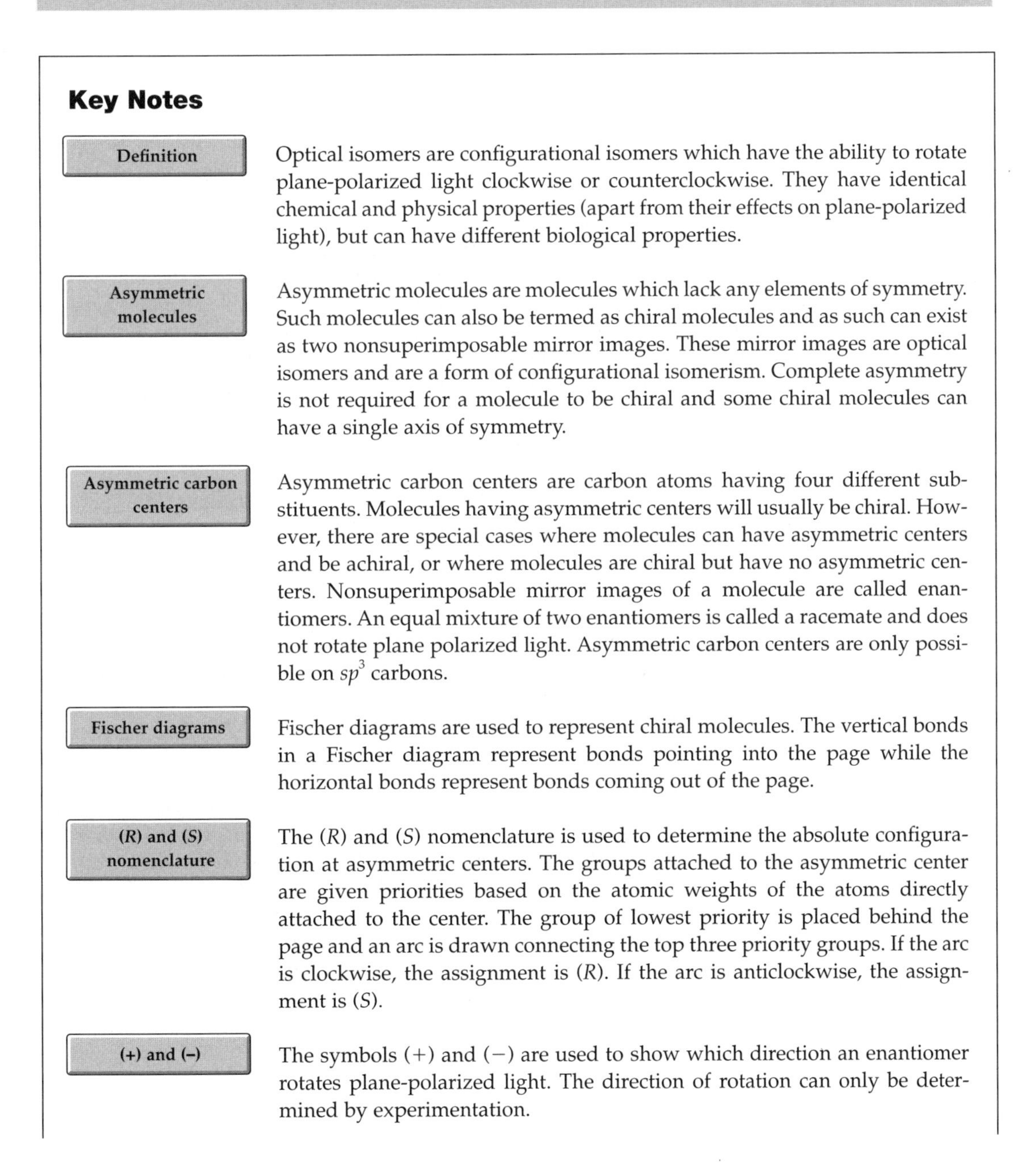

Key Notes

Definition

Optical isomers are configurational isomers which have the ability to rotate plane-polarized light clockwise or counterclockwise. They have identical chemical and physical properties (apart from their effects on plane-polarized light), but can have different biological properties.

Asymmetric molecules

Asymmetric molecules are molecules which lack any elements of symmetry. Such molecules can also be termed as chiral molecules and as such can exist as two nonsuperimposable mirror images. These mirror images are optical isomers and are a form of configurational isomerism. Complete asymmetry is not required for a molecule to be chiral and some chiral molecules can have a single axis of symmetry.

Asymmetric carbon centers

Asymmetric carbon centers are carbon atoms having four different substituents. Molecules having asymmetric centers will usually be chiral. However, there are special cases where molecules can have asymmetric centers and be achiral, or where molecules are chiral but have no asymmetric centers. Nonsuperimposable mirror images of a molecule are called enantiomers. An equal mixture of two enantiomers is called a racemate and does not rotate plane polarized light. Asymmetric carbon centers are only possible on sp^3 carbons.

Fischer diagrams

Fischer diagrams are used to represent chiral molecules. The vertical bonds in a Fischer diagram represent bonds pointing into the page while the horizontal bonds represent bonds coming out of the page.

(*R*) and (*S*) nomenclature

The (*R*) and (*S*) nomenclature is used to determine the absolute configuration at asymmetric centers. The groups attached to the asymmetric center are given priorities based on the atomic weights of the atoms directly attached to the center. The group of lowest priority is placed behind the page and an arc is drawn connecting the top three priority groups. If the arc is clockwise, the assignment is (*R*). If the arc is anticlockwise, the assignment is (*S*).

(+) and (–)

The symbols (+) and (–) are used to show which direction an enantiomer rotates plane-polarized light. The direction of rotation can only be determined by experimentation.

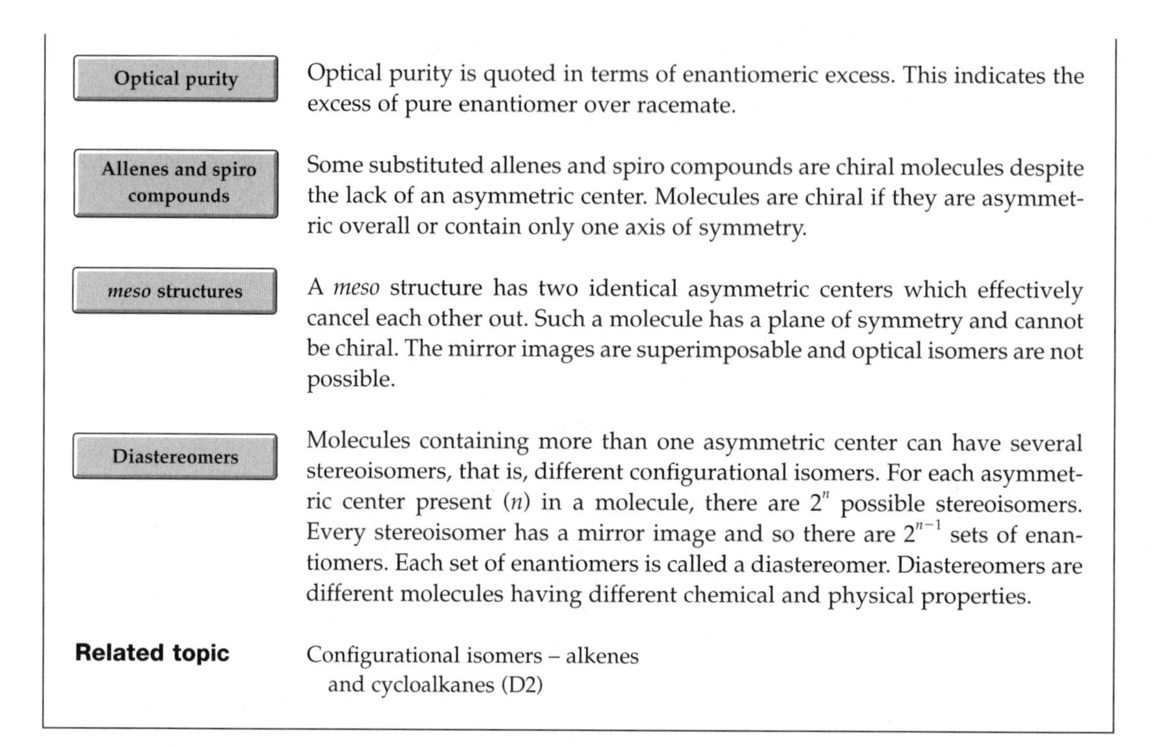

Optical purity

Optical purity is quoted in terms of enantiomeric excess. This indicates the excess of pure enantiomer over racemate.

Allenes and spiro compounds

Some substituted allenes and spiro compounds are chiral molecules despite the lack of an asymmetric center. Molecules are chiral if they are asymmetric overall or contain only one axis of symmetry.

***meso* structures**

A *meso* structure has two identical asymmetric centers which effectively cancel each other out. Such a molecule has a plane of symmetry and cannot be chiral. The mirror images are superimposable and optical isomers are not possible.

Diastereomers

Molecules containing more than one asymmetric center can have several stereoisomers, that is, different configurational isomers. For each asymmetric center present (n) in a molecule, there are 2^n possible stereoisomers. Every stereoisomer has a mirror image and so there are 2^{n-1} sets of enantiomers. Each set of enantiomers is called a diastereomer. Diastereomers are different molecules having different chemical and physical properties.

Related topic

Configurational isomers – alkenes and cycloalkanes (D2)

Definition

Optical isomerism is another example of configurational isomerism and is so named because of the ability of optical isomers to rotate plane-polarized light clockwise or counterclockwise. The existence of optical isomers has very important consequences for life, since optical isomers often have significant differences in their biological activity. Apart from their biological activity and their effects on plane-polarized light, optical isomers have identical chemical and physical properties.

Asymmetric molecules

A molecule such as chloroform ($CHCl_3$) is tetrahedral and there is only one way of fitting the atoms together. This is not the case for a molecule such as lactic acid. There are two ways of constructing a model of lactic acid, such that the two structures obtained are nonsuperimposable and cannot be interconverted without breaking covalent bonds. As such, they represent two different molecules which are configurational isomers. The difference between the two possible molecules lies in the way the substituents are attached to the central carbon. This can be represented by the following drawings (*Fig. 1*) where the bond to the hydroxyl group comes out of the page in one isomer but goes into the page in the other isomer.

The two isomers of lactic acid are mirror images (*Fig. 2*). A molecule which

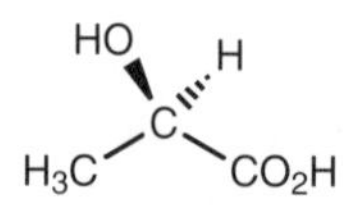

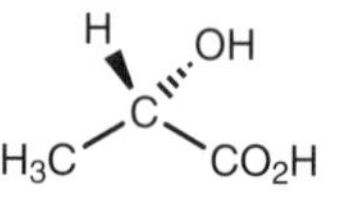

Fig. 1. Lactic acid.

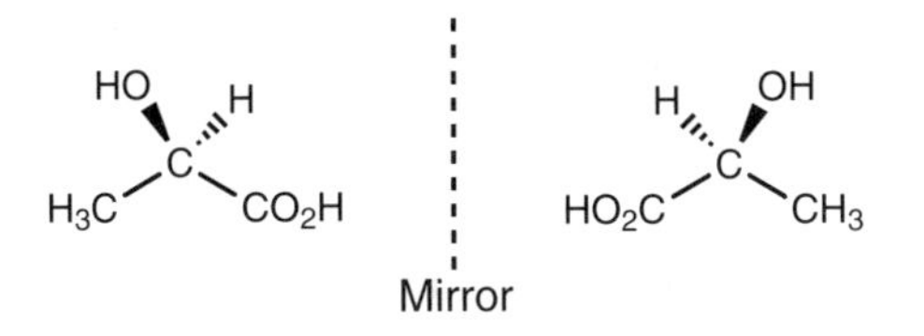

Fig. 2. Nonsuperimposable mirror images of lactic acid.

exists as two nonsuperimposable mirror images has optical activity if only one of the mirror images is present.

Lactic acid exists as two nonsuperimposable mirror images because it is **asymmetric** – in other words has no element of symmetry present. Asymmetric molecules can also be termed as **chiral**, and the ability of molecules to exist as two optical isomers is called **chirality**. In fact, a molecule does not have to be totally asymmetric to be chiral. Molecules containing a single axis of symmetry can also be chiral.

Asymmetric carbon centers

A simple method of identifying most chiral molecules involves identifying what are known as **asymmetric carbon centers**. This works for most chiral molecules, but it is important to realize that it is not foolproof and that there are several cases where it will not work. For example, some chiral molecules have no asymmetric carbon centers, and some molecules having more than one asymmetric carbon center are not chiral.

Usually, a compound will have optical isomers if there are four different substituents attached to a central carbon (*Fig. 3*). In such cases, the mirror images are nonsuperimposable and the structure will exist as two configurational isomers called **enantiomers**. The carbon center which contains these four different substituents is known as a **stereogenic** or an asymmetric center. A solution of each enantiomer or optical isomer is capable of rotating plane-polarized light. One enantiomer will rotate plane-polarized light clockwise while the other (the mirror image) will rotate it counterclockwise by the same amount. A mixture of the two isomers (a **racemate**) will not rotate plane-polarized light at all. In all other respects, the two isomers are identical in physical and chemical properties and are therefore indistinguishable. The asymmetric centers in the molecules shown (*Fig. 4*) have been identified with an asterisk. The structure lacking the asymmetric center is symmetric or **achiral** and does not have optical isomers. A structure can also have more than one asymmetric center.

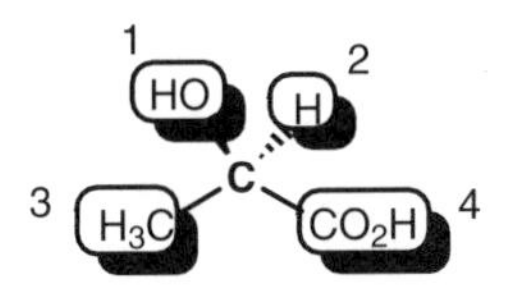

Fig. 3. Four different substituents of lactic acid.

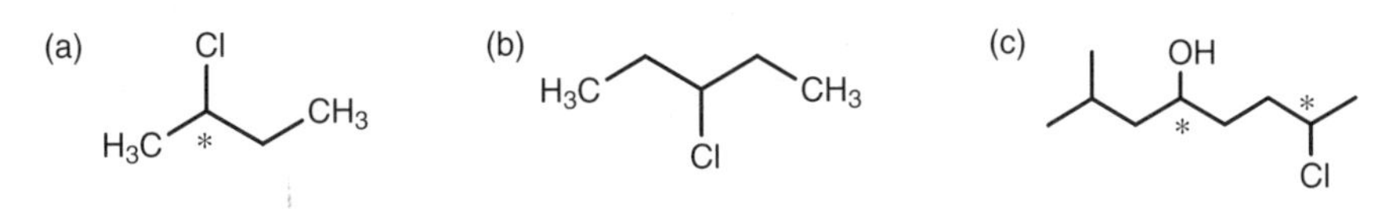

Fig. 4. Chiral and achiral structures: (a) chiral; (b) achiral; (c) chiral.

Asymmetric centers are only possible on sp^3 carbons. An sp^2 center is planar and cannot be asymmetric. Similarly, an *sp* center cannot be asymmetric.

Fischer diagrams

A chiral molecule can be represented by a Fischer diagram (*Fig. 5*). The molecule is drawn such that the carbon chain is vertical with the functional group positioned at the top. The vertical C–C bonds from the asymmetric center point into the page while the horizontal bonds from the asymmetric center come out of the page. This is usually drawn without specifying the wedged and hatched bonds.

The Fischer diagrams of alanine allow the structures to be defined as L- or D- from the position of the amino group. If the amino group is to the left, then it is the L- enantiomer. If it is positioned to the right, it is the D-enantiomer. This is an old fashioned nomenclature which is only used for amino acids and sugars. The L- and D- nomenclature depends on the absolute configuration at the asymmetric center and not the direction in which the enantiomer rotates plane-polarized light. It is not possible to predict which way a molecule will rotate plane-polarized light and this can only be found out by experimentation.

(*R*) and (*S*) nomenclature

The structure of an enantiomer can be specified by the (*R*) and (*S*) nomenclature, determined by the **Cahn–Ingold–Prelog** rules. The example (*Fig. 6*) shows how the nomenclature is worked out. First of all, the atoms directly attached to the asymmetric center and their atomic numbers are identified. Next, you give the attached atoms a priority based on their atomic numbers. In this example, there are two carbon atoms with the same atomic numbers and so they cannot be given a priority. When this happens, the next stage is to move to the next atom of highest atomic number (*Fig. 7a*). This means moving to an oxygen for one of the carbons and to a hydrogen for the other. The oxygen has the higher priority and so this substituent takes priority over the other.

Once the priorities have been settled, the structure is redrawn such that the group of lowest priority is positioned 'behind the page'. In this example (*Fig. 7b*),

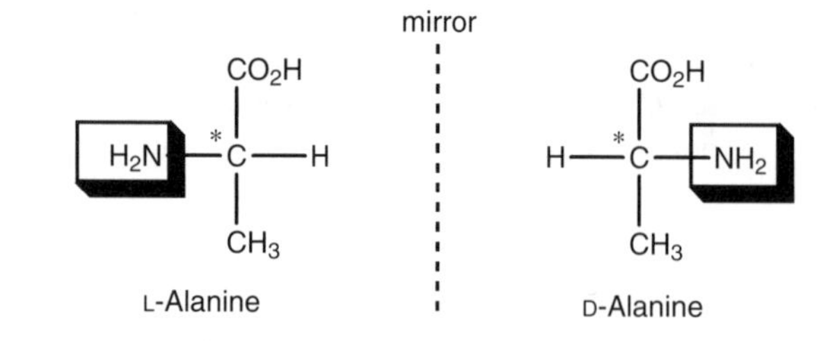

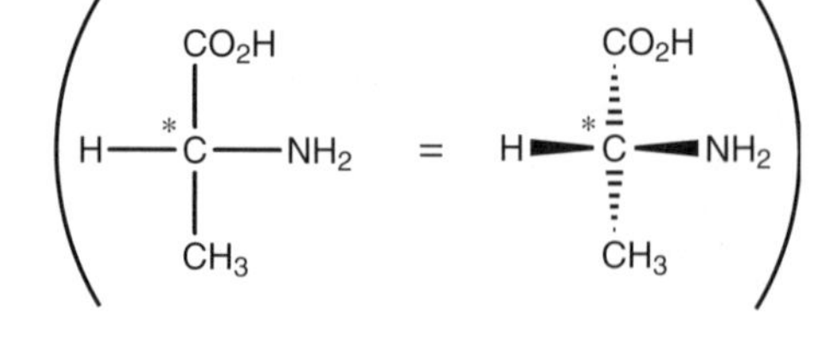

Fig. 5. Fischer diagrams of L-alanine and D-alanine.

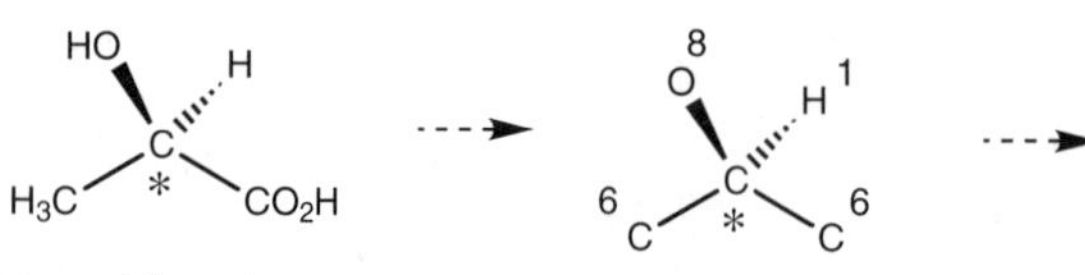

Fig. 6. Assigning priorities to substituents of an asymmetric center.

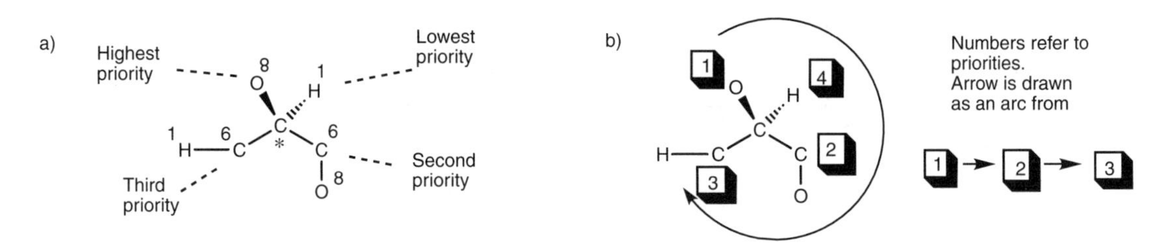

Fig. 7. (a) Assigning priorities to substituents; (b) assigning an asymmetric center as (R) *or* (S).

the group of lowest priority (the hydrogen) is already positioned behind the page (note the hatched bond indicating the bond going away from you). An arc is now drawn connecting the remaining groups, starting from the group of highest priority and finishing at the group of third priority. If the arc is drawn clockwise, the assignment is (*R*) (rectus). If the arc is drawn counterclockwise, the assignment is (*S*) (sinister). In this example the arrow is drawn clockwise. Therefore, the molecule is (*R*)-lactic acid.

A second example (*Fig. 8*) illustrates another rule involving substituents with double bonds. The asymmetric center is marked with an asterisk. The atoms directly attached to the asymmetric center are shown on the right with their atomic numbers. At this stage, it is possible to define the group of highest priority (the oxygen) and the group of lowest priority (the hydrogen). There are two identical carbons attached to the asymmetric center so we have to move to the next stage and identify the atom with the highest atomic number joined to each of the identical carbons (*Fig. 9*). This still does not distinguish between the CHO and CH_2OH groups since both carbon atoms have an oxygen atom attached. The next stage is to look at the second most important atom attached to the two carbon atoms. However, if there is a double bond present, you are allowed to 'visit' the same atom twice. The next most important atom in the CH_2OH group is the hydrogen. In the CHO group, the oxygen can be 'revisited' since there is a double bond. Therefore, this group takes priority over the CH_2OH group. The priorities

Fig. 8. Assigning priority to substituents of an asymmetric center.

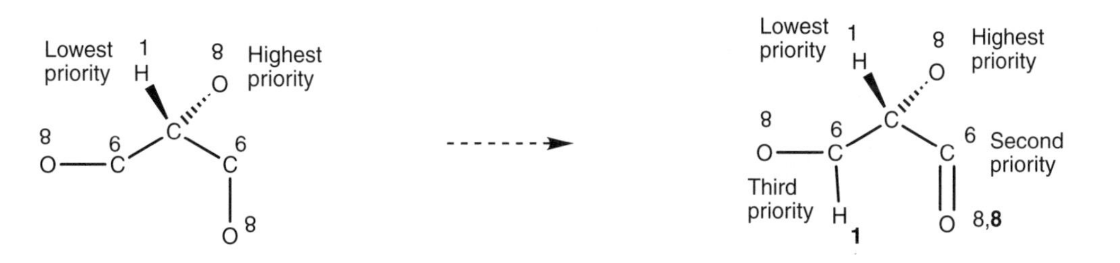

Fig. 9. Assigning priority to substituents of an asymmetric center.

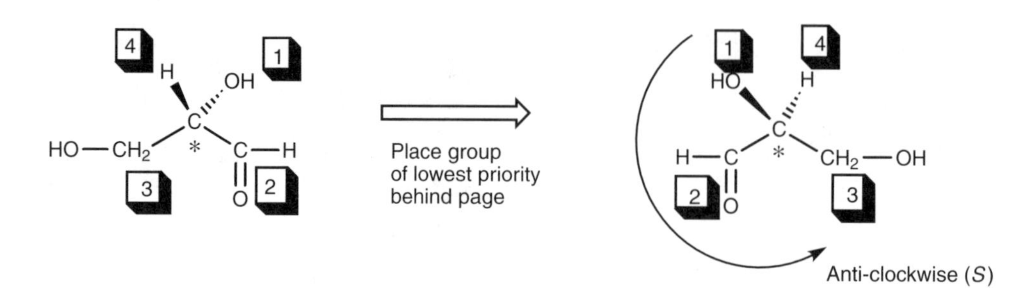

Fig. 10. Assigning an asymmetric center as (R) or (S).

have been determined, and so the group of lowest priority is placed behind the page and the three most important groups are connected to see if they are clockwise or counterclockwise (*Fig. 10*).

(+) and (–)

The assignment of an asymmetric center as (*R*) or (*S*) has nothing to do with whichever direction the molecule rotates plane-polarized light. Optical rotation can only be determined experimentally. By convention, molecules which rotate plane-polarized light clockwise are defined as (+) or *d*. Molecules which rotate plane-polarized light counterclockwise are defined as (−) or *l*. The (*R*) enantiomer of lactic acid is found to rotate plane-polarized light counterclockwise and so this molecule is defined as (*R*)-(−)-lactic acid.

Optical purity

The optical purity of a compound is a measure of its enantiomeric purity and is given in terms of its **enantiomeric excess** (ee). A pure enantiomer would have an optical purity and enantiomeric excess of 100%. A fully racemized compound would have an optical purity of 0%. If the enantiomeric excess is 90%, it signifies that 90% of the sample is pure enantiomer and the remaining 10% is a racemate containing equal amounts of each enantiomer (i.e. 5% + 5%). Therefore the ratio of enantiomers in a sample having 90% optical purity is 95:5.

Allenes and spiro compounds

Not all chiral molecules have asymmetric centers. For example, some substituted allenes and spiro structures have no asymmetric center but are still chiral (*Fig. 11*). The substituents at either end of the allene are in different planes, and the rings in the spiro structure are at right angles to each other. The mirror images of the allene and the spiro structures are nonsuperimposable and are enantiomers.

A better rule for determining whether a molecule is chiral or not is to study the symmetry of the molecule. A molecule will be chiral if it is asymmetric (i.e. has no elements of symmetry) **or** if it has no more than one axis of symmetry.

meso structures

The molecule in *Fig. 12a* has two identical asymmetric centers but is not chiral. The mirror images of this structure are superimposable and so the compound cannot be chiral. This is because the molecule contains a plane of symmetry as

a)

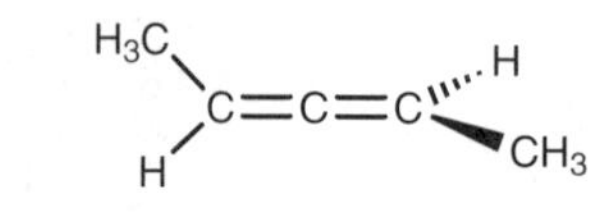

b) H3C CH3

Fig. 11. (a) Allene; (b) spiro structure.

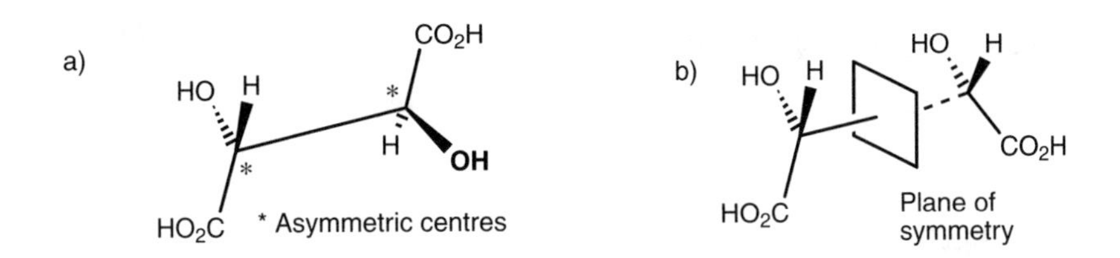

Fig. 12. (a) Meso structure showing asymmetric centers; (b) plane of symmetry in meso structure.

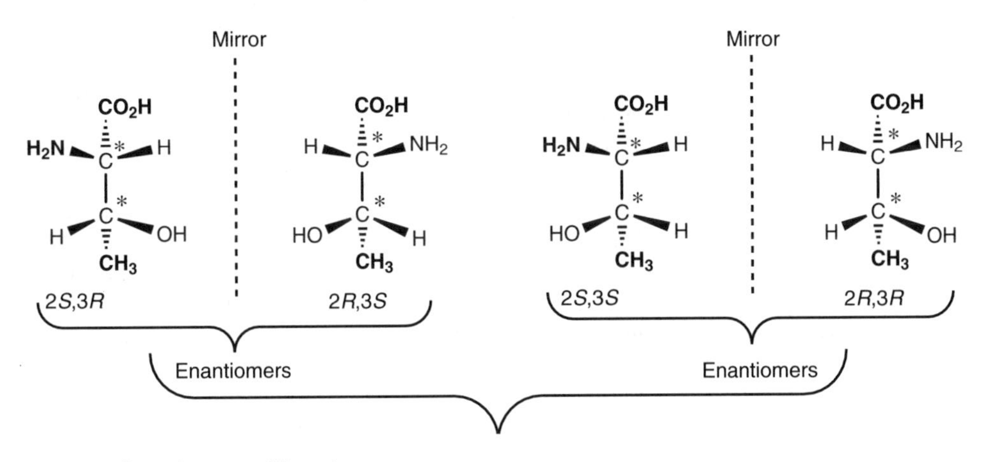

Fig. 13. Stereoisomers of threonine.

demonstrated in *Fig. 12b* where the molecule has been rotated around the central C–C bond. A structure such as this is called a ***meso*** structure.

Diastereomers

Whenever a molecule has two or more asymmetric centers, there are several possible structures which are possible and we need to use terms such as **stereoisomers, diastereomers** and enantiomers in order to discuss them. To illustrate the relative meaning of these terms, we shall look at the possible structures of the amino acid threonine (*Fig. 13*). This molecule has two asymmetric centers. As a result, four different structures are possible arising from the two different possible configurations at each center. These are demonstrated with the asymmetric centers at positions 2 and 3 defined as (*R*) or (*S*). The four different structures are referred to as stereoisomers. The (2*S*,3*R*) stereoisomer is a nonsuperimposable mirror image of the (2*R*,3*S*) stereoisomer and so these structures are enantiomers having the same chemical and physical properties. The (2*S*,3*S*) stereoisomer is the nonsuperimposable mirror image of the (2*R*,3*R*) stereoisomer and so these structures are also enantiomers having the same chemical and physical properties. Each set of enantiomers is called a diastereomer. Diastereomers are not mirror images of each other and are completely different compounds having different physical and chemical properties. To conclude, threonine has two asymmetric centers which means that there are two possible diastereomers consisting of two enantiomers each, making a total of four stereoisomers. As the number of asymmetric centers increases, the number of possible stereoisomers and diastereomers increases. For a molecule having n asymmetric centers, the number of possible stereoisomers is 2^n and the number of diastereomers is 2^{n-1}.

D4 Conformational isomers

Key Notes

Definition

Conformational isomers are different shapes of the same molecule resulting from rotation round C–C single bonds. Conformational isomers are not different compounds and are freely interconvertable.

Alkanes

Alkanes can take up different shapes or conformations due to rotation around the C–C bonds. The most stable conformations are those where the bonds are staggered, rather than eclipsed. The torsional angle in butane is the angle between the first and third C–C bonds when viewed along the middle C–C bond. The most stable conformation of butane has a torsional angle of 180° where the carbon atoms and the C–C bonds are as far apart from each other as possible. The other possible staggered conformation has a torsional angle of 60° which results in some steric and electronic strain – called a gauche interaction. The most stable conformation for a straight chain alkane is zigzag shaped where all the torsional angles are at 180°.

Cycloalkanes

Cycloalkanes can adopt different conformations or shapes. The most stable conformation for cyclohexane is the chair. Each carbon in the chair has two C–H bonds, one of which is equatorial and one of which is axial. A chair structure can invert through a high energy boat intermediate such that the equatorial bonds become axial and the axial bonds become equatorial. If a substituent is present, the most stable chair conformation is where the substituent is equatorial. In the axial position, the substituent experiences two gauche interactions with C–C bonds in the ring.

Definition

Conformational isomers are essentially different shapes of the same molecule resulting from rotation round C–C single bonds. Since rotation round a single bond normally occurs easily at room temperature, conformational isomers are not different compounds and are freely interconvertable. Unlike constitutional and configurational isomers, conformational isomers cannot be separated.

Alkanes

Conformational isomers arise from the rotation of C–C single bonds. There are many different shapes which a molecule like ethane could adopt by rotation around the C–C bond. However, it is useful to concentrate on the most distinctive ones (*Fig. 1*). The two conformations I and II are called 'staggered' and 'eclipsed' respectively. In conformation I, the C–H bonds on carbon 1 are staggered with respect to the C–H bonds on carbon 2. In conformation II, they are eclipsed. **Newman projections** (*Fig. 2*) represent the view along the C1–C2 bond and emphasize the difference. Carbon 1 is represented by the small black circle and carbon 2 is represented by the larger

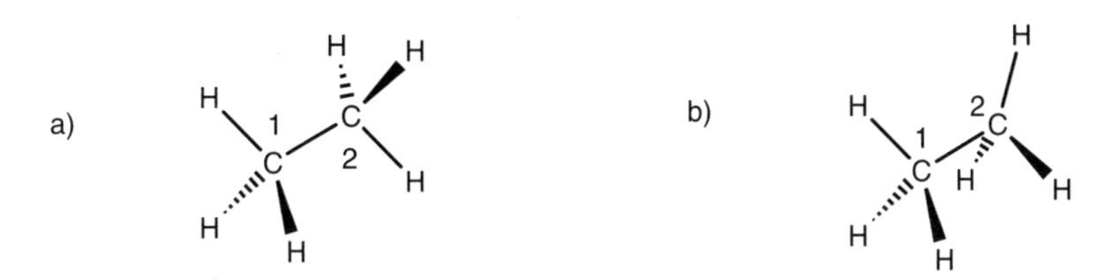

Fig. 1. (a) 'Staggered' conformation of ethane ; (b) 'eclipsed' conformation of ethane.

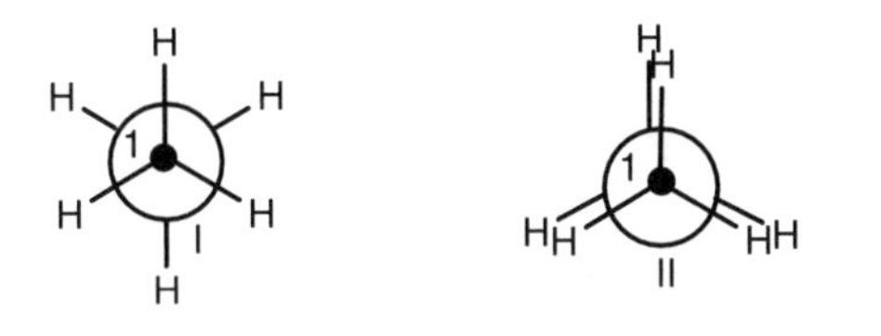

Fig. 2. Newman projections of the staggered (I) and eclipsed (II) conformations of ethane.

sphere. Viewed in this way, it can be seen that the C–H bonds on carbon 1 are eclipsed with the C–H bonds on carbon 2 in conformation II.

Of these two conformations, the staggered conformation is the more stable since the C–H bonds and hydrogen atoms are as far apart from each other as possible. In the eclipsed conformation, both the bonds and the atoms are closer together and this can cause strain due to electron repulsion between the eclipsed bonds and between the eclipsed atoms. Therefore, the vast majority of ethane molecules are in the staggered conformation at any one time. However, it is important to realize that the energy difference between the staggered and eclipsed conformations is still small enough to allow each ethane molecule to pass through an eclipsed conformation (*Fig. 3*) – otherwise C–C bond rotation would not occur.

Ethane has only one type of staggered conformation, but different staggered conformations are possible with larger molecules such as butane (*Fig. 4*). The first

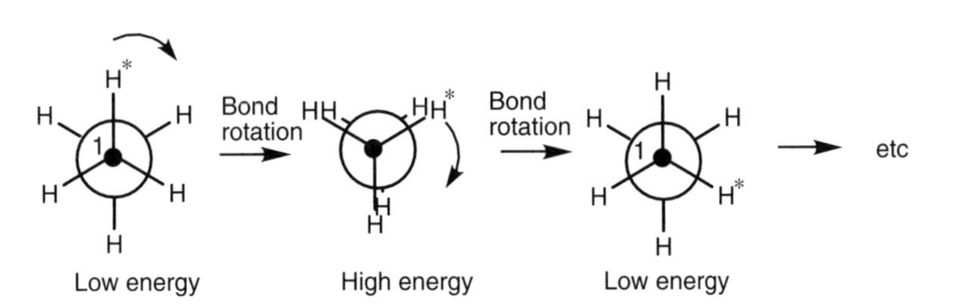

Fig. 3. Bond rotation of ethane.

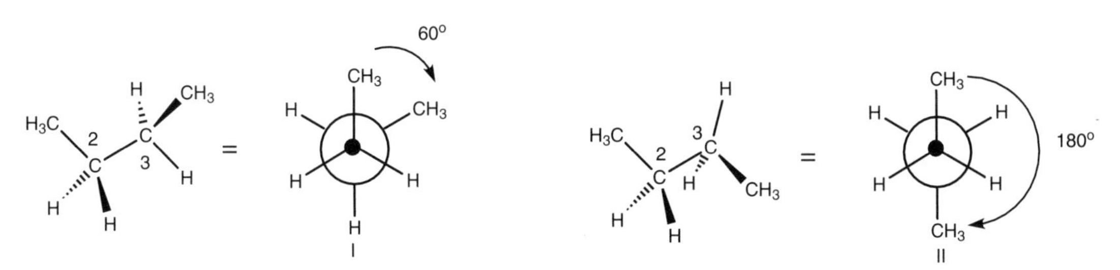

Fig. 4. Gauche conformation (I) and anti conformation (II) of butane.

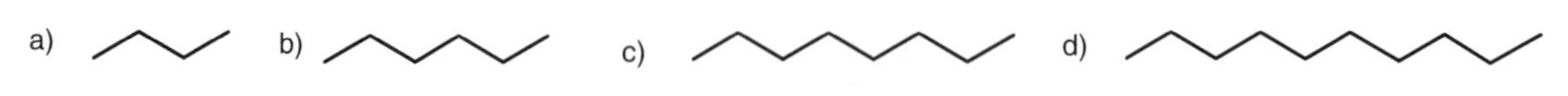

Fig. 5. Zigzag conformations of (a) butane; (b) hexane; (c) octane; (d) decane.

and the third C–C bonds in isomer I are at an angle of 60° with respect to each other when viewed along the middle C–C bond. In isomer II, these bonds are at an angle of 180°. This angle is known as the **torsional angle** or **dihedral angle**. Isomer II is more stable than isomer I. This is because the methyl groups and the C–C bonds in this conformer are as far apart from each other as possible. The methyl groups are bulky and in conformation I they are close enough to interact with each other and lead to some strain. There is also an interaction between the C–C bonds in isomer I since a torsional angle of 60° is small enough for some electronic repulsion to exist between the C–C bonds. When C–C bonds have a torsional angle of 60°, the steric and electronic repulsions which arise are referred to as a **gauche interaction**.

As a result, the most stable conformation for butane is where the C–C bonds are at torsional angles of 180° which results in a 'zigzag' shape. In this conformation, the carbon atoms and C–C bonds are as far apart from each other as possible. The most stable conformations for longer chain hydrocarbons will also be zigzag (*Fig. 5*). However, since bond rotation is occurring all the time for all the C–C bonds, it is unlikely that many molecules will be in a perfect zigzag shape at any one time.

Cycloalkanes

Cyclopropane (*Fig. 6*) is a flat molecule as far as the carbon atoms are concerned, with the hydrogen atoms situated above and below the plane of the ring. There are no conformational isomers. Cyclobutane (*Fig. 7*) on the other hand can form three distinct shapes – a planar shape and two 'butterfly' shapes. Cyclopentane (*Fig. 8*) can also form a variety of shapes or conformations. The planar structures for cyclobutane and cyclopentane are too strained to exist in practice due to eclipsed C–H bonds.

The two main conformational shapes for cyclohexane are known as the chair and the boat (*Fig. 9*). The chair is more stable than the boat since the latter has eclipsed C–C and C–H bonds. This can be seen better in the Newman projections (*Fig. 10*) which have been drawn such that we are looking along two bonds at the same time – bonds 2–3 and 6–5. In the chair conformation, there are no eclipsed

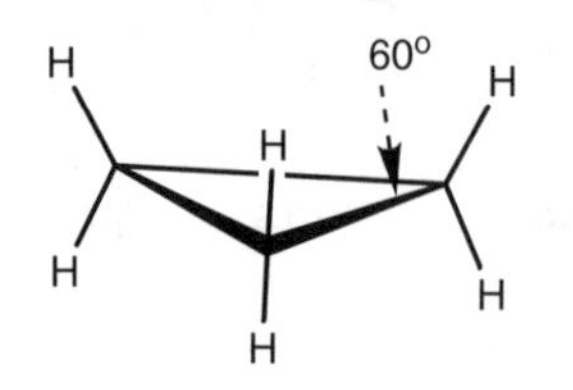

Fig. 6. Cyclopropane.

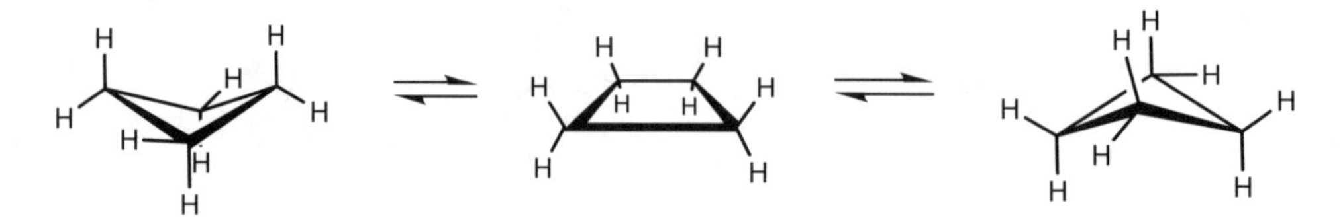

Fig. 7. Cyclobutane.

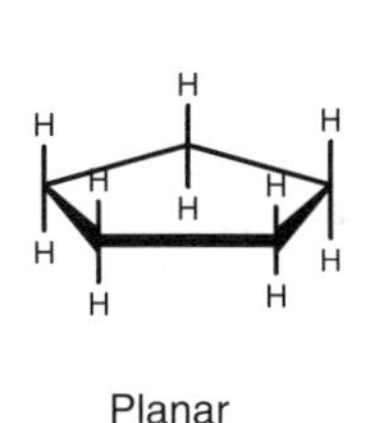

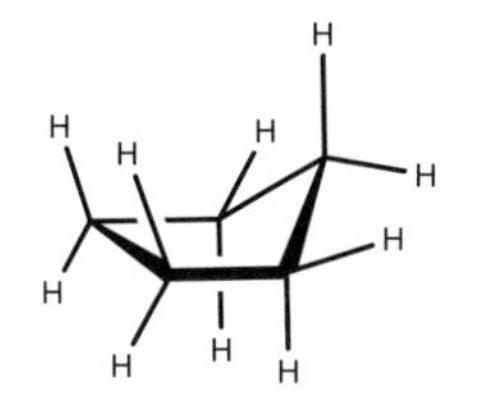

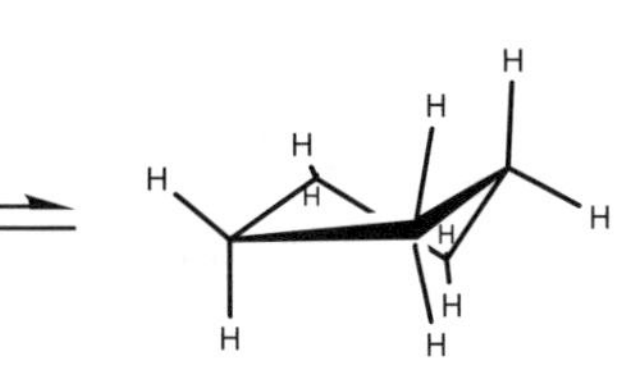

Fig. 8. Cyclopentane.

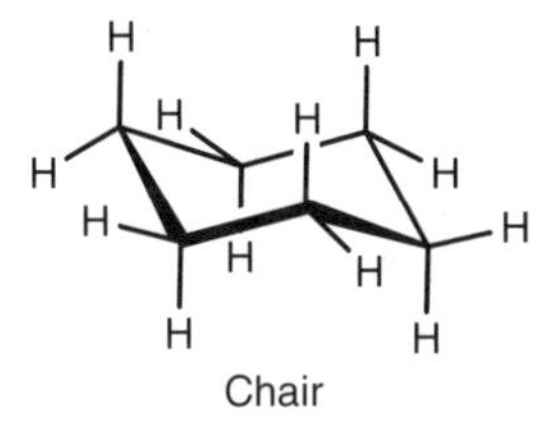

Boat

Fig. 9. Cyclohexane.

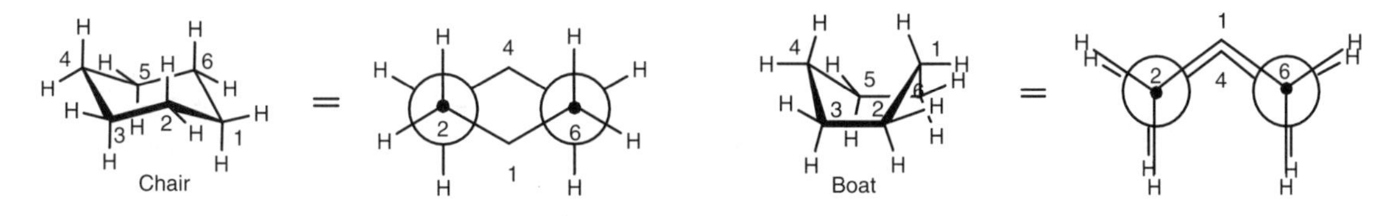

Fig. 10. Newman projections of the chair and boat conformations of cyclohexane.

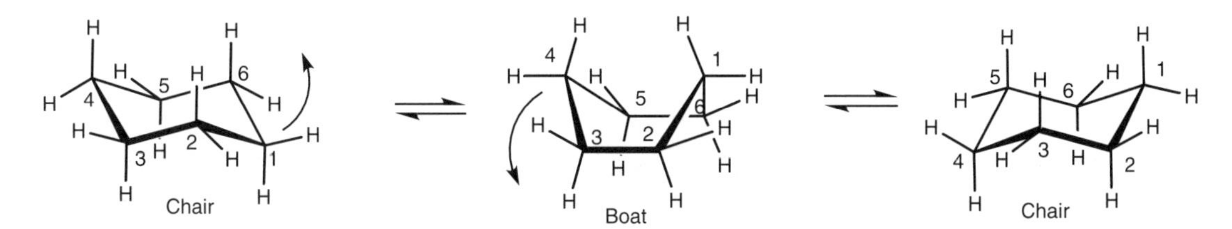

Fig. 11. Ring flipping of cyclohexane.

C–C bonds. However, in the boat conformation, bond 1–2 is eclipsed with bond 3–4, and bond 1–6 is eclipsed with bond 5–4. This means that the boat conformation is less stable than the chair conformation and the vast majority of cyclohexane molecules exist in the chair conformation. However, the energy barrier is small enough for the cyclohexane molecules to pass through the boat conformation in a process called 'ring flipping' (*Fig. 11*). The ability of a cyclohexane molecule to ring-flip is important when substituents are present. Each carbon atom in the chair structure has two C–H bonds, but these are not identical (*Fig. 12*). One of these bonds is termed **equatorial** since it is roughly in the plane of the ring. The other C–H bond is vertical to the plane of the ring and is called the **axial** bond.

When ring flipping takes place from one chair to another, all the axial bonds become equatorial bonds and all the equatorial bonds become axial bonds. This does not matter for cyclohexane itself, but it becomes important when there is a

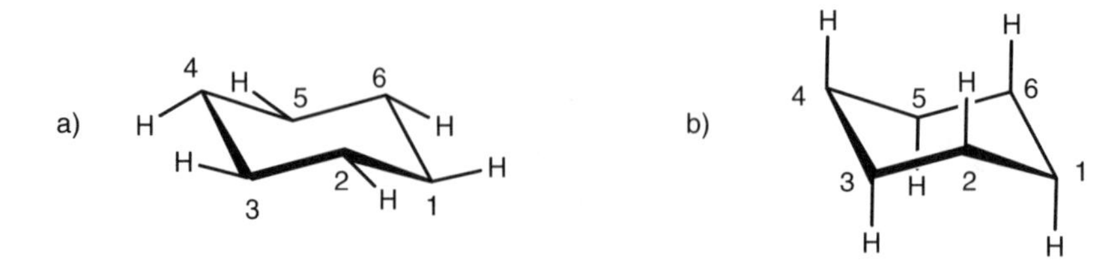

Fig. 12. (a) Equatorial C–H bonds; (b) axial C–H bonds.

substituent present in the ring. For example, methylcyclohexane can have two chair structures where the methyl group is either on an equatorial bond or on an axial bond (*Fig. 13*).

These are different shapes of the same molecule which are interconvertable due to rotation of C–C single bonds (the ring flipping process). The two chair structures are conformational isomers but they are not of equal stability. The more stable conformation is the one where the methyl group is in the equatorial position. In this position, the C–C bond connecting the methyl group to the ring has a torsional angle of 180° with respect to bonds 5–6 and 3–2 in the ring. In the axial position, however, the C–C bond has a torsional angle of 60° with respect to these same

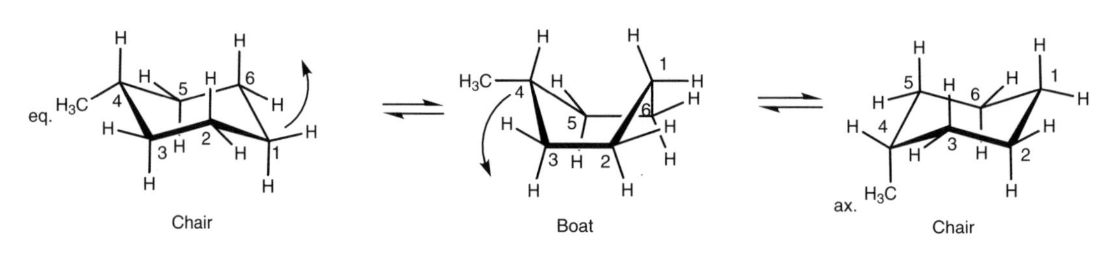

Fig. 13. Ring flipping of methylcyclohexane.

Equatorial methyl

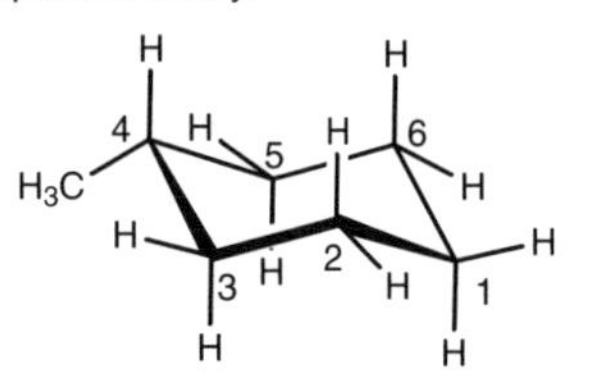

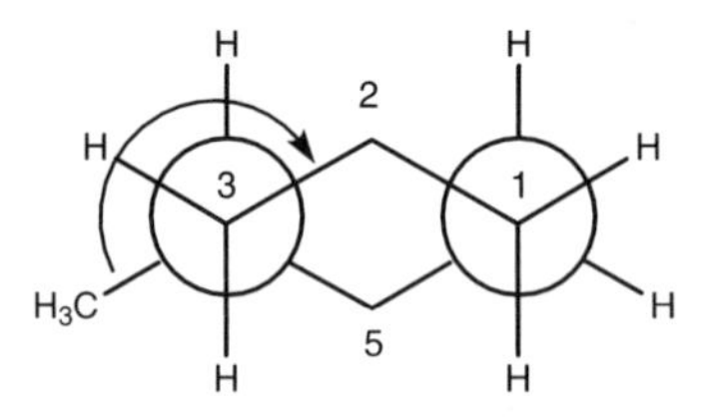

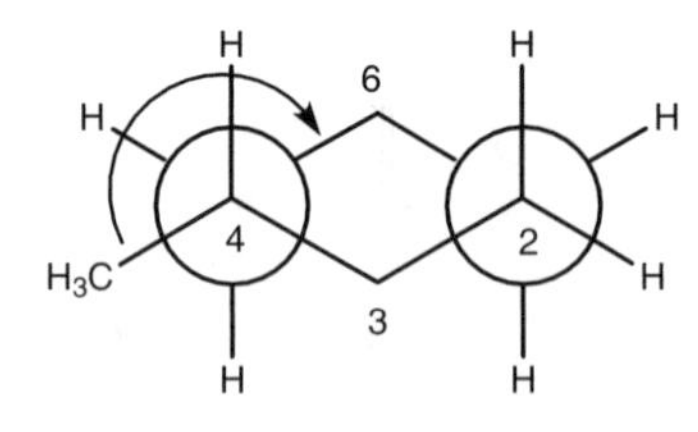

Torsion angle = 180°

Torsion angle = 180°

Axial methyl

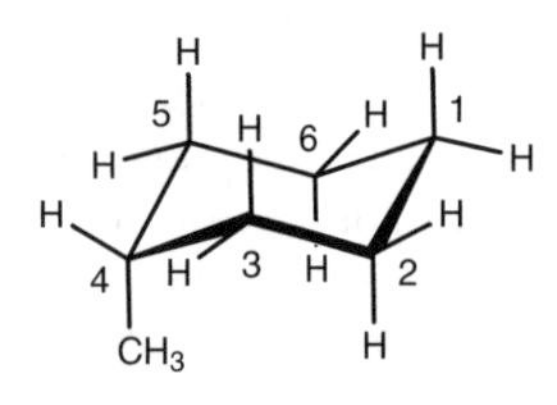

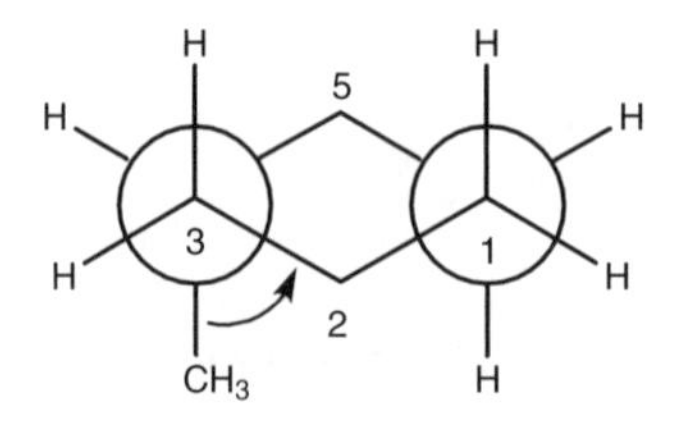

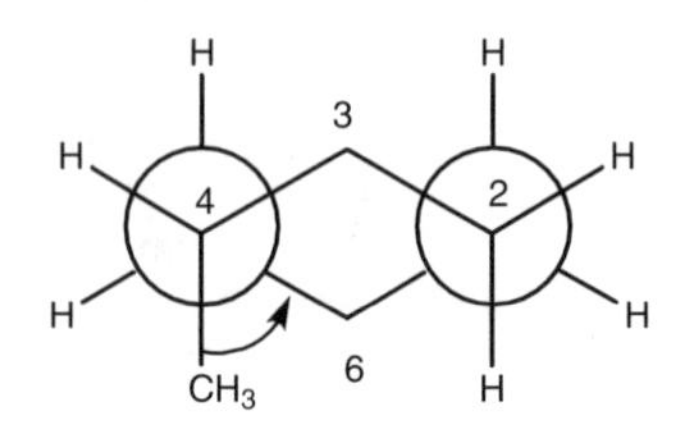

Torsion angle = 60°

Torsion angle = 60°

Fig. 14. Newman projections of the chair conformations of methylcyclohexane.

two bonds. This can be illustrated by comparing Newman diagrams of the two methylcyclohexane conformations (*Fig. 14*).

A torsion angle of 60° between C–C bonds represents a gauche interaction and so an axial methyl substituent experiences two gauche interactions with the cyclohexane ring whereas the equatorial methyl substituent experiences none. As a result, the latter chair conformation is preferred and about 95% of methylcyclohexane molecules are in this conformation at any one time, compared to 5% in the other conformation.

E1 DEFINITION

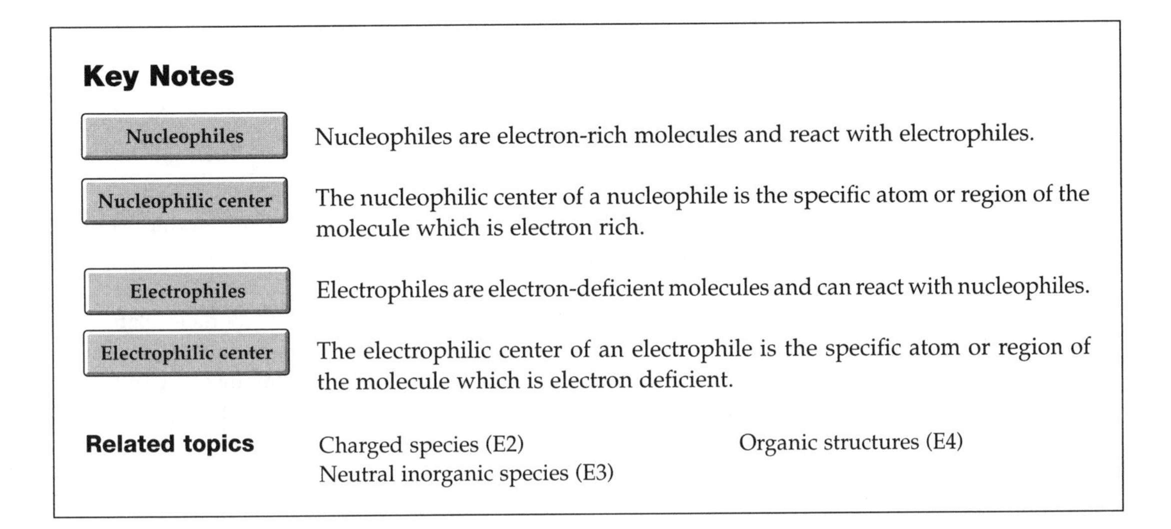

Key Notes

Nucleophiles	Nucleophiles are electron-rich molecules and react with electrophiles.
Nucleophilic center	The nucleophilic center of a nucleophile is the specific atom or region of the molecule which is electron rich.
Electrophiles	Electrophiles are electron-deficient molecules and can react with nucleophiles.
Electrophilic center	The electrophilic center of an electrophile is the specific atom or region of the molecule which is electron deficient.

Related topics Charged species (E2) Organic structures (E4)
Neutral inorganic species (E3)

Nucleophiles

Most organic reactions involve the reaction between a molecule which is rich in electrons and a molecule which is deficient in electrons. The reaction involves the formation of a new bond where the electrons are provided by the electron-rich molecule. Electron-rich molecules are called **nucleophiles** (meaning nucleus-loving). The easiest nucleophiles to identify are negatively charged ions with lone pairs of electrons (e.g. the hydroxide ion), but neutral molecules can also act as nucleophiles if they contain electron-rich functional groups (e.g. an amine).

Nucleophilic center

Nucleophiles have a specific atom or region of the molecule which is electron rich. This is called the **nucleophilic center**. The nucleophilic center of an ion is the atom bearing a lone pair of electrons and the negative charge. The nucleophilic center of a neutral molecule is usually an atom with a lone pair of electrons (e.g. nitrogen or oxygen), or a multiple bond (e.g. alkene, alkyne, aromatic ring).

Electrophiles

Electron-deficient molecules are called **electrophiles** (electron-loving) and react with nucleophiles. Positively charged ions can easily be identified as electrophiles (e.g. a carbocation), but neutral molecules can also act as electrophiles if they contain certain types of functional groups (e.g. carbonyl groups or alkyl halides).

Electrophilic center

Electrophiles have a specific atom or region of the molecule which is electron deficient. This region is called the **electrophilic center**. In a positively charged ion, the electrophilic center is the atom bearing the positive charge (e.g. the carbon atom of a carbocation). In a neutral molecule, the electrophilic center is an electron-deficient atom within a functional group (e.g. a carbon or hydrogen atom linked to an electronegative atom such as oxygen or nitrogen).

E2 Charged species

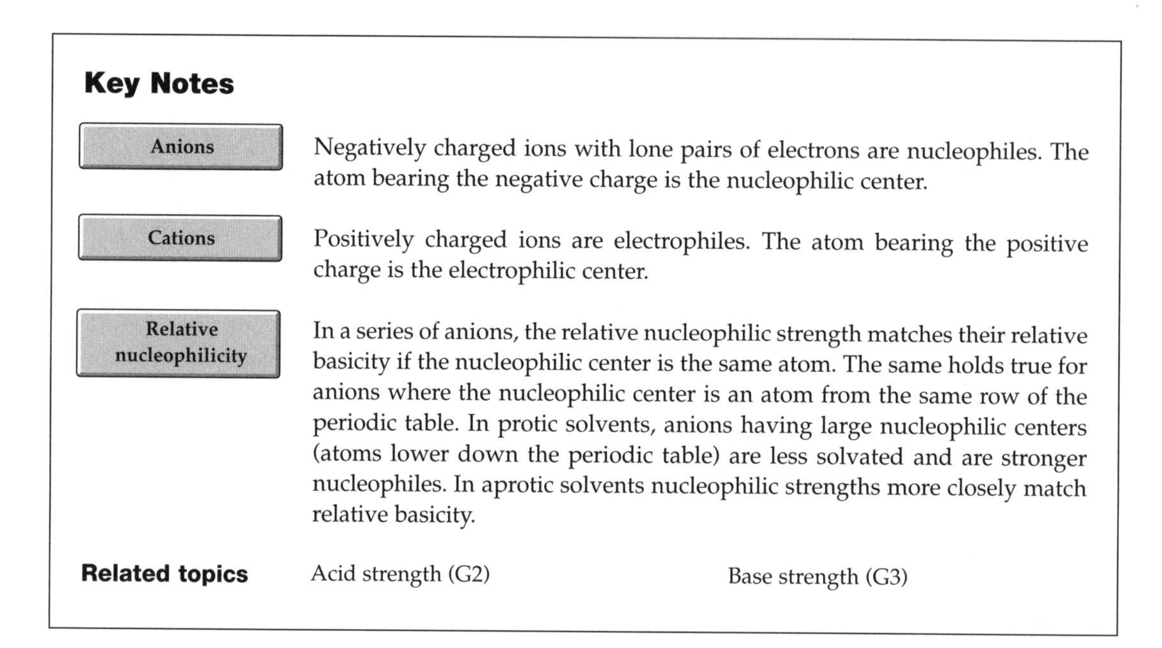

Key Notes

Anions	Negatively charged ions with lone pairs of electrons are nucleophiles. The atom bearing the negative charge is the nucleophilic center.
Cations	Positively charged ions are electrophiles. The atom bearing the positive charge is the electrophilic center.
Relative nucleophilicity	In a series of anions, the relative nucleophilic strength matches their relative basicity if the nucleophilic center is the same atom. The same holds true for anions where the nucleophilic center is an atom from the same row of the periodic table. In protic solvents, anions having large nucleophilic centers (atoms lower down the periodic table) are less solvated and are stronger nucleophiles. In aprotic solvents nucleophilic strengths more closely match relative basicity.

Related topics Acid strength (G2) Base strength (G3)

Anions

A negatively charged molecule such as the hydroxide ion (*Fig. 1*) is electron rich and acts as a nucleophile. The atom which bears the negative charge and a lone pair of electrons is the nucleophilic center, which in the case of the hydroxide ion is the oxygen atom. Some ions (e.g. the carboxylate ion) are able to share the negative charge between two or more atoms through a process known as **delocalization**. In this case, the negative charge is shared between both oxygen atoms and so both of these atoms are nucleophilic centers (*Fig. 1*).

Cations

A positively charged ion is electron deficient and acts as an electrophile. The atom which bears the positive charge is the electrophilic center. In the case of a carbocation (*Fig. 2*), this is the carbon atom. Some molecules (e.g. the allylic cation) are able to delocalize their positive charge between two or more atoms in which case all the atoms capable of sharing the charge are electrophilic centers (*Fig. 2*).

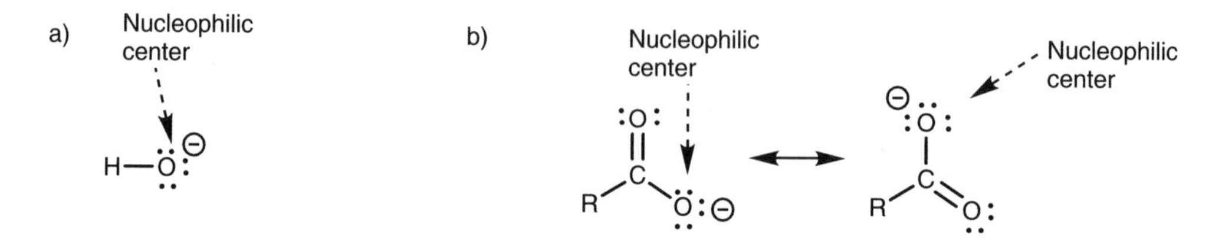

Fig. 1. Examples of nucleophiles; (a) hydroxide ion; (b) carboxylate ion.

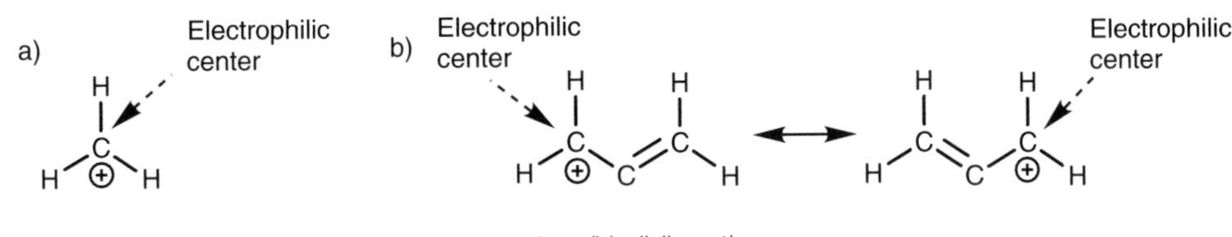

Fig. 2. *Examples of electrophiles: (a) carbocation; (b) allylic cation.*

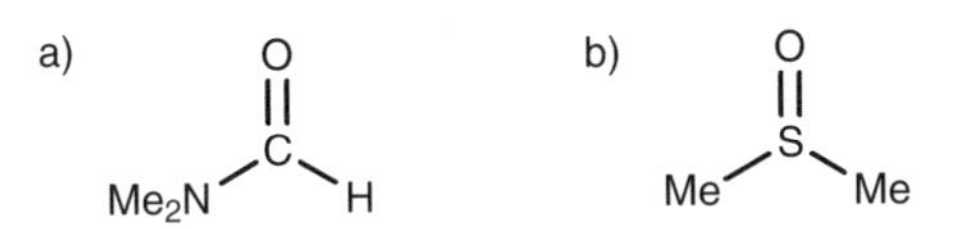

Fig. 3. *(a) Dimethylformamide (DMF); (b) dimethylsulfoxide (DMSO).*

Relative nucleophilicity

In a series of anions, nucleophilic strength parallels basicity if the nucleophilic center is the same atom. For example the nucleophilic strengths of the following oxygen compounds ($RO^- > HO^- >> RCO_2^-$) matches their order of basicity.

The same holds true for anions where the nucleophilic center is an element in the same row of the periodic table (e.g. C,N,O,F). Thus, the order of nucleophilicity of the following anions ($R_3C^- > R_2N^- > RO^- > F^-$) is the same as their order of basicity. This trend is related to the electronegativities of these atoms. The more electronegative the atom (e.g. F), the more tightly it holds on to its electrons and the less available these electrons are for forming new bonds (less nucleophilic).

The story becomes more complex if we compare anions having nucleophilic centers from different parts of the periodic table. Here, relative nucleophilicity does not necessarily match relative basicity. This is because the solvent used in a reaction has an important effect. In protic solvents such as water or alcohol, the stronger nucleophiles are those which have a large nucleophilic center, that is, an atom lower down the periodic table (e.g. S^- is more nucleophilic than O^- but is less basic). This is because protic solvents can form hydrogen bonds to the anion. The smaller the anion, the stronger the solvation and the more difficult it is for the anion to react as a nucleophile.

The order of nucleophilicity of some common anions in protic solvents is as follows: $SH^- > CN^- > I^- > OH^- > N_3^- > Br^- > CH_3CO_2^- > Cl^- > F^-$.

When an organic solvent is used which is incapable of forming hydrogen bonds to the anion (e.g. DMF or DMSO; *Fig. 3*), the order of nucleophilicity changes to more closely match that of basicity. For example, the order of nucleophilicity of the halides in DMSO is $F^- > Cl^- > Br^- > I^-$.

E3 NEUTRAL INORGANIC SPECIES

Key Notes

Polar bonds

Polar bonds are polarized such that the more electronegative atom has the greater share of the bonding electrons. As a result, it will be electron rich and will be a nucleophilic center. The less electronegative atom will be electron deficient and will be an electrophilic center. Electronegative atoms are to the right of the periodic table and have lone pairs of electrons.

Nucleophilic strength

The relative strengths of neutral nucleophilic centers are determined by how well they can accommodate a positive charge. The more electronegative the atom, the less nucleophilic it will be. Therefore, nitrogen is more nucleophilic than oxygen, and oxygen is more nucleophilic than fluorine.

Electrophilic strength

The relative electrophilic strengths of hydrogen atoms in different molecules is determined by the stability of the ions formed. Hydrogen atoms attached to nitrogen are only weakly electrophilic, whereas hydrogens attached to halogen atoms are strongly electrophilic. Therefore, the electrophilic strength of hydrogen atoms depends on the electronegativity of the neighboring atom.

Properties

It is possible to predict whether a molecule is likely to react as an electrophile or as a nucleophile, based on the strength of the nucleophilic or electrophilic centers present.

Related topics Acid strength (G2) Base strength (G3)

Polar bonds

If two atoms of quite different electronegativities are linked together, then the bond connecting them will be polar covalent such that the bonding electrons are biased towards the more electronegative atom. This will give the latter a slightly negative charge and make it a nucleophilic center. Conversely, the less electronegative atom will gain a slightly positive charge and be an electrophilic center (*Fig. 1*). The further right an element is in the periodic table, the more electronegative it is. Thus, fluorine is more electronegative than oxygen, which in turn is more electronegative than nitrogen. Note also that all the nucleophilic

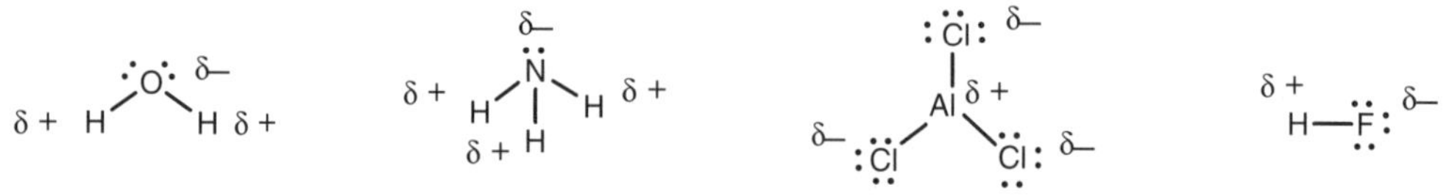

Fig. 1. Nucleophilic (δ–) and electrophilic (δ+) centers in neutral inorganic molecules.

atoms identified above have lone pairs of electrons. This is another way of identifying nucleophilic atoms.

Nucleophilic strength

The molecules above have both nucleophilic and electrophilic centers and could react as nucleophiles or as electrophiles. However, it is usually found that there is a preference to react as one rather than the other. This is explained by considering the relative strengths of nucleophilic and electrophilic centers. First of all, let us consider the relative strengths of nucleophilic centers by comparing N, O, and F. If we compare the relative positions of these atoms in the periodic table, we find that fluorine is more electronegative than oxygen, which in turn is more electronegative than nitrogen. However, when we compare the nucleophilic strengths of these atoms, we find that the nitrogen is more nucleophilic than oxygen, which in turn is more nucleophilic than fluorine.

The relative nucleophilic strengths of these atoms is explained by looking at the products which would be formed if these atoms **were** to act as nucleophiles. Let us compare the three molecules HF, H_2O, and NH_3 and see what happens if they were to form a bond to a proton (*Fig. 2*). Since the proton has no electrons, both electrons for the new bond must come from the nucleophilic centers (i.e. the F, O, and N). As a result, these atoms will gain a positive charge. If hydrogen fluoride acts as a nucleophile, then the fluorine atom gains a positive charge. Since the fluorine atom is strongly electronegative, it does not tolerate a positive charge. Therefore, this reaction does not take place. Oxygen is less electronegative and is able to tolerate the positive charge slightly better, such that an equilibrium is possible between the charged and uncharged species. Nitrogen is the least electronegative of the three atoms and tolerates the positive charge so well that the reaction is irreversible and a salt is formed.

Thus, nitrogen is strongly nucleophilic and will usually react as such, whereas halogens are weakly nucleophilic and will rarely react as such.

Lastly, it is worth noting that all these molecules are weaker nucleophiles than their corresponding anions, i.e. HF, H_2O, and NH_3 are weaker nucleophiles than F^-, OH^- and NH_2^- respectively.

Electrophilic strength

The same argument can be used in reverse when looking at the relative electrophilic strengths of atoms in different molecules. Let us compare the electrophilic strengths of the hydrogens in HF, H_2O, and NH_3. In this case, reaction with a strong nucleophile or base would generate anions (*Fig. 3*). Fluorine being the most electronegative atom is best able to stabilize a negative charge and so the fluoride ion is the most stable ion of the three. Oxygen is also able to stabilize a

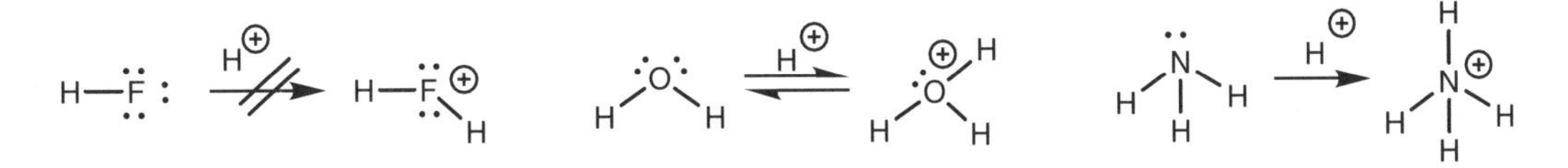

Fig. 2. Cations formed if HF, H_2O, and NH_3 act as nucleophiles.

Fig. 3. Anions generated when HF, H_2O, and NH_3 act as electrophiles.

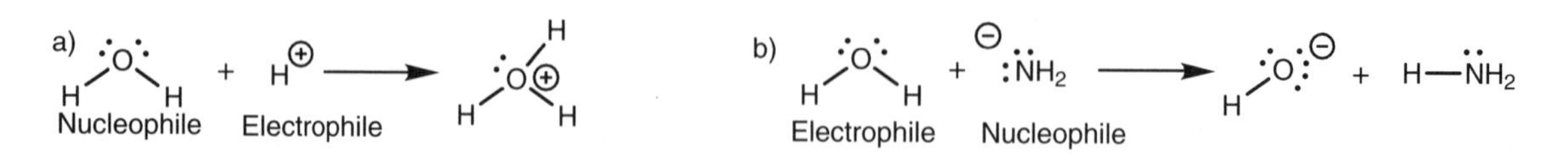

Fig. 4. Water acting as (a) a nucleophile and (b) an electrophile.

negative charge, though not as well as fluorine. Nitrogen is the least electronegative of the three atoms and has the least stabilizing influence on a negative charge and so the NH_2^- ion is unstable. The more stable the anion, the more easily it is formed and hence the hydrogen which is lost will be strongly electrophilic. This is the case for HF. In contrast, the hydrogen in ammonia is a very weak electrophilic center since the anion formed is unstable. As a result, nitrogen anions are only formed with very strong bases.

Properties

It is possible to predict whether molecules are more likely to react as nucleophiles or electrophiles depending on the strength of the nucleophilic and electrophilic centers present. For example, ammonia has both electrophilic and nucleophilic centers. However, it usually reacts as a nucleophile since the nitrogen atom is a strong nucleophilic center and the hydrogen atom is a weak electrophilic center. By contrast, molecules such as hydrogen fluoride or aluminum chloride prefer to react as electrophiles. This is because the nucleophilic centers in both these molecules (halogen atoms) are weak, whereas the electrophilic centers (H or Al) are strong. Water is a molecule which can react equally well as a nucleophile or as an electrophile. For example, water reacts as a nucleophile with a proton and as an electrophile with an anion (*Fig. 4*).

E4 ORGANIC STRUCTURES

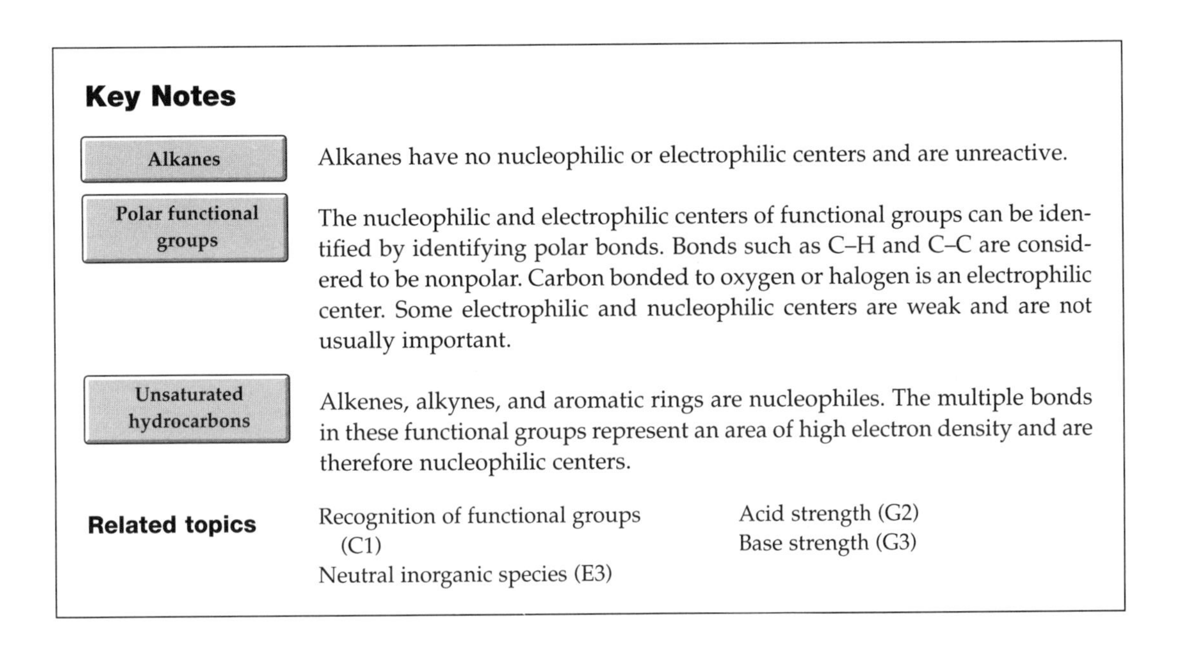

Key Notes

Alkanes — Alkanes have no nucleophilic or electrophilic centers and are unreactive.

Polar functional groups — The nucleophilic and electrophilic centers of functional groups can be identified by identifying polar bonds. Bonds such as C–H and C–C are considered to be nonpolar. Carbon bonded to oxygen or halogen is an electrophilic center. Some electrophilic and nucleophilic centers are weak and are not usually important.

Unsaturated hydrocarbons — Alkenes, alkynes, and aromatic rings are nucleophiles. The multiple bonds in these functional groups represent an area of high electron density and are therefore nucleophilic centers.

Related topics

Recognition of functional groups (C1)
Neutral inorganic species (E3)
Acid strength (G2)
Base strength (G3)

Alkanes

Alkanes are made up of carbon–carbon and carbon–hydrogen single bonds and are unreactive compounds. This is because C–C and C–H bonds are covalent in nature and so there are no electrophilic or nucleophilic centers present. Since most reagents react with nucleophilic or electrophilic centers, alkanes are unreactive molecules.

Polar functional groups

It is possible to identify the nucleophilic and electrophilic centers in common functional groups, based on the relative electronegativities of the atoms present. The following guidelines are worth remembering:

- C–H and C–C bonds are covalent. Therefore, neither carbon nor hydrogen is a nucleophilic or electrophilic center;
- nitrogen is immediately to the right of carbon in the periodic table. The nitrogen is more electronegative but the difference in electronegativity between these two atoms is small and so the N–C bond is not particularly polar. Therefore, the carbon atom can usually be ignored as an electrophilic center;
- N–H and O–H bonds are polar covalent. Nitrogen and oxygen are strong nucleophilic centers. Hydrogen is a weak electrophilic center;
- C=O, C=N and C≡N bonds are polar covalent. The O and N are nucleophilic centers and the carbon is an electrophilic center;
- C–O and C–X bonds (X=halogen) are polar covalent. The oxygen atom is moderately nucleophilic whereas the halogen atom is weakly nucleophilic. The carbon atom is an electrophilic center.

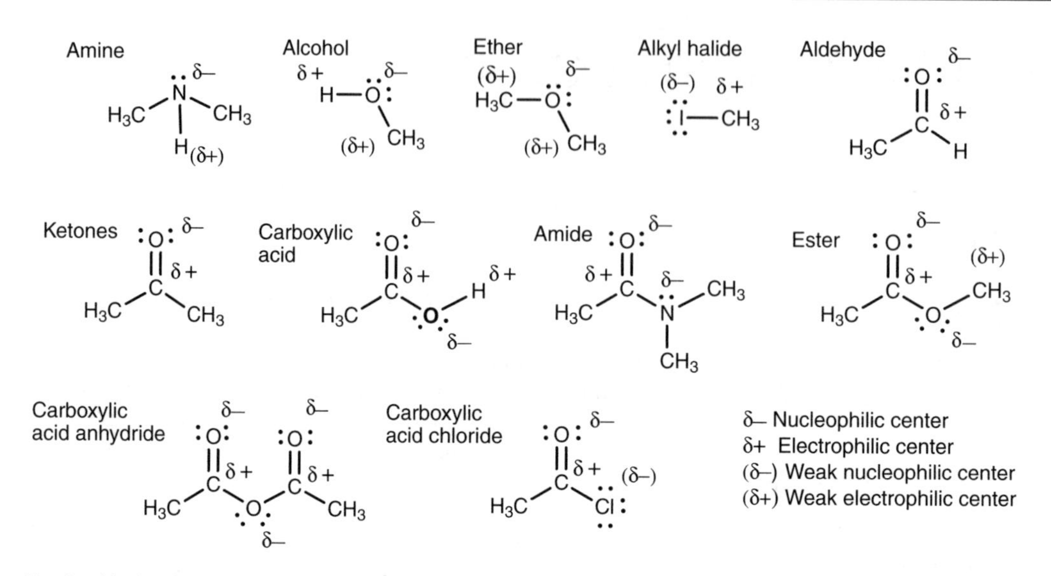

Fig. 1. Nucleophilic and electrophilic centers of common functional groups.

Using the above guidelines, the nucleophilic and electrophilic centers of the common functional groups can be identified, where atoms having a slightly negative charge are nucleophilic centers and atoms having a slightly positive charge are electrophilic centers (*Fig. 1*).

Not all the nucleophilic and electrophilic centers are of equal importance. For example, a nitrogen atom is more nucleophilic than an oxygen atom. Also halogen atoms are very weakly nucleophilic and will not usually react with electrophiles if there is a stronger nucleophilic center present. Hydrogen atoms attached to halogens are more electrophilic than hydrogen atoms attached to oxygen. Hydrogen atoms attached to nitrogen are very weakly electrophilic.

Taking this into account, some functional groups are more likely to react as nucleophiles while some functional groups are more likely to react as electrophiles. For example, amines, alcohols and ethers are more likely to react as nucleophiles, since they have strong nucleophilic centers and weak electrophilic centers. Alkyl halides are more likely to react as electrophiles since they have strong electrophilic centers and weak nucleophilic centers. Aldehydes and ketones can react as nucleophiles or electrophiles since both electrophilic and nucleophilic centers are strong.

Some functional groups contain several nucleophilic and electrophilic centers. For example, carboxylic acids and their derivatives fall into this class and so there are several possible centers where a nucleophile or an electrophile could react.

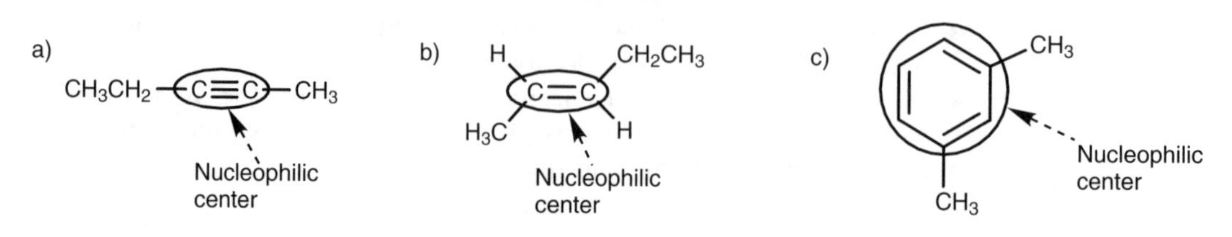

Fig. 2. Nucleophilic centers in (a) an alkyne; (b) an alkene; (c) an aromatic compound.

Unsaturated hydrocarbons

Not all functional groups have polar bonds. Alkenes, alkynes, and aromatic compounds are examples of functional groups which have covalent multiple bonds. The space between the multiple bonded carbons is rich in electrons and is therefore nucleophilic. Thus, the nucleophilic center in these molecules is not a specific atom, but the multiple bond (*Fig. 2*)!

F1 Reactions

Key Notes

Bond formation

Most organic reactions take place between nucleophiles and electrophiles, where the nucleophilic center of the nucleophile forms a bond to the electrophilic center of the electrophile.

Classification of reactions

Reactions can be classified as acid/base reactions, functional group transformations or as carbon–carbon bond formations. Reactions can also be classified according to the process or mechanism taking place and these are specific for particular functional groups.

Related topics

Definition (E1)
Organic structures (E4)
Nucleophilic substitution (L2)

Bond formation

Synthetic organic chemistry is about creating complex molecules from simple starting materials – a process which may involve many different reactions. Designing a synthesis is a bit like chess. A grand master has to know the pieces and the moves that can be made before planning a game strategy. As far as an organic chemist is concerned, he/she has to know the molecules and the sort of reactions which can be carried out before planning a synthetic 'game strategy'.

Inevitably, there is a lot of memory work involved in knowing reactions, but there is a logic involved as well. Basically, most reactions involve electron-rich molecules forming bonds to electron deficient molecules (i.e. nucleophiles forming bonds to electrophiles). The bond will be formed specifically between the nucleophilic center of the nucleophile and the electrophilic center of the electrophile.

Classification of reactions

There are a large number of reactions in organic chemistry, but we can simplify the picture by grouping these reactions into various categories. To begin with, we can classify reactions as being:

- acid/base reactions;
- functional group transformations;
- carbon–carbon bond formations.

The first category of reaction is relatively simple and involves the reaction of an acid with a base to give a salt. These reactions are covered in Section G. The second category of reaction is where one functional group can be converted into another. Normally these reactions are relatively straightforward and proceed in high yield. The third category of reactions is extremely important to organic chemistry since these are the reactions which allow the chemist to construct complex molecules from simple starting materials. In general, these reactions are the most difficult and temperamental to carry out. Some of these

Table 1. Different categories of reaction undergone by functional groups.

Reaction category	Functional group	Section
Electrophilic addition	Alkenes and alkynes	H
Electrophilic substitution	Aromatic	I
Nucleophilic addition	Aldehydes and ketones	J
Nucleophilic substitution	Carboxylic acid derivatives	K
	Alkyl halides	L
Elimination	Alcohols and alkyl halides	M,L
Reduction	Alkenes, alkynes, aromatic, aldehydes, ketones, nitriles, carboxylic acids, and carboxylic acid derivatives,	H–N
Oxidation	Alkenes, alcohols, aldehydes	H–N
Acid/base reactions	Carboxylic acids, phenols, amines	G

reactions are so important that they are named after the scientists who developed them (e.g. Grignard and Aldol reactions).

Another way of categorizing reactions is to group similar types of reactions together, depending on the process or mechanism involved. This is particularly useful since specific functional groups will undergo certain types of reaction category. *Table 1* serves as a summary of the types of reactions which functional groups normally undergo.

F2 Mechanisms

Key Notes

Definition

A mechanism describes how a reaction takes place by showing what is happening to valence electrons during the formation and breaking of bonds.

Curly arrows

Curly arrows are used to show what happens to valence electrons during the making and breaking of bonds. They always start from the source of two electrons (i.e. a lone pair of electrons on an atom or the middle of a bond about to be broken). They always point to where the valence electrons will end up. If the electrons end up as a lone pair of electrons on an atom, the arrow points to that specific atom. If the electrons are being used to form a new bond, the arrow points to where the center of the new bond will be formed.

Half curly arrows

Half curly arrows are used to show the movement of single electrons during radical reactions. Bond breaking during a radical reaction involves homolytic cleavage where the bonding electrons move to different atoms. However, most reactions in organic chemistry involve heterolytic cleavage where the bonding electrons move as a pair onto one atom and not the other.

Related topic

Acid strength (G2)

Definition

An understanding of electrophilic and nucleophilic centers allows a prediction of **where** reactions might occur but not what **sort** of reaction will occur. In order to understand and predict the outcome of reactions, it is necessary to understand what goes on at the electronic level. This process is known as a **mechanism**.

A mechanism is the 'story' of how a reaction takes place. It explains how molecules react together to give the final product. The mechanism tells us how bonds are formed and how bonds are broken and in what order. It explains what is happening to the valence electrons in the molecule since it is the movement of these electrons which result in a reaction. Take as a simple example the reaction between a hydroxide ion and a proton to form water (*Fig. 1*). The hydroxide ion is a nucleophile and the proton is an electrophile. A reaction takes place between the nucleophilic center (the oxygen) and the electrophilic center (the hydrogen) and water is formed. A new bond has been formed between the oxygen of the hydroxide ion and the proton. The mechanism looks at what happens to the electrons. In

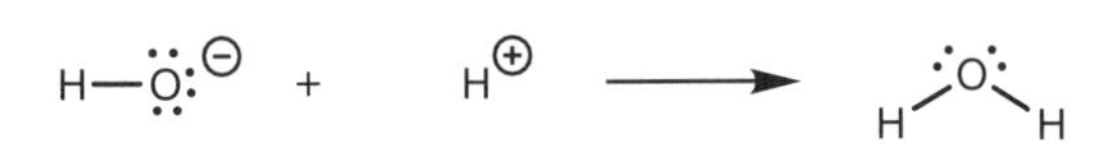

Fig. 1. Reaction of a hydroxide ion and a proton to form water.

this example, a lone pair of electrons from oxygen is used to form a bond to the proton. By doing so, the oxygen effectively 'loses' one electron and the proton effectively gains one electron. As a result, the oxygen loses its negative charge and the proton loses its positive charge.

Curly arrows

Explaining what happens to all the valence electrons during a reaction mechanism can be rather long-winded if you are trying to explain it all in words. Fortunately, there is a diagrammatic way of showing the same thing – using curly arrows. For example, the mechanism described above can be explained by using a curly arrow to show what happens to the lone pair of electrons (*Fig. 2*). In this case, the arrow starts from a lone pair of electrons on the oxygen (the source of the two electrons) and points to where the **center** of the new bond will be formed.

In some textbooks, you may see the arrow written directly to the proton (*Fig. 3*). Formally, this is incorrect. Arrows should only be drawn directly to an atom if the electrons are going to end up on that atom as a lone pair of electrons.

The following rules are worth remembering when drawing arrows:

- curly arrows show the movement of electrons, **not** atoms;
- curly arrows start from the source of two electrons (i.e. a lone pair of electrons on an atom or the middle of a bond which is about to be broken);
- curly arrows point to an **atom** if the electrons are going to end up as a lone pair on that atom;
- curly arrows point to where a new bond will be formed if the electrons are being used to form a new bond.

The mechanism (*Fig. 4*) explains what happens when a hydroxide ion reacts with a carboxylic acid and is a demonstration of how arrows should be drawn. One of the lone pairs of electrons on the hydroxide ion is used to form a bond to the acidic proton of the carboxylic acid. The curly arrow representing this starts from a lone pair of electrons and points to the space between the two atoms to show that a bond is being formed.

Fig. 2. Mechanism for the reaction of a hydroxide ion with a proton.

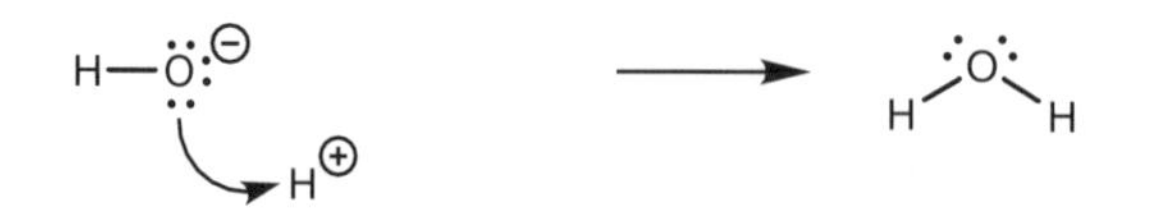

Fig. 3. Incorrect way of drawing a curly arrow.

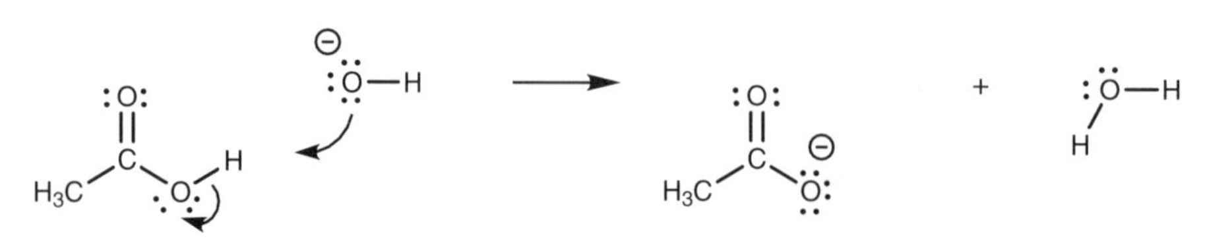

Fig. 4. Mechanism for the reaction of a hydroxide ion with ethanoic acid.

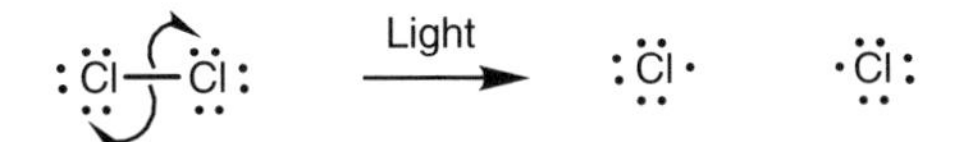

Fig. 5. Use of half curly arrows in a mechanism (homolytic cleavage).

Fig. 6. Heterolytic cleavage of a bond.

At the same time as this new bond is being formed, the O–H bond of the carboxylic acid has to break. This is because the hydrogen atom is only allowed one bond. The electrons in this bond end up on the carboxylate oxygen as a third lone pair of electrons. The arrow representing this starts from the **center** of the bond being broken and points directly to the atom where the electrons will end up as a lone pair.

Notice also what happens to the charges. The negatively charged oxygen of the hydroxide ion ends up as a neutral oxygen in water. This is because one of the oxygen's lone pairs is used to form the new bond. Both electrons are now shared between two atoms and so the oxygen effectively loses one electron and its negative charge. The oxygen in the carboxylate ion (which was originally neutral in the carboxylic acid) becomes negatively charged since it now has three lone pairs of electrons and has effectively gained an extra electron.

Half curly arrows

Occasionally reactions occur which involve the movement of single electrons rather than pairs of electrons. Such reactions are known as **radical** reactions. For example, a chlorine molecule can be split into two chlorine radicals on treatment with light. One of the original bonding electrons ends up on one chlorine radical and the second bonding electron ends up on the other chlorine radical. The movement of these single electrons can be illustrated by using half curly arrows rather than full curly arrows (*Fig. 5*).

This form of bond breaking is known as a **homolytic cleavage**. The radical atoms obtained are neutral but highly reactive species since they have an unpaired valence electron.

There are some important radical reactions in organic chemistry, but the majority of organic reactions involve the **heterolytic cleavage** of covalent bonds where electrons move together as a pair (*Fig. 6*).

G1 Brønsted–Lowry acids and bases

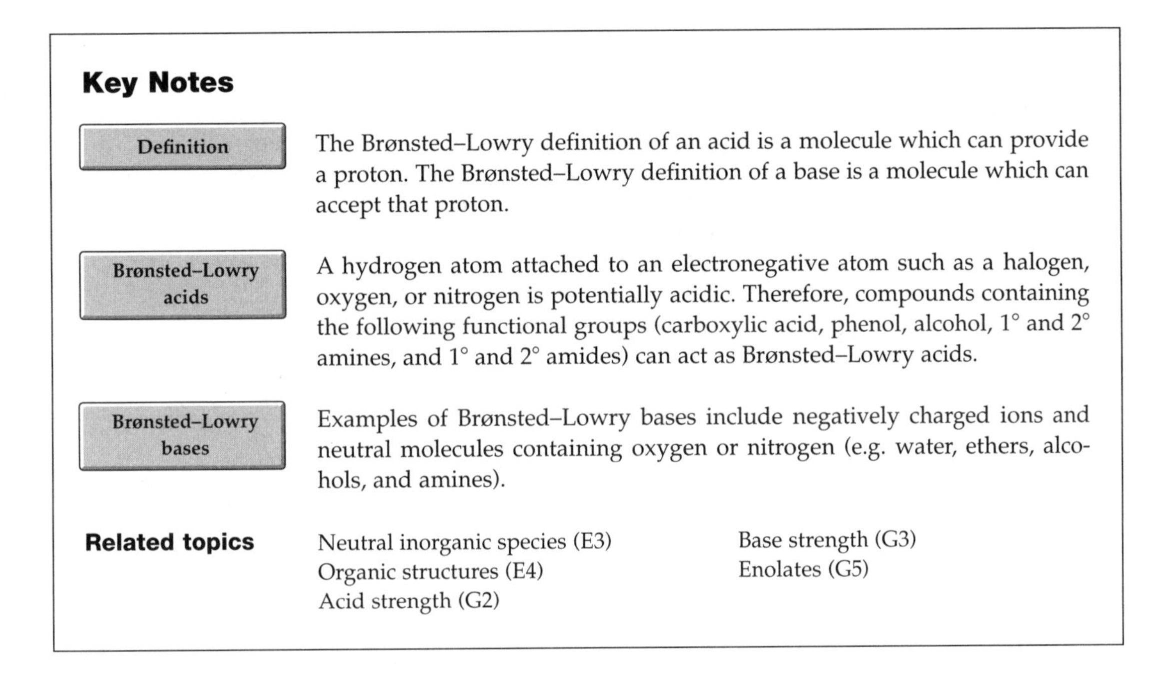

Key Notes

Definition — The Brønsted–Lowry definition of an acid is a molecule which can provide a proton. The Brønsted–Lowry definition of a base is a molecule which can accept that proton.

Brønsted–Lowry acids — A hydrogen atom attached to an electronegative atom such as a halogen, oxygen, or nitrogen is potentially acidic. Therefore, compounds containing the following functional groups (carboxylic acid, phenol, alcohol, 1° and 2° amines, and 1° and 2° amides) can act as Brønsted–Lowry acids.

Brønsted–Lowry bases — Examples of Brønsted–Lowry bases include negatively charged ions and neutral molecules containing oxygen or nitrogen (e.g. water, ethers, alcohols, and amines).

Related topics

Neutral inorganic species (E3)
Organic structures (E4)
Acid strength (G2)
Base strength (G3)
Enolates (G5)

Definition

Put at its simplest, the Brønsted–Lowry definition of an acid is a molecule which can provide a proton. The Brønsted–Lowry definition of a base is a molecule which can accept that proton.

An example of a simple acid/base reaction is the reaction of ammonia with water (*Fig. 1*). Here, water loses a proton and is an acid. Ammonia accepts that proton and is the base.

Fig. 1. Reaction of ammonia with water.

As far as the mechanism of the reaction is concerned, the ammonia uses its lone pair of electrons to form a new bond to the proton and is therefore acting as a nucleophile. This means that the water is acting as an electrophile.

As the nitrogen uses its lone pair of electrons to form the new bond, the bond between hydrogen and oxygen must break since hydrogen is only allowed one bond. The electrons making up the O–H bond will move onto oxygen to produce a third lone pair of electrons, thus giving the oxygen a negative charge (*Fig. 2*). Since the nitrogen atom on ammonia has used its lone pair of electrons to form a new bond, it now has to share the electrons with hydrogen and so nitrogen gains a positive charge.

Fig. 2. Mechanism for the reaction of ammonia with water.

Brønsted–Lowry acids

A Brønsted–Lowry acid is a molecule which contains an acidic hydrogen. In order to be acidic, the hydrogen must be slightly positive or electrophilic. This is possible if hydrogen is attached to an electronegative atom such as a halogen, oxygen, or nitrogen. The following mineral acids and functional groups contain hydrogens which are potentially acidic (*Fig. 3*).

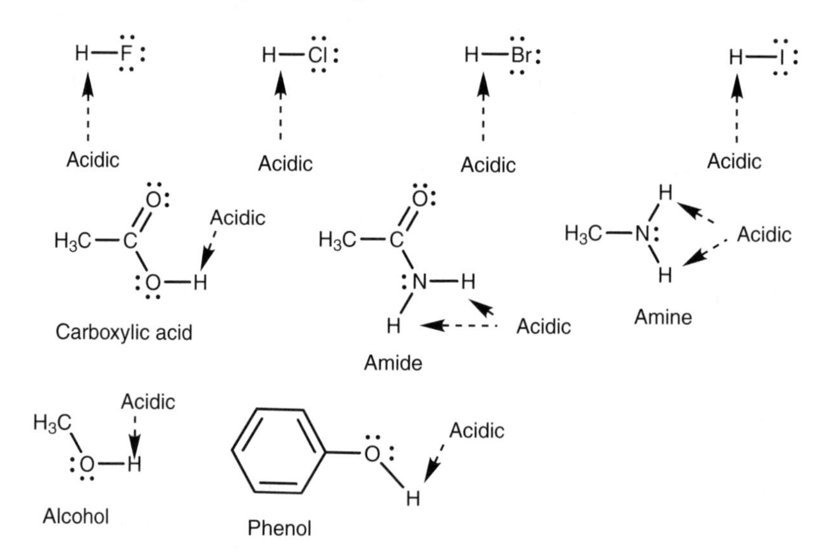

Fig. 3. Acidic protons in mineral acids and common functional groups.

Hydrogens attached to carbon are not normally acidic. However, in topic G5, we shall look at special cases where hydrogens attached to carbon **are** acidic.

Brønsted–Lowry bases

A Brønsted–Lowry base is a molecule which can form a bond to a proton. Examples include negatively charged ions with a lone pair of electrons (*Fig. 4*).

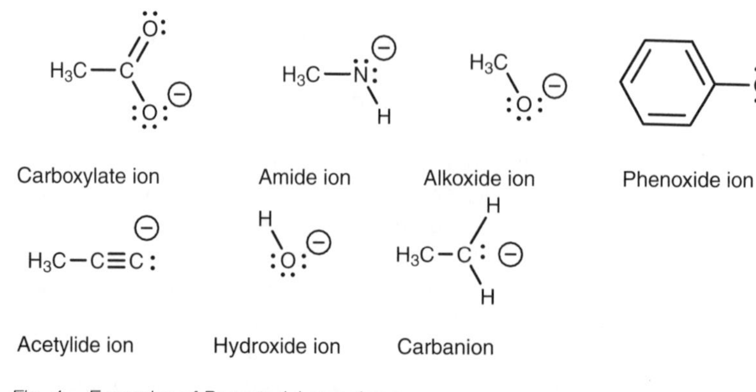

Fig. 4. Examples of Brønsted–Lowry bases.

Neutral molecules can also act as bases if they contain an oxygen or nitrogen atom. The most common examples are amines. However, water, ethers and alcohols are also capable of acting as bases (*Fig. 5*).

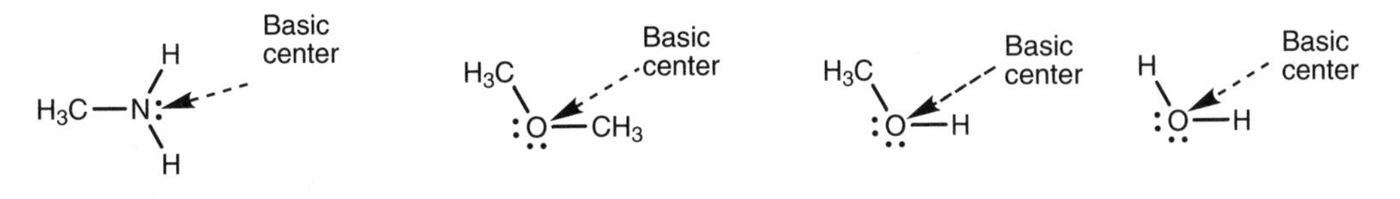

Fig. 5. Examples of neutral Brønsted–Lowry bases.

G2 ACID STRENGTH

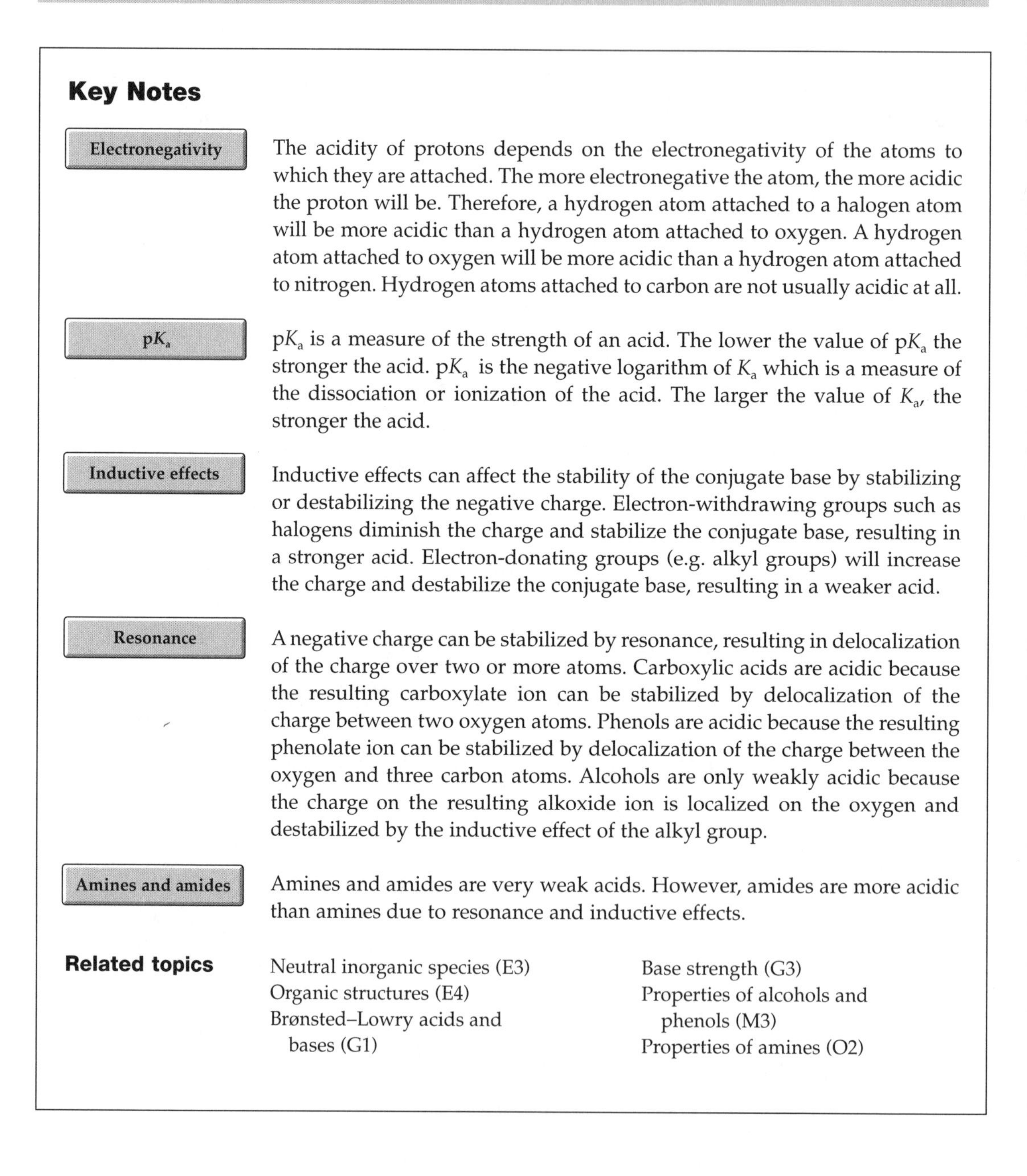

Key Notes

Electronegativity The acidity of protons depends on the electronegativity of the atoms to which they are attached. The more electronegative the atom, the more acidic the proton will be. Therefore, a hydrogen atom attached to a halogen atom will be more acidic than a hydrogen atom attached to oxygen. A hydrogen atom attached to oxygen will be more acidic than a hydrogen atom attached to nitrogen. Hydrogen atoms attached to carbon are not usually acidic at all.

pK_a pK_a is a measure of the strength of an acid. The lower the value of pK_a the stronger the acid. pK_a is the negative logarithm of K_a which is a measure of the dissociation or ionization of the acid. The larger the value of K_a, the stronger the acid.

Inductive effects Inductive effects can affect the stability of the conjugate base by stabilizing or destabilizing the negative charge. Electron-withdrawing groups such as halogens diminish the charge and stabilize the conjugate base, resulting in a stronger acid. Electron-donating groups (e.g. alkyl groups) will increase the charge and destabilize the conjugate base, resulting in a weaker acid.

Resonance A negative charge can be stabilized by resonance, resulting in delocalization of the charge over two or more atoms. Carboxylic acids are acidic because the resulting carboxylate ion can be stabilized by delocalization of the charge between two oxygen atoms. Phenols are acidic because the resulting phenolate ion can be stabilized by delocalization of the charge between the oxygen and three carbon atoms. Alcohols are only weakly acidic because the charge on the resulting alkoxide ion is localized on the oxygen and destabilized by the inductive effect of the alkyl group.

Amines and amides Amines and amides are very weak acids. However, amides are more acidic than amines due to resonance and inductive effects.

Related topics

Neutral inorganic species (E3)
Organic structures (E4)
Brønsted–Lowry acids and bases (G1)
Base strength (G3)
Properties of alcohols and phenols (M3)
Properties of amines (O2)

Electronegativity

The acidic protons of various molecules are not equally acidic and their relative acidity depends on a number of factors, one of which is the electronegativity of the atom to which they are attached. For example, consider hydrofluoric acid, ethanoic acid, and methylamine (*Fig. 1*). Hydrofluoric acid has the most acidic proton since the hydrogen is attached to a strongly electronegative fluorine. The fluorine strongly polarizes the H–F bond such that the hydrogen becomes highly electron deficient and is easily lost. Once the proton is lost, the fluoride ion can stabilize the resulting negative charge.

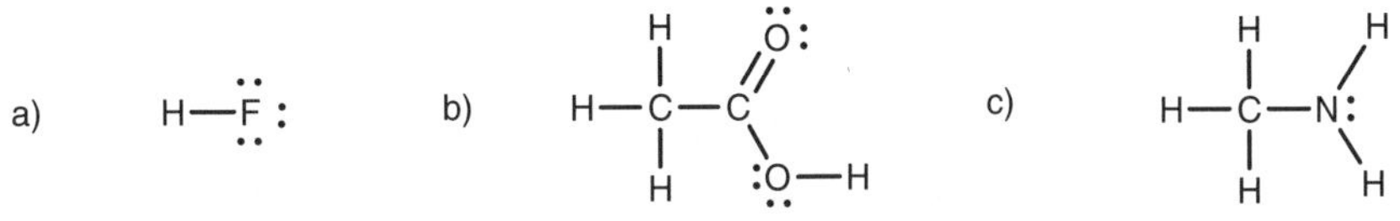

Fig. 1. *(a) Hydrofluoric acid; (b) ethanoic acid; (c) methylamine.*

The acidic protons on methylamine are attached to nitrogen which is less electronegative than fluorine. Therefore, the N–H bonds are less polarized, and the protons are less electron deficient. If one of the protons is lost, the nitrogen is left with a negative charge which it cannot stabilize as efficiently as a halide ion. All of this means that methylamine is a much weaker acid than hydrogen fluoride. Ethanoic acid is more acidic than methylamine but less acidic than hydrofluoric acid. This is because the electronegativity of oxygen lies between that of a halogen and that of a nitrogen atom.

These differences in acid strength can be demonstrated if the three molecules above are placed in water. Mineral acids such as HF, HCl, HBr, and HI are strong acids and **dissociate** or ionize completely (*Fig. 2*).

Fig. 2. *Ionization of hydrochloric acid.*

Ethanoic acid (acetic acid) partially dissociates in water and an equilibrium is set up between the carboxylic acid (termed the **free acid**) and the carboxylate ion (*Fig. 3*). An acid which only partially ionizes in this manner is termed a **weak acid**.

Fig. 3. *Partial ionization of ethanoic acid.*

If methylamine is dissolved in water, none of the acidic protons are lost at all and the amine behaves as a weak base instead of an acid, and is in equilibrium with its protonated form (*Fig. 4*).

Fig. 4. *Equilibrium acid–base reaction of methylamine with water.*

Methylamine can act as an acid but it has to be treated with a strong base such as butyl lithium (*Fig. 5*).

Lastly, hydrogen atoms attached to carbon are not usually acidic since carbon atoms are not electronegative. There are exceptions to this rule as described in Topic G5.

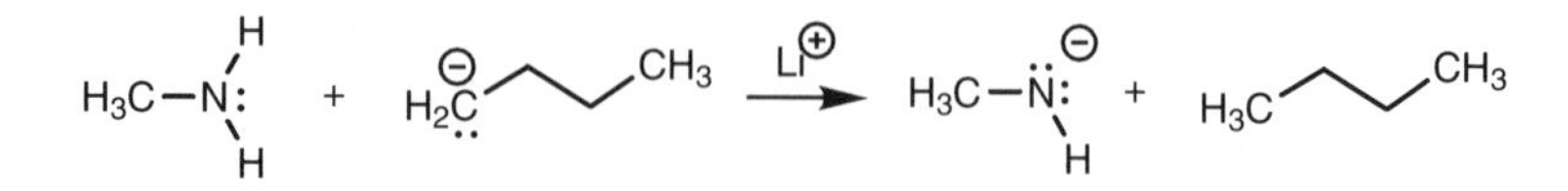

Fig. 5. Methylamine acting as an acid with a strong base (butyl lithium).

pK_a

Acids can be described as being weak or strong and the pK_a is a measure of this. Dissolving acetic acid in water, results in an equilibrium between the carboxylic acid and the carboxylate ion (*Fig. 6*).

$$H_3C{-}C(=\ddot{O}{:}){-}\ddot{O}{-}H \; + \; H_2\ddot{O}{:} \; \rightleftharpoons \; H_3C{-}C(=\ddot{O}{:}){-}\ddot{O}{:}^{\ominus} \; + \; H_3O{:}^{\oplus}$$

Fig. 6. Equilibrium acid–base reaction of ethanoic acid with water.

Ethanoic acid on the left hand of the equation is termed the **free acid**, while the carboxylate ion formed on the right hand side is termed its **conjugate base**. The extent of ionization or dissociation is defined by the **equilibrium constant** (K_{eq});

$$K_{eq} = \frac{[\text{PRODUCTS}]}{[\text{REACTANTS}]} = \frac{[CH_3CO_2^-]\,[H_3O^+]}{[CH_3CO_2H]\,[H_2O]}.$$

K_{eq} is normally measured in a dilute aqueous solution of the acid and so the concentration of water is high and assumed to be constant. Therefore, we can rewrite the equilibrium equation in a simpler form where K_a is the acidity constant and includes the concentration of pure water (55.5 M).

$$K_a = K_{eq}\,[H_2O] = \frac{[CH_3CO_2^-]\,[H_3O^+]}{[CH_3CO_2H]}.$$

The acidity constant is also a measure of dissociation and of how acidic a particular acid is. The stronger the acid, the more it is ionized and the greater the concentration of products in the above equation. This means that a strong acid has a high K_a value. The K_a values for the following ethanoic acids are in brackets and demonstrate that the strongest acid in the series is trichloroacetic acid.

$$Cl_3CCO_2H\ (23\,200 \times 10^{-5}) > Cl_2CHCO_2H\ (5530 \times 10^{-5}) > ClCH_2CO_2H\ (136 \times 10^{-5}) > CH_3CO_2H\ (1.75 \times 10^{-5}).$$

K_a values are awkward to work with and so it is more usual to measure the acidic strength as a pK_a value rather than K_a. The pK_a is the negative logarithm of K_a (p$K_a = -\log_{10} K_a$) and results in more manageable numbers. The pK_a values for each of the above ethanoic acids is shown in brackets below. The strongest acid (trichloroacetic acid) has the lowest pK_a value.

$$Cl_3CCO_2H\ (0.63) < Cl_2CHCO_2H\ (1.26) < ClCH_2CO_2H\ (2.87) < CH_3CO_2H\ (4.76).$$

Therefore the stronger the acid, the **higher** the value of K_a, and the **lower** the value of pK_a. An amine such as ethylamine ($CH_3CH_2NH_2$) is an extremely weak acid (pK_a = 40) compared to ethanol (pK_a = 16). This is due to the relative electronegativities of oxygen and nitrogen as described above. However, the electronegativity of neighboring atoms is not the only influence on acidic strength. For example, the pK_a values of ethanoic acid (4.76), ethanol (16), and phenol (10) show that ethanoic acid is more acidic than phenol, and that phenol is more acidic than ethanol. The difference in acidity is quite marked, yet hydrogen is attached to oxygen in all three structures.

Similarly, the ethanoic acids Cl_3CCO_2H (0.63), Cl_2CHCO_2H (1.26), $ClCH_2CO_2H$ (2.87), and CH_3CO_2H (4.76) have significantly different pK_a values and yet the acidic hydrogen is attached to an oxygen in each of these structures. Therefore, factors other than electronegativity have a role to play in determining acidic strength.

Inductive effects

Stabilizing the negative charge of the conjugate base is important in determining the strength of the acid and so any effect which stabilizes the charge will result in a stronger acid. Substituents can help to stabilize a negative charge and do so by an **inductive effect**. This is illustrated by comparing the pK_a values of the alcohols CF_3CH_2OH and CH_3CH_2OH (12.4 and 16, respectively) where CF_3CH_2OH is more acidic than CH_3CH_2OH. This implies that the anion $CF_3CH_2O^-$ is more stable than $CH_3CH_2O^-$ (*Fig. 7*).

a) $F_3C-CH_2-\ddot{O}:^{\ominus}$ b) $H_3C-CH_2-\ddot{O}:^{\ominus}$

Fig. 7. (a) 2,2,2-Trifluoroethoxy ion; (b) ethoxy ion.

Fluorine atoms are strongly electronegative and this means that each C–F bond is strongly polarized such that the carbon bearing the fluorine atoms becomes strongly electropositive. Since this carbon atom is now electron deficient, it will 'demand' a greater share of the electrons in the neighboring C–C bond. This results in electrons being withdrawn from the neighboring carbon, making it electron deficient too. This inductive effect will continue to be felt through the various bonds of the structure. It will decrease through the bonds but it is still significant enough to be felt at the negatively charged oxygen. Since the inductive effect is electron withdrawing it will decrease the negative charge on the oxygen and help to stabilize it. This means that the original fluorinated alcohol will lose its proton more readily and will be a stronger acid.

This inductive effect explains the relative acidities of the chlorinated ethanoic acids Cl_3CCO_2H (0.63), Cl_2CHCO_2H (1.26), $ClCH_2CO_2H$ (2.87), and CH_3CO_2H (4.76). Trichloroethanoic acid is the strongest acid since its conjugate base (the carboxylate ion) is stabilized by the inductive effect created by three electronegative chlorine atoms. As the number of chlorine atoms decrease, so does the inductive effect.

Inductive effects also explain the difference between the acid strengths of ethylamine (pK_a ~ 40) and ammonia (pK_a ~ 33). The pK_a values demonstrate that ammonia is a stronger acid than ethylamine. In this case, the inductive effect is electron donating. The alkyl group of ethylamine enhances the negative charge of the conjugate base and so destabilizes it, making ethylamine a weaker acid than ammonia (*Fig. 8*).

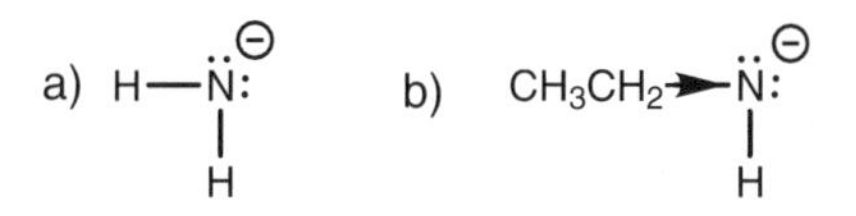

Fig. 8. Conjugate bases of (a) ammonia and (b) ethylamine.

Resonance

The negative charge on some conjugate bases can be stabilized by **resonance**. Resonance involves the movement of valence electrons around a structure, resulting in the sharing of charge between different atoms – a process called **delocalization**. The effects of resonance can be illustrated by comparing the acidities of ethanoic acid (pK_a 4.76), phenol (pK_a 10.0) and ethanol (pK_a 12.4). The pK_a values illustrate that ethanoic acid is a stronger acid than phenol, and that phenol is a stronger acid than ethanol.

The differing acidic strengths of ethanoic acid, phenol and ethanol can be explained by considering the relative stabilities of their conjugate bases (*Fig. 9*).

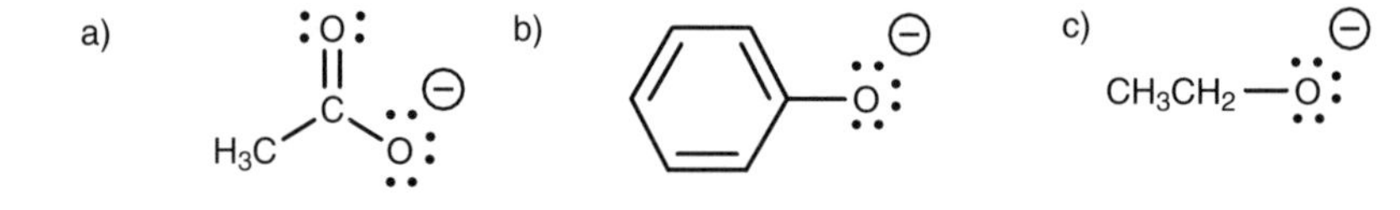

Fig. 9. Conjugate bases of (a) ethanoic acid; (b) phenol; (c) ethanol.

The charge of the carboxylate ion is on an oxygen atom, and since oxygen is electronegative, the charge is stabilized. However, the charge can be shared with the other oxygen leading to delocalization of the charge. This arises by a resonance interaction between a lone pair of electrons on the negatively charged oxygen and the π electrons of the carbonyl group (*Fig. 10*). A lone pair of electrons on the 'bottom' oxygen forms a new π bond to the neighboring carbon. At the same time as this takes place, the weak π bond of the carbonyl group breaks. This is essential or else the carbonyl carbon would end up with five bonds and that is not permitted. Both electrons in the original π bond now end up on the 'top' oxygen which means that this oxygen ends up with three lone pairs and gains a negative charge. Note that the π bond and the charge have effectively 'swapped places'. Both the structures involved are called resonance structures and are easily interconvertible. The negative charge is now shared or delocalized equally between both oxygens and is stabilized. Therefore, ethanoic acid is a stronger acid than one would expect based on the electronegativity of oxygen alone.

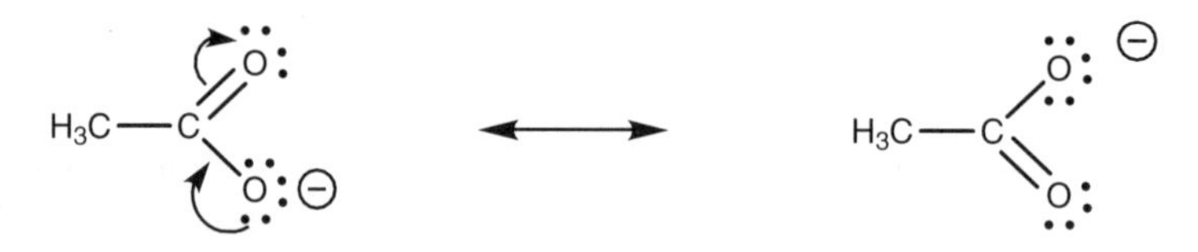

Fig. 10. Resonance interaction for the carboxylate ion.

Phenol is less acidic than ethanoic acid but is more acidic than ethanol. Once again, resonance can explain these differences. The conjugate base of phenol is called the phenolate ion. In this case, the resonance process can be carried out several times to place the negative charge on four separate atoms – the oxygen atom and three of the aromatic carbon atoms (*Fig. 11*). The fact that the negative charge can be spread over four atoms might suggest that the phenolate anion should be more stable than the carboxylate anion, since the charge is spread over more atoms. However, with the phenolate ion, three of the resonance structures place the charge on a carbon atom which is much less electronegative than an oxygen atom. These resonance structures will be far less important than the resonance structure having the charge on oxygen. As a result, delocalization is weaker for the phenolate ion than it is for the ethanoate ion. Nevertheless, a certain amount of delocalization still takes place which is why a phenolate ion is more stable than an ethoxide ion.

Lastly, we turn to ethanol. The conjugate base is the ethoxide ion which cannot

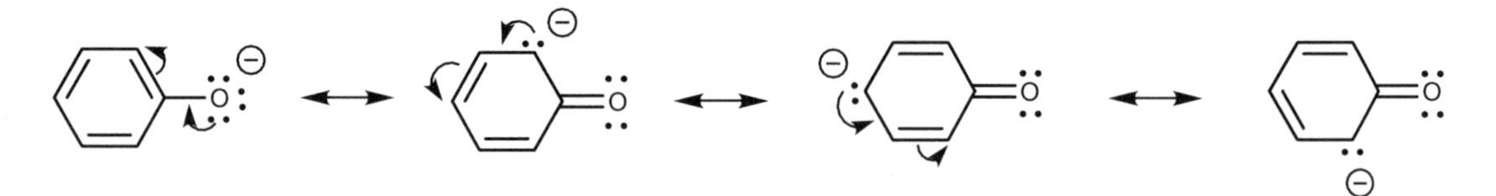

Fig. 11. *Resonance interactions for the phenolate ion.*

be stabilized by delocalizing the charge, since resonance is not possible. There is no π bond available to participate in resonance. Therefore, the negative charge is localized on the oxygen. Furthermore, the inductive donating effect of the neighboring alkyl group (ethyl) enhances the charge and destabilizes it (*Fig. 12*). This makes the ethoxide ion the least stable (or most reactive) of the three anions we have studied. As a result, ethanol is the weakest acid.

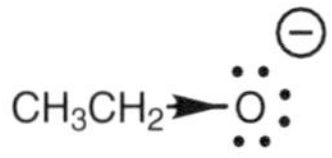

$CH_3CH_2 \rightarrow O^{\ominus}$

Fig. 12. *Destabilizing inductive effect of the ethoxide ion.*

Amines and amides

Amines and amides are very weak acids and only react with very strong bases. The pK_a values for ethanamide and ethylamine are 15 and 40, respectively, which means that ethanamide has the more acidic proton (*Fig. 13*). This can be explained by resonance and inductive effects (*Fig. 14*).

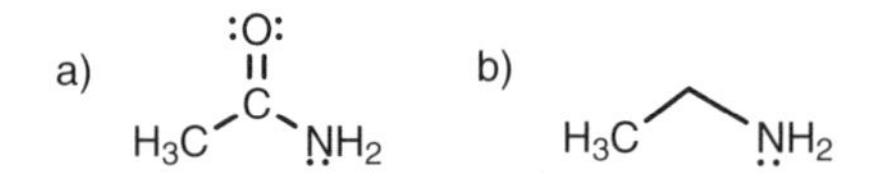

Fig. 13. *(a) Ethanamide; (b) ethylamine.*

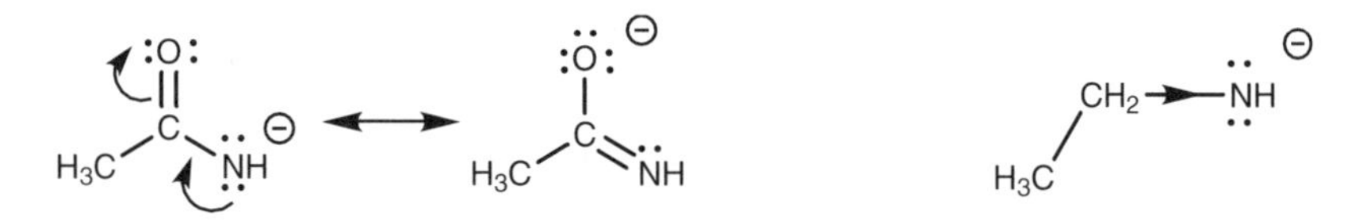

Fig. 14. *(a) Resonance stabilization for the conjugate base of ethanamide; (b) inductive destabilization for the conjugate base of ethylamine.*

G3 BASE STRENGTH

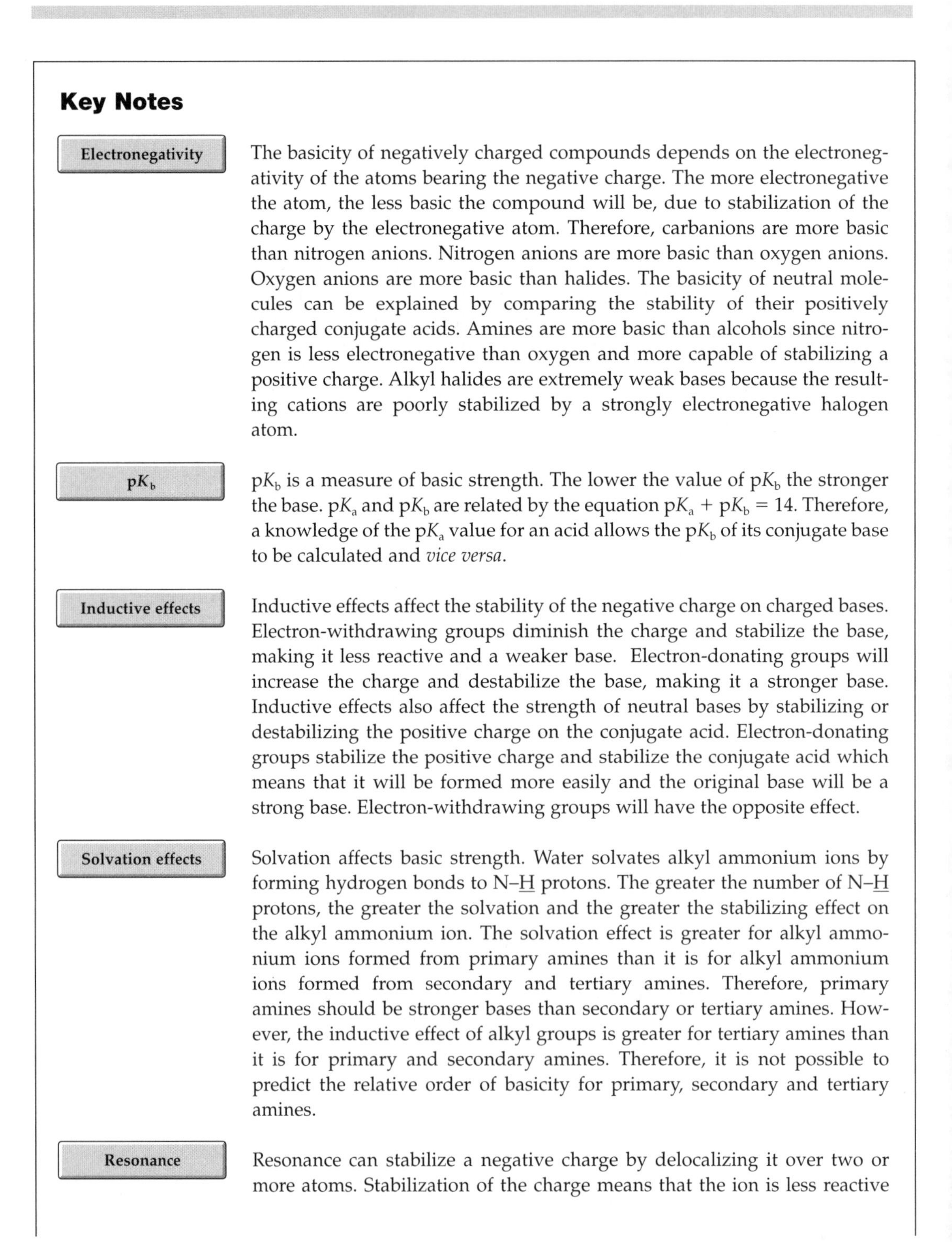

Key Notes

Electronegativity

The basicity of negatively charged compounds depends on the electronegativity of the atoms bearing the negative charge. The more electronegative the atom, the less basic the compound will be, due to stabilization of the charge by the electronegative atom. Therefore, carbanions are more basic than nitrogen anions. Nitrogen anions are more basic than oxygen anions. Oxygen anions are more basic than halides. The basicity of neutral molecules can be explained by comparing the stability of their positively charged conjugate acids. Amines are more basic than alcohols since nitrogen is less electronegative than oxygen and more capable of stabilizing a positive charge. Alkyl halides are extremely weak bases because the resulting cations are poorly stabilized by a strongly electronegative halogen atom.

pK_b

pK_b is a measure of basic strength. The lower the value of pK_b the stronger the base. pK_a and pK_b are related by the equation pK_a + pK_b = 14. Therefore, a knowledge of the pK_a value for an acid allows the pK_b of its conjugate base to be calculated and *vice versa*.

Inductive effects

Inductive effects affect the stability of the negative charge on charged bases. Electron-withdrawing groups diminish the charge and stabilize the base, making it less reactive and a weaker base. Electron-donating groups will increase the charge and destabilize the base, making it a stronger base. Inductive effects also affect the strength of neutral bases by stabilizing or destabilizing the positive charge on the conjugate acid. Electron-donating groups stabilize the positive charge and stabilize the conjugate acid which means that it will be formed more easily and the original base will be a strong base. Electron-withdrawing groups will have the opposite effect.

Solvation effects

Solvation affects basic strength. Water solvates alkyl ammonium ions by forming hydrogen bonds to N–H protons. The greater the number of N–H protons, the greater the solvation and the greater the stabilizing effect on the alkyl ammonium ion. The solvation effect is greater for alkyl ammonium ions formed from primary amines than it is for alkyl ammonium ions formed from secondary and tertiary amines. Therefore, primary amines should be stronger bases than secondary or tertiary amines. However, the inductive effect of alkyl groups is greater for tertiary amines than it is for primary and secondary amines. Therefore, it is not possible to predict the relative order of basicity for primary, secondary and tertiary amines.

Resonance

Resonance can stabilize a negative charge by delocalizing it over two or more atoms. Stabilization of the charge means that the ion is less reactive and is a weaker base. Carboxylate ions are weaker bases than phenolate ions, and phenolate ions are weaker bases than alkoxide ions. Aromatic amines are weaker bases than alkylamines since the lone pair of electrons on an aromatic amine interacts with the aromatic ring through resonance and is less available for bonding to a proton.

Amines and amides

Amines are weak bases. They have a lone pair of electrons which can bind to a proton and are in equilibrium with their conjugate acid in aqueous solution. Amides are not basic because the lone pair of electrons on the nitrogen is involved in a resonance mechanism which involves the neighboring carbonyl group.

Related topics

Neutral inorganic species (E3)
Organic structures (E4)
Brønsted–Lowry acids and bases (G1)
Acid strength (G2)
Properties of amines (O2)

Electronegativity

Electronegativity has an important influence to play on basic strength. If we compare the fluoride ion, hydroxide ion, amide ion and the methyl carbanion, then the order of basicity is as shown (*Fig. 1*).

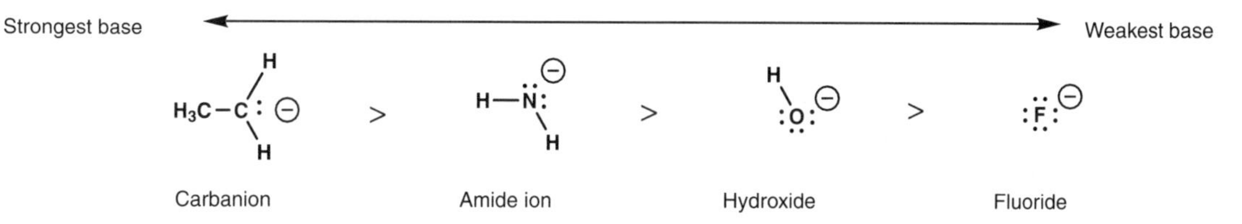

Fig. 1. Comparison of basic strength.

The strongest base is the carbanion since this has the negative charge situated on the least electronegative atom – the carbon atom. The weakest base is the fluoride ion which has the negative charge situated on the most electronegative atom – the fluorine atom. Strongly electronegative atoms such as fluorine are able to stabilize a negative charge making the ion less reactive and less basic. The order of basicity of the anions formed from alkanes, amines, and alcohols follows a similar order for the same reason (*Fig. 2*).

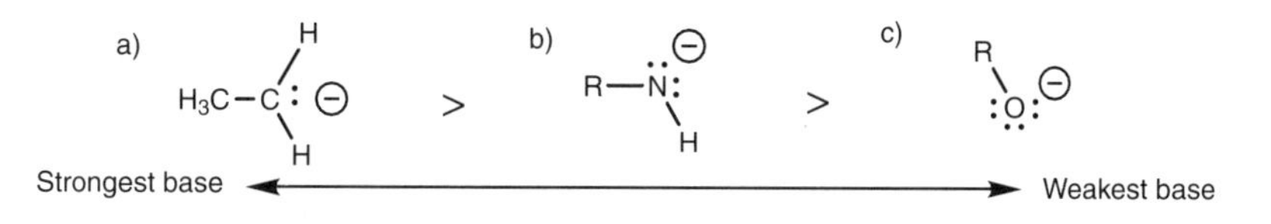

Fig. 2. Comparison of basic strengths: (a) a carbanion; (b) an amide ion; (c) an alkoxide ion.

Electronegativity also explains the order of basicity for neutral molecules such as amines, alcohols, and alkyl halides (*Fig. 3*).

These neutral molecules are much weaker bases than their corresponding anions, but the order of basicity is still the same and can be explained by

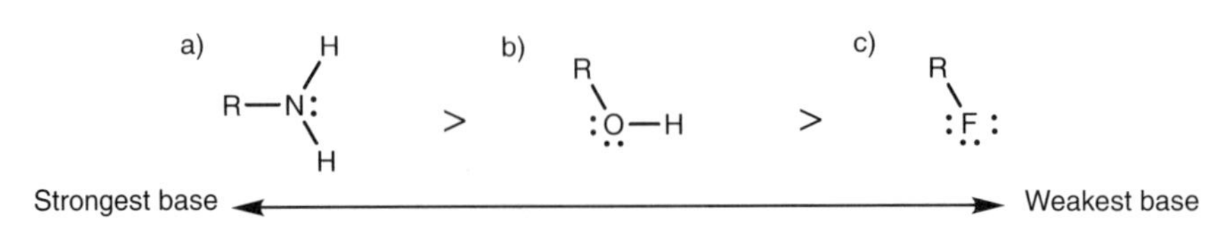

Fig. 3. Comparison of basic strengths: (a) an amine; (b) an alcohol; (c) an alkyl fluoride.

considering the relative stability of the cations which are formed when these molecules bind a proton (*Fig. 4*).

A nitrogen atom can stabilize a positive charge better than a fluorine atom since the former is less electronegative. Electronegative atoms prefer to have a negative charge rather than a positive charge. Fluorine is so electronegative that its basicity is negligible. Therefore, amines act as weak bases in aqueous solution and are partially ionized. Alcohols only act as weak bases in acidic solution. Alkyl halides are essentially nonbasic even in acidic solutions.

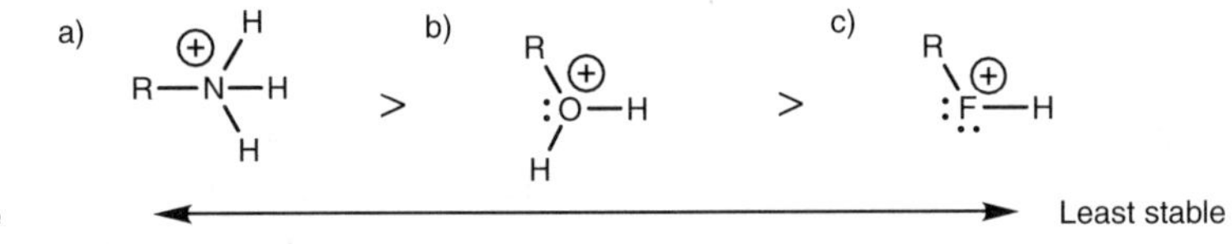

Fig. 4. Relative stability of the cations formed from (a) an amine; (b) an alcohol; (c) an alkyl fluoride

pK_b

pK_b is a measure of basic strength. If methylamine is dissolved in water, an equilibrium is set up (*Fig. 5*).

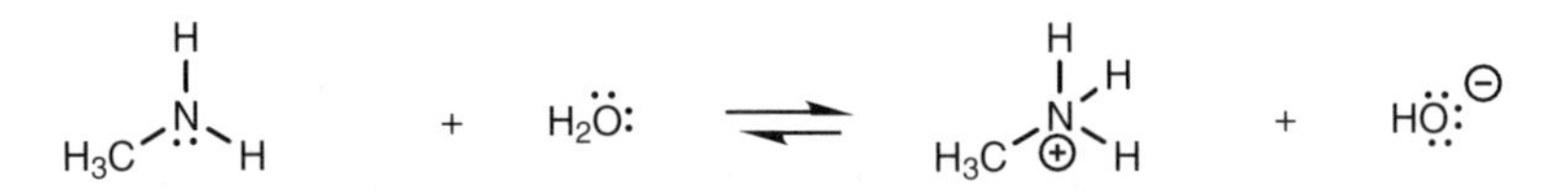

Fig. 5. Acid–base equilibrium of methylamine and water.

Methylamine on the left hand side of the equation is termed the **free base**, while the methyl ammonium ion formed on the right hand side is termed the **conjugate acid.** The extent of ionization or dissociation in the equilibrium reaction is defined by the **equilibrium constant** (K_{eq});

$$K_{eq} = \frac{[\text{Products}]}{[\text{Reactants}]} = \frac{[CH_3NH_3^+][HO^-]}{[CH_3NH_2][H_2O]} \qquad K_b = K_{eq}[H_2O] = \frac{[CH_3NH_3^+][HO^-]}{[CH_3NH_2]}$$

K_{eq} is normally measured in a dilute aqueous solution of the base and so the concentration of water is high and assumed to be constant. Therefore, we can rewrite the equilibrium equation in a simpler form where K_b is the basicity constant and includes the concentration of pure water (55.5 M). pK_b is the negative logarithm of K_b and is used as a measure of basic strength (pK_b = $-\text{Log}_{10}K_b$).

A large pK_b indicates a weak base. For example, the pK_b values of ammonia and methylamine are 4.74 and 3.36, respectively, which indicates that ammonia is a weaker base than methylamine.

pK_b and pK_a are related by the equation pK_a + pK_b = 14. Therefore, if one knows the pK_a of an acid, the pK_b of the conjugate base can be calculated and *vice versa*.

Inductive effects

Inductive effects affect the strength of a charged base by influencing the negative charge. For example, an electron-withdrawing group helps to stabilize a negative charge, resulting in a weaker base. An electron-donating group will destabilize a negative charge resulting in a stronger base. We discussed this in Topic G2 when we compared the relative acidities of the chlorinated ethanoic acids Cl_3CCO_2H, Cl_2CHCO_2H, $ClCH_2CO_2H$, and CH_3CO_2H. Trichloroacetic acid is a strong acid because its conjugate base (the carboxylate ion) is stabilized by the three electronegative chlorine groups (*Fig. 6*).

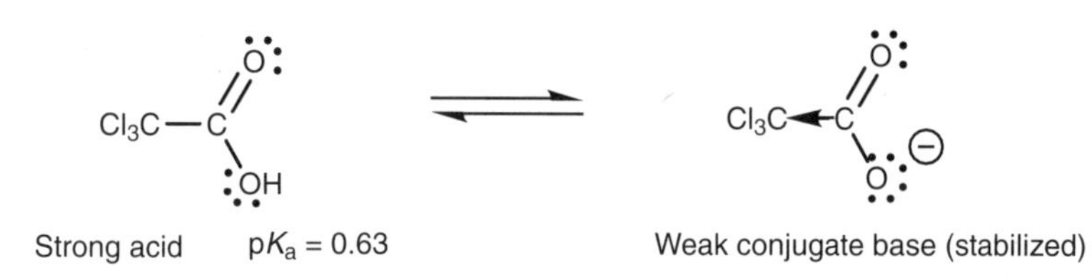

Fig. 6. Inductive effect on the conjugate base of trichloroacetic acid.

The chlorine atoms have an electron-withdrawing effect on the negative charge which helps to stabilize it. If the negative charge is stabilized, it makes the conjugate base less reactive and a weaker base. Note that the conjugate base of a strong acid is weak, while the conjugate base of a weak acid is strong. Therefore, the order of basicity for the ethanoate ions $Cl_3CCO_2^-$, $Cl_2CHCO_2^-$, $ClCH_2CO_2^-$, and $CH_3CO_2^-$ is the opposite to the order of acidity for the corresponding carboxylic acids, that is, the ethanoate ion is the strongest base, while the trichlorinated ethanoate ion is the weakest base.

Inductive effects also influence the basic strength of neutral molecules (e.g. amines). The pK_b for ammonia is 4.74, which compares with pK_b values for methylamine, ethylamine, and propylamine of 3.36, 3.25 and 3.33 respectively. The alkylamines are stronger bases than ammonia because of the inductive effect of an alkyl group on the **alkyl ammonium ion** (RNH_3^+; *Fig. 7*). Alkyl groups donate electrons towards a neighboring positive center and this helps to stabilize the ion since some of the positive charge is partially dispersed over the alkyl group. If the ion is stabilized, the equilibrium of the acid–base reaction will shift to the ion, which means that the amine is more basic. The larger the alkyl group, the more significant this effect.

If one alkyl group can influence the basicity of an amine, then further alkyl groups should have an even greater inductive effect. Therefore, one might expect secondary and tertiary amines to be stronger bases than primary amines. In fact, this is not necessarily the case. There is no easy relationship between basicity and the number of alkyl groups attached to nitrogen. Although the inductive effect of more alkyl groups is certainly greater, this effect is counterbalanced by a solvation effect.

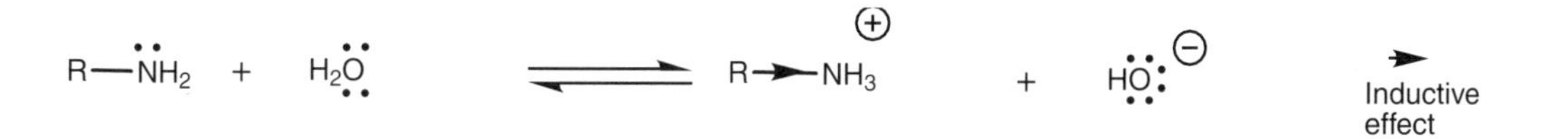

Fig. 7. Inductive effects of an alkyl group on the alkyl ammonium ion.

Solvation effects

Once the alkyl ammonium ion is formed, it is solvated by water molecules – a process which involves hydrogen bonding between the oxygen atom of water and any N–H̲ group present in the alkyl ammonium ion (*Fig. 8*). Water solvation is a stabilizing factor which is as important as the inductive effect of alkyl substituents and the more hydrogen bonds which are possible, the greater the stabilization. Solvation is stronger for the alkyl ammonium ion formed from a primary amine than for the alkyl ammonium ion formed from a tertiary amine. This is because the former ion has three N–H̲ hydrogens available for H-bonding, compared with only one such N–H̲ hydrogen for the latter. As a result, there is more solvent stabilization experienced for the alkyl ammonium ion of a primary amine compared to that experienced by the alkyl ammonium ion of a tertiary amine. This means that tertiary amines are generally weaker bases than primary or secondary amines.

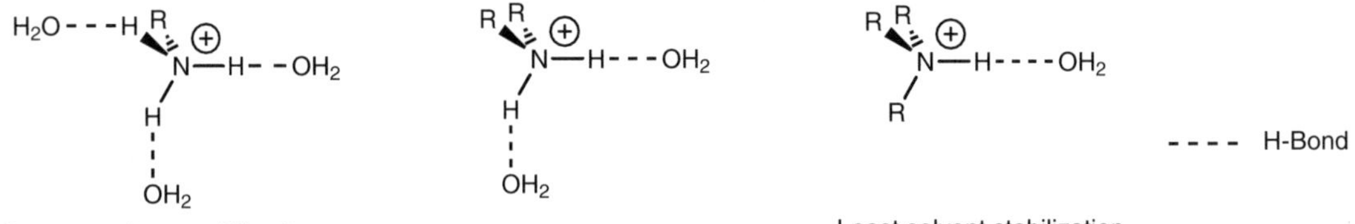

Fig. 8. Solvent effect on alkyl ammonium ions from primary, secondary, and tertiary amines.

Resonance

We have already discussed how resonance can stabilize a negative charge by delocalizing it over two or more atoms (Topic G2). This explains why a carboxylate ion is more stable than an alkoxide ion. The negative charge in the former can be delocalized between two oxygens whereas the negative charge on the former is localized on the oxygen. We used this argument to explain why a carboxylic acid is a stronger acid than an alcohol. We can use the same argument in reverse to explain the difference in basicities between a carboxylate ion and an alkoxide ion (*Fig. 9*). Since the latter is less stable, it is more reactive and is therefore a stronger base.

Resonance effects also explain why aromatic amines (arylamines) are weaker bases than alkylamines. The lone pair of electrons on nitrogen can interact with the π system of the aromatic ring, resulting in the possibility of three **zwitterionic** resonance structures (*Fig. 10*). (A zwitterion is a neutral molecule containing a positive and a negative charge.) Since nitrogen's lone pair of electrons is involved in this interaction, it is less available to form a bond to a proton and so the amine is less basic.

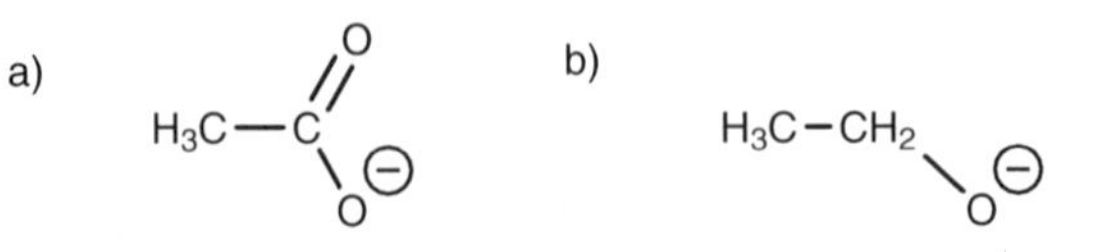

Fig. 9. (a) Carboxylate ion; (b) alkoxide ion.

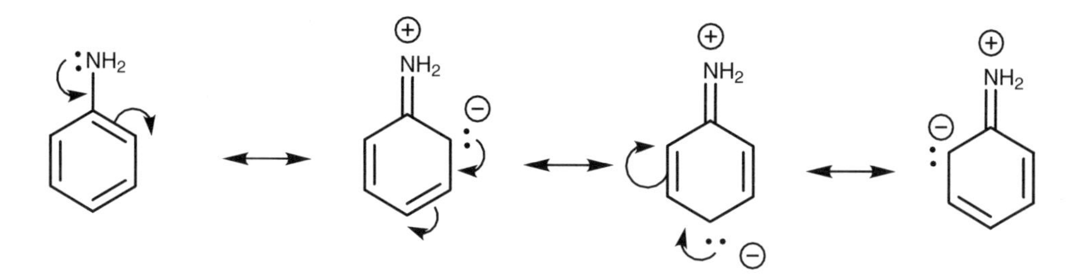

Fig. 10. Resonance structures for aniline.

Amines and amides

Amines are weak bases. They form water soluble salts in acidic solutions (*Fig. 11a*), and in aqueous solution they are in equilibrium with their conjugate acid (*Fig. 11b*).

Amines are basic because they have a lone pair of electrons which can form a bond to a proton. Amides also have a nitrogen with a lone pair of electrons, but unlike amines they are not basic. This is because a resonance takes place within the amide structure which involves the nitrogen lone pair (*Fig. 12*). The driving force behind this resonance is the electronegative oxygen of the neighboring carbonyl group which is 'hungry' for electrons. The lone pair of electrons on nitrogen forms a π bond to the neighboring carbon atom. As this takes place, the π bond of the carbonyl group breaks and both electrons move onto the oxygen to give it a total of three lone pairs and a negative charge. Since the nitrogen's lone pair is involved in this resonance, it is unavailable to bind to a proton and therefore amides are not basic.

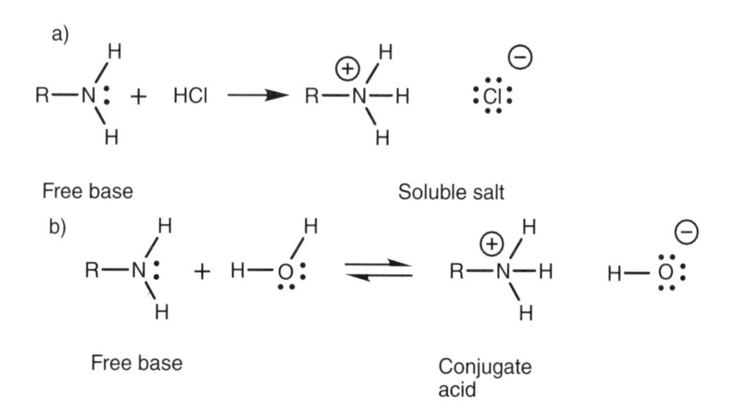

Fig. 11. (a) Salt formation; (b) acid–base equilibrium.

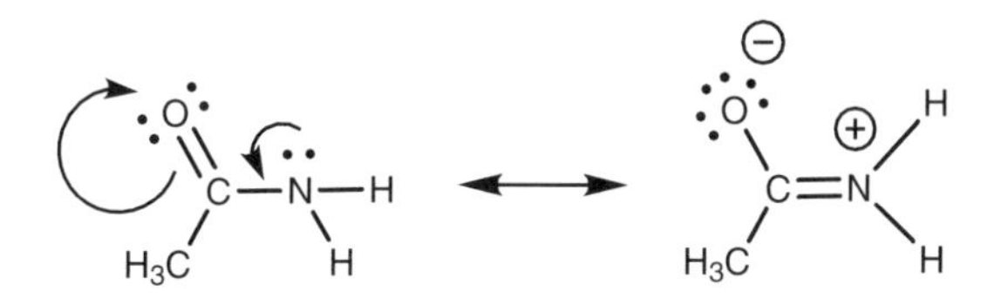

Fig. 12. Resonance interaction of an amide.

G4 LEWIS ACIDS AND BASES

Key Notes

Lewis acids

Lewis acids are electron deficient molecules which are termed acidic because they will accept a pair of electrons (in the form of a bond) from an electron-rich species in order to fill up their valence shell. BF_3, $AlCl_3$, $TiCl_4$, and $SnCl_4$ are examples of Lewis acids.

Lewis bases

Lewis bases are ions or neutral molecules containing an atom with a lone pair of electrons. Lewis bases use a lone pair of electrons to form a bond to a Lewis acid.

Related topics

Brønsted–Lowry acids and bases (G1)

Electrophilic substitutions of benzene (I3)

Lewis acids

Lewis acids are ions or electron deficient molecules with an unfilled valence shell. They are classed as acids because they can accept a lone pair of electrons from another molecule to fill their valence shell. Lewis acids include all the Brønsted–Lowry acids we have already discussed, as well as ions (e.g. H^+, Mg^{2+}), and neutral species such as BF_3 and $AlCl_3$.

Both Al and B are in Group 3A of the periodic table and have three valence electrons in their outer shell. This means that these elements can form three bonds. However, there is still room for a fourth bond. For example in BF_3, boron is surrounded by six electrons (three bonds containing two electrons each). However, boron's valence shell can accommodate eight electrons and so a fourth bond is possible if the fourth group can provide both electrons for the new bond. Since both boron and aluminum are in Group 3A of the periodic table, they are electropositive and will react with electron-rich molecules in order to obtain this fourth bond. Many transition metal compounds can also act as Lewis acids (e.g. $TiCl_4$ and $SnCl_4$).

Lewis bases

A Lewis base is a molecule which can provide a lone pair of electrons to fill the valence shell of a Lewis acid (*Fig. 1*). The base can be a negatively charged group such as a halide, or a neutral molecule such as water, an amine, or an ether, as long as there is an atom present with a lone pair of electrons (i.e. O, N, or a halogen).

All the Brønsted–Lowry bases discussed earlier can also be defined as Lewis bases. The crucial feature is the presence of a lone pair of electrons which is available for bonding. Therefore, all negatively charged ions and all functional groups containing a nitrogen, oxygen, or halogen atom can act as Lewis bases.

(a) $:\ddot{F}:^{-}$ (Lewis base) + BF_3 (Lewis acid) → $:\ddot{F}-BF_3^{-}$

(b) $R-\ddot{O}(R):$ (Lewis base) + BF_3 (Lewis acid) → $R-\overset{+}{O}(R)-BF_3^{-}$

Fig. 1. Reactions between Lewis acids and Lewis bases.

G5 ENOLATES

Key Notes

Acidic C–H protons

Hydrogen atoms attached to a carbon are not usually acidic, but if the carbon is next to a carbonyl group, any attached α protons are potentially acidic.

Stabilization

The negative charge resulting from loss of an α proton can be stabilized by a resonance process which places it onto an electronegative oxygen atom. The ion formed is called an enolate ion.

Mechanism

The mechanism is a concerted process whereby the acidic proton is lost at the same time as the C=C double bond is formed and the C=O π bond is broken. The electronegative carbonyl oxygen is the reason why the α proton is acidic.

Enolate ion

The enolate ion is a hybrid of two resonance structures where the negative charge is delocalized over three sp^2 hybridized atoms. Orbital diagrams can be used to predict whether an acidic proton is in the correct orientation to be lost.

Related topics

Reactions of enolate ions (J8)
α-Halogenation (J9)
Enolate reactions (K7)

Acidic C–H protons

Most acidic protons are attached to heteroatoms such as halogen, oxygen, and nitrogen. Protons attached to carbon are not normally acidic but there are exceptions. One such exception occurs with aldehydes or ketones when there is a C<u>H</u>R_2, C<u>H</u>$_2$R or C<u>H</u>$_3$ group next to the carbonyl group (*Fig. 1*). The protons indicated are acidic and are attached to what is known as the α (alpha) carbon. They are therefore termed as α protons.

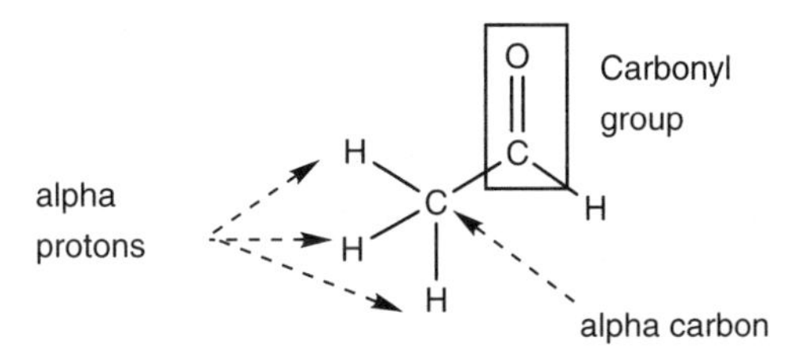

Fig. 1. Acidic α protons.

Treatment with a base results in loss of one of the acidic α protons (*Fig. 2*).

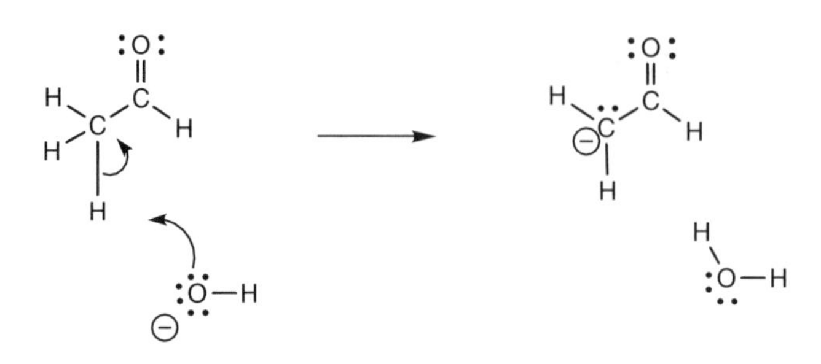

Fig. 2. Loss of an α proton and formation of a carbanion.

A lone pair on the hydroxide oxygen forms a new bond to an α proton. At the same time as this happens, the C–H bond breaks. Both electrons of that bond end up on the carbon atom and give it a lone pair of electrons and a negative charge (a **carbanion**). However, carbanions are usually very reactive, unstable species which are not easily formed. Therefore, some form of stabilization is involved here.

Stabilization

Since carbon is not electronegative, it cannot stabilize the charge. However, stabilization is possible through resonance (*Fig. 3*). The lone pair of electrons on the carbanion form a new π bond to the carbonyl carbon. As this bond is formed, the weak π bond of the carbonyl group breaks and both these electrons move onto the oxygen. This results in the negative charge ending up on the electronegative oxygen where it is more stable. This mechanism is exactly the same as the one described for the carboxylate ion (Topic G2). However, whereas both resonance structures are equally stable in the carboxylate ion, this is not the case here. The resonance structure having the charge on the oxygen atom (an **enolate ion**) is more stable than the original carbanion resonance structure. Therefore, the enolate ion will predominate over the carbanion.

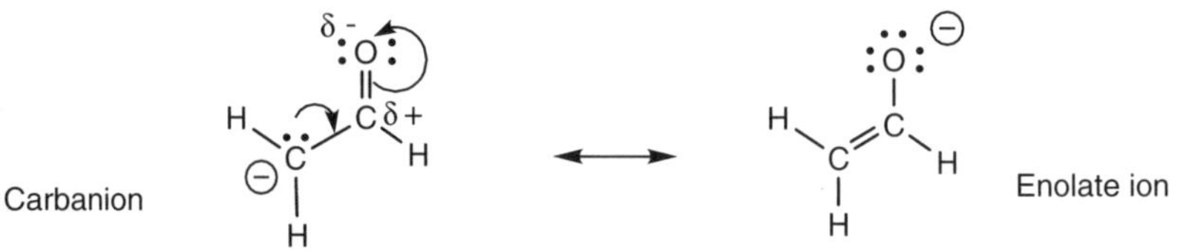

Fig. 3. Resonance interaction between carbanion and enolate ion.

Mechanism

Since the enolate ion is the preferred resonance structure, a better mechanism for the acid base reaction shows the enolate ion being formed at the same time as the acidic proton is lost (*Fig. 4*). As the hydroxide ion forms its bond to the acidic proton, the C–H bond breaks, and the electrons in that bond form a π bond to the carbonyl carbon atom. At the same time, the carbonyl π bond breaks such that both electrons move onto the oxygen. Note that it is the electronegative oxygen which is responsible for making the α proton acidic.

Enolate ion

Resonance structures represent the extreme possibilities for a particular molecule and the true structure is really a hybrid of both (*Fig. 5*). The 'hybrid' structure shows that the negative charge is 'smeared' or delocalized between three sp^2 hybridized atoms. Since these atoms are sp^2 hybridized, they are planar and have

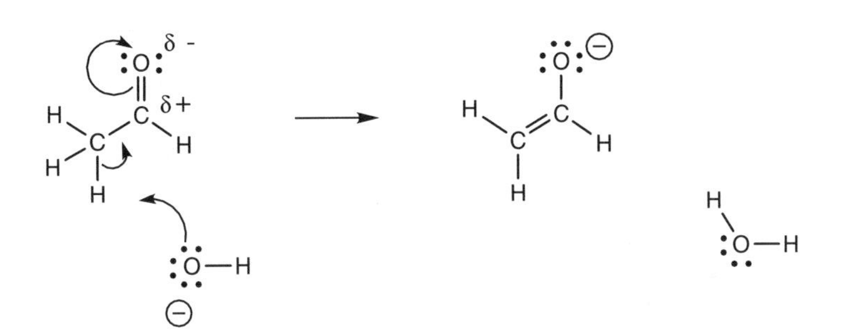

Fig. 4. Mechanism for the formation of the enolate ion.

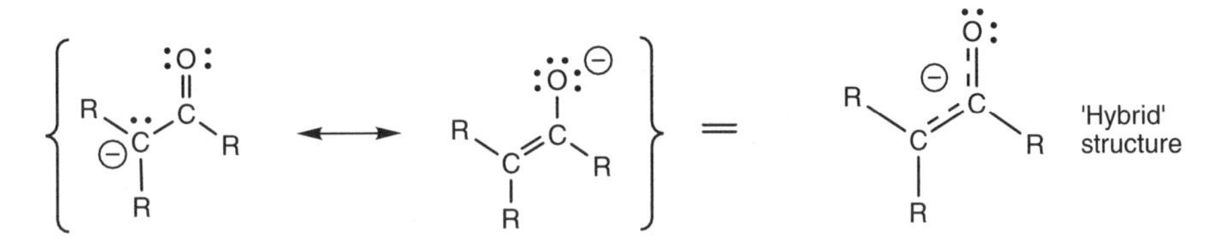

Fig. 5. Resonance structures and 'hybrid' structure for the enolate ion.

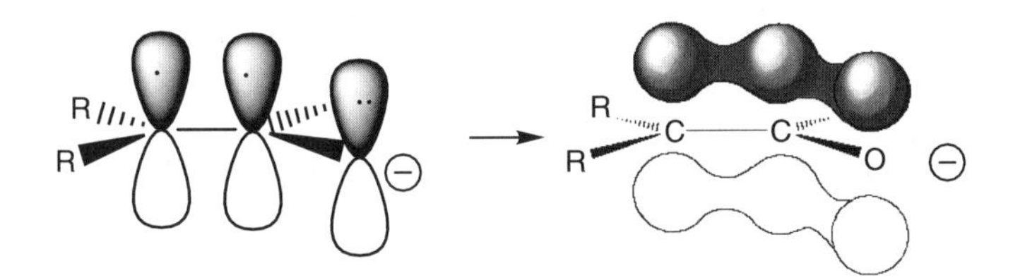

Fig. 6. Interaction of 2p orbitals to form a molecular orbital.

a $2p$ orbital which can interact with its neighbors to form one molecular orbital, thus spreading the charge between the three atoms (*Fig. 6*). Bearing this in mind, it is possible to state which of the methyl hydrogens is most likely to be lost in the formation of an enolate ion. The hydrogen circled (*Fig. 7a*) is the one which will be lost since the σ C–H bond is correctly orientated to interact with the π orbital of the carbonyl bond. The orbital diagram (*Fig. 7b*) illustrates this interaction. A Newman diagram can also be drawn by looking along the C–C bond to indicate the relative orientation of the α hydrogen which will be lost (*Fig. 7c*). In this particular example, there is no difficulty in the proton being in the correct orientation since there is free rotation around the C–C single bond. However, in cyclic systems, the hydrogen atoms are locked in space and the relative stereochemistry is important if the α proton is to be acidic.

Enolate ions formed from ketones or aldehydes are extremely important in the synthesis of more complex organic molecules (Topics J8 and J9). The ease with which an enolate ion is formed is related to the acidity of the α proton. The pK_a of propanone (acetone) is ≈19.3 which means that it is a stronger acid compared to ethane (pK_a ≈60) and a much weaker acid than acetic acid (pK_a 4.7). This means that strong bases such as sodium hydride, sodium amide, and lithium diisopropylamide ($LiN(i\text{-}C_3H_7)_2$) are required to form an enolate ion.

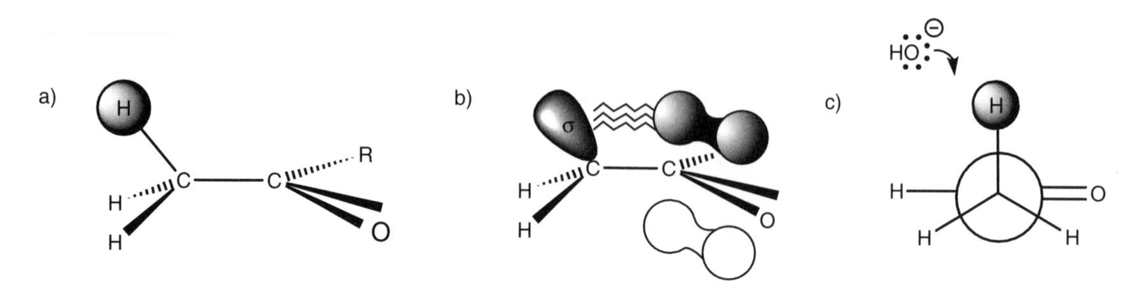

Fig. 7. *(a) α proton; (b) orbital diagram illustrating orbital interactions; (c) Newman projection.*

However, the acidity of the α proton is increased if it is flanked by two carbonyl groups rather than one, for example, 1,3-diketones (β-diketones) or 1,3-diesters (β-keto esters). This is because the negative charge of the enolate ion can be stabilized by both carbonyl groups resulting in three resonance structures (*Fig. 8*). For example, the pK_a of 2,4-pentanedione is 9.

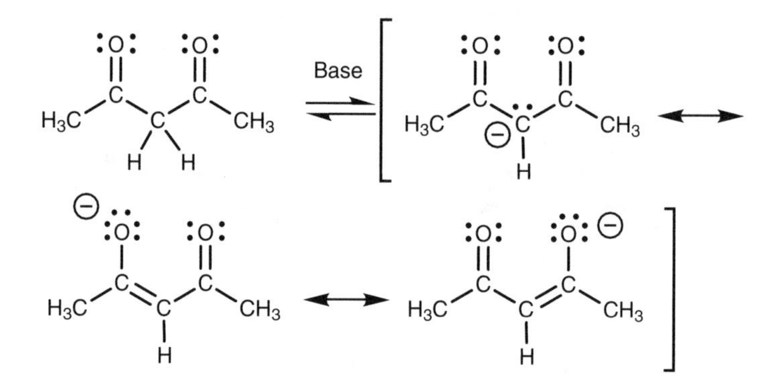

Fig. 8. *Resonance structures for the conjugate base of a 1,3-diketone.*

H1 PREPARATION OF ALKENES AND ALKYNES

Key Notes

Alkenes

Alkenes can be synthesized by the reduction of alkynes or by the elimination of alkyl halides and alcohols. Vicinal dibromides can be debrominated by treatment with zinc dust in acetic acid or with sodium iodide in acetone.

Alkynes

Alkenes can be treated with bromine to give a vicinal dibromide. Treatment of the dibromide with a strong base such as sodium amide results in the loss of two molecules of hydrogen bromide (dehydrohalogenation) and the formation of an alkyne.

Related topics

Reduction of alkynes (H9)
Elimination (L4)
Reactions of alcohols (M4)

Alkenes

Alkenes can be obtained by the transformation of various functional groups such as the reduction of alkynes (Topic H9), the elimination of alkyl halides (Topic L6), or the elimination of alcohols (Topic M4).

Alkenes can also be synthesized from vicinal dibromides, that is, molecules which have bromine atoms on neighboring carbon atoms. This reaction is called a **debromination** reaction and is carried out by treating the dibromide with sodium iodide in acetone or with zinc dust in acetic acid (*Fig. 1*).

The dibromide itself is usually prepared from the same alkene (Topic H3) and so the reaction is not particularly useful for the synthesis of alkenes. It is useful, however, in protection strategy. During a lengthy synthesis, it may be necessary to protect a double bond so that it does not undergo any undesired reactions. Bromine can be added to form the dibromide and removed later by debromination in order to restore the functional group.

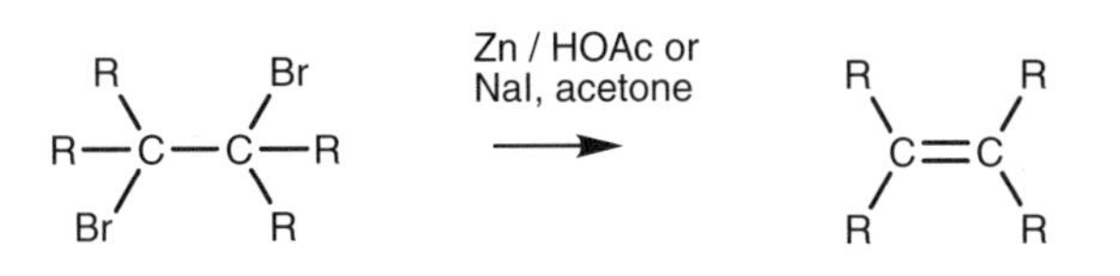

Fig. 1. Synthesis of an alkene from a vicinal dibromide.

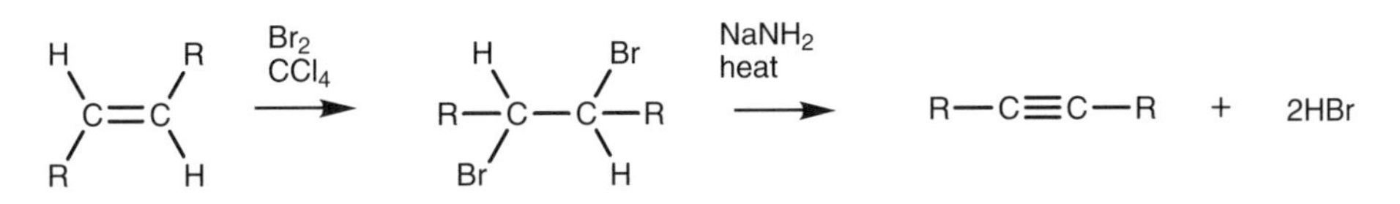

Fig. 2. Synthesis of an alkyne from an alkene.

Alkynes

Alkynes can be synthesized from alkenes through a two-step process which involves the **electrophilic addition** of bromine to form a vicinal dibromide (Topic H3) then **dehydrohalogenation** with strong base (*Fig. 2*). The second stage involves the loss of two molecules of hydrogen bromide and so two equivalents of base are required.

H2 PROPERTIES OF ALKENES AND ALKYNES

Key Notes

Structure

Alkenes are planar with bond angles of 120°. The carbon atoms of the C=C bond are sp^2 hybridized and the double bond is made up of one σ bond and one π bond. Alkynes are linear with the triple bond carbons being *sp* hybridized. The triple bond is made up of one σ bond and two π bonds.

C=C Bond

The C=C bond is stronger and shorter than a C–C single bond. However, the two bonds making up the C=C bond are not of equal strength. The π bond is weaker than the σ bond. Bond rotation round a C=C bond is not possible and isomers are possible depending on the substituents present. The more substituents which are present on an alkene, the more stable the alkene is.

C≡C Bond

An alkyne triple bond is stronger than a C–C single bond or a C=C double bond. The two π bonds present in the triple bond are weaker and more reactive than the σ bond.

Properties

Alkenes and alkynes are nonpolar compounds which dissolve in nonpolar solvents and are very poorly soluble in water. They have low boiling points since only weak van der Waals interactions are possible between the molecules.

Nucleophilicity

Alkenes and alkynes act as nucleophiles and react with electrophiles by a reaction known as electrophilic addition. The nucleophilic centers are the multiple bonds which are areas of high electron density.

Related topics

sp^2 Hybridization (A4)
sp Hybridization (A5)
Bonds and hybridized centers (A6)
Configurational isomers – alkenes and cycloalkanes (D2)
Organic structures (E4)

Structure

The alkene functional group ($R_2C{=}CR_2$) is planar in shape with bond angles of 120°. The two carbon atoms involved in the double bond are both sp^2 hybridized. Each carbon has three sp^2 hybridized orbitals which are used for σ bonds while the *p* orbital is used for a π bond. Thus, the double bond is made up of one σ bond and one π bond (*Fig. 1*).

The alkyne functional group consists of a carbon carbon triple bond and is linear in shape with bond angles of 180° (*Fig. 2*).

The two carbon atoms involved in the triple bond are both *sp* hybridized, such that each carbon atom has two *sp* hybridized orbitals and two *p* orbitals. The *sp* hybridized orbitals are used for two σ bonds while the *p* orbitals are used for two π bonds. Thus, the triple bond is made up of one σ bond and two π bonds.

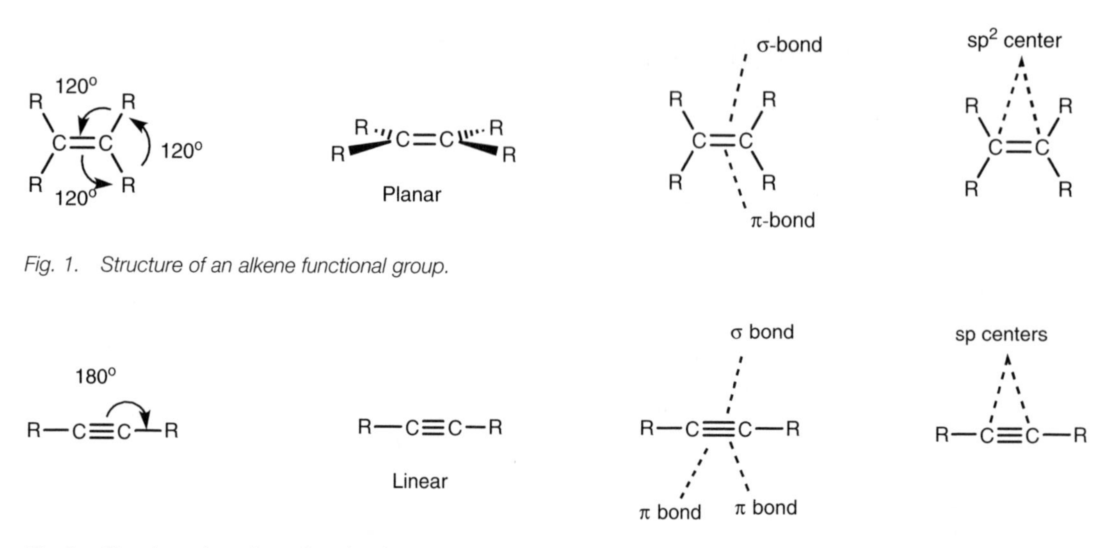

Fig. 1. Structure of an alkene functional group.

Fig. 2. Structure of an alkyne functional group.

C=C Bond

The C=C bond is stronger (152 kcal mol^{-1}) and shorter (1.33 Å) than a C–C single bond (88 kcal mol^{-1} and 1.54 Å respectively). A C=C bond contains one σ bond and one π bond, with the π bond being weaker than the σ bond. This is important with respect to the reactivity of alkenes.

Bond rotation is not possible for a C=C double bond since this would require the π bond to be broken. Therefore, isomers of alkenes are possible depending on the relative position of the substituents. These can be defined as *cis* or *trans*, but are more properly defined as (*Z*) or (*E*) (see Topic D2).

Alkenes are defined as mono-, di-, tri-, or tetrasubstituted depending on the number of substituents which are present. The more substituents which are present, the more stable the alkene.

C≡C Bond

The bond length of a carbon carbon triple bond is 1.20 Å and the bond strength is 200 kcal mol^{-1}. The π bonds are weaker than the σ bond. The presence of the π bonds explains why alkynes are more reactive than alkanes.

Properties

Alkenes and alkynes have physical properties similar to alkanes. They are relatively nonpolar, dissolve in nonpolar solvents and are not very soluble in water. Only weak van der Waals interactions are possible between unsaturated molecules such as alkene and alkynes, and so these structures have low boiling points compared to other functional groups.

Fig. 3. Nucleophilic centers of an alkene and an alkyne.

Nucleophilicity

Alkenes and alkynes are nucleophilic and commonly react with electrophiles in a reaction known as electrophilic addition. The nucleophilic center of the alkene or alkyne is the double bond or triple bond (*Fig. 3*). These are areas of high electron density due to the bonding electrons. The specific electrons which are used to form bonds to attacking electrophiles are those involved in π bonding.

H3 Electrophilic addition to symmetrical alkenes

Key Notes

Reactions

Alkenes readily undergo electrophilic addition reactions. The π bond is involved in the reaction and new substituents are added to either end of the original alkene.

Symmetrical and unsymmetrical alkenes

Symmetrical alkenes have the same substituents at each end of the double bond. Unsymmetrical alkenes do not.

Hydrogen halide addition

Treating an alkene with a hydrogen halide results in the formation of an alkyl halide. The proton from the hydrogen halide adds to one end of the double bond and the halogen atom to the other. The mechanism of electrophilic addition is a two stage process which goes through a carbocation intermediate. In the first stage, the alkene acts as a nucleophile and the hydrogen halide acts as an electrophile. In the second stage of the mechanism, the halide ion acts as a nucleophile and the carbocation intermediate is an electrophile.

Halogen addition

Alkenes react with bromine or chlorine to produce vicinal dihalides with the halogen atoms adding to each end of the double bond. The reaction is useful in the protection or purification of alkenes or as a means of synthesizing alkynes. The halogen molecule is polarized as it approaches the alkene double bond thus generating the required electrophilic center. The intermediate formed is called a bromonium ion intermediate in the case of bromine and is stabilized since it is possible to share or delocalize the positive charge between three atoms. If the reaction is carried out in water, water can act as a nucleophile and intercept the reaction intermediate to form a halohydrin where a halogen atom is added to one end of the double bond and a hydroxyl group is added to the other.

Alkenes to alcohols

Alkenes can be converted to alcohols by treatment with aqueous acid (e.g. sulfuric acid). Milder conditions can be used if mercuric acetate is used to produce an organomercury intermediate which is reduced with sodium borohydride.

Alkenes to ethers

A similar reaction to the mercuric acetate/sodium borohydride synthesis of alcohols allows the conversion of alkenes to ethers. In this case, mercuric trifluoracetate is used.

Alkenes to arylalkanes

Alkenes can be reacted with aromatic rings to give arylalkanes. The reaction is known as a Friedel–Crafts alkylation of the aromatic ring but can also be viewed as another example of an electrophilic addition to an alkene.

Related topics

Organic structures (E4)
Electrophilic addition to unsymmetrical alkenes (H4)
Carbocation stabilization (H5)
Hydroboration of alkenes (H7)
Electrophilic substitutions of benzene (I3)

Reactions

Many of the reactions which alkenes undergo take place by a mechanism known as **electrophilic addition** (*Fig. 1*). In these reactions, the π bond of the double bond has been used to form a bond to an incoming electrophile and is no longer present in the product. Furthermore, a new substituent has been added to each of the carbon atoms.

Symmetrical and unsymmetrical alkenes

In this section we shall look at the electrophilic addition of symmetrical alkenes. A symmetrical alkene is an alkene which has the same substituents at each end of the double bond (*Fig. 2a*). Unsymmetrical alkenes have different substituents at each end of the double bond (*Fig. 2b*). Electrophilic addition to unsymmetrical alkenes will be covered in Topic H4.

Hydrogen halide addition

Alkenes react with hydrogen halides (HCl, HBr, and HI) to give an alkyl halide. The hydrogen halide molecule is split and the hydrogen atom adds to one end of the double bond while the halogen atom adds to the other. The reaction of HBr with 2,3-dimethyl-2-butene is an example of this reaction (*Fig. 3*). In this reaction, the alkene acts as a nucleophile. It has an electron-rich double bond containing four

Fig. 1. Electrophilic additions.

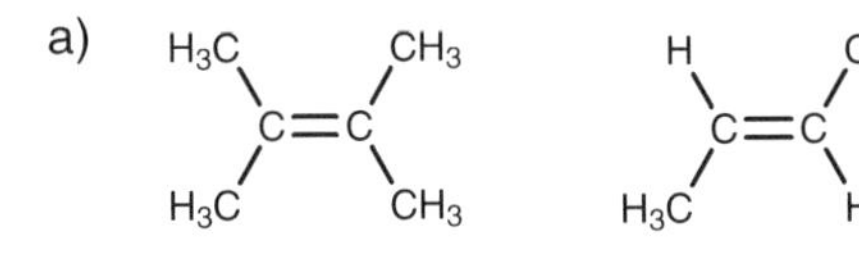

Fig. 2. (a) Symmetrical alkenes; (b) unsymmetrical alkenes.

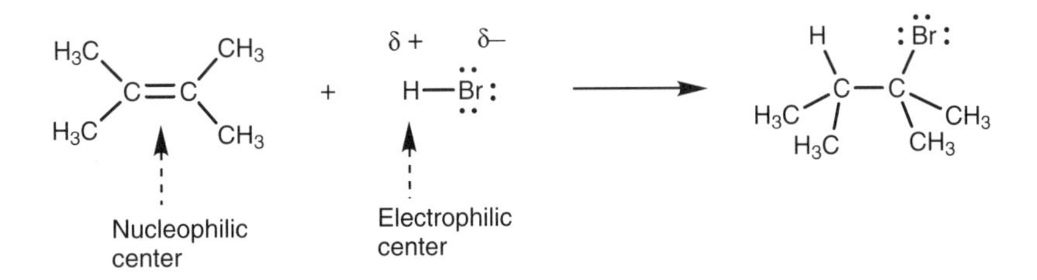

Fig. 3. Reaction of HBr with 2,3-dimethyl-2-butene.

electrons, two of which make up a strong σ bond and two of which make up a weaker π bond. The double bond can be viewed as a nucleophilic center. Hydrogen bromide has a polar H–Br bond and so the hydrogen is an electrophilic center and the bromine is a nucleophilic center. However, halogen atoms are extremely weak nucleophilic centers (see Topic E4) and so this molecule is more likely to react as an electrophile through its electrophilic hydrogen.

In the first step of electrophilic addition (*Fig. 4*), the alkene acts as a nucleophile and uses its two π electrons to form a new bond to the hydrogen of HBr. As this new bond is formed, the H–Br bond breaks since hydrogen is only allowed one bond. Both electrons in that bond end up on the bromine atom to produce a bromide ion. Since the electrons from the π bond have been used for the formation of a new σ bond, the π bond is no longer present. As a result, the 'left hand' carbon has been left with only three bonds and becomes positively charged. This is known as a **carbocation** since the positive charge is on a carbon atom.

This structure is known as a **reaction intermediate**. It is a reactive species and will not survive very long with the bromide ion in the vicinity. The carbocation is an electrophile since it is positively charged. The bromide ion is a nucleophile since it is negatively charged. Therefore, the bromide ion uses one of its lone pairs of electrons to form a new σ bond to the carbocation and the final product is formed.

The mechanism involves the addition of HBr to the alkene. It is an electrophilic addition since the first step of the mechanism involves the addition of the electrophilic hydrogen to the alkene. Note that the second step involves a nucleophilic addition of the bromide ion to the carbocation intermediate, but it is the first step which defines this reaction.

In the mechanism shown (*Fig. 4*), the π electrons of the alkene provided the electrons for a new bond between the right hand carbon and hydrogen. They could equally well have been used to form a bond between the left hand carbon and hydrogen (*Fig. 5*).

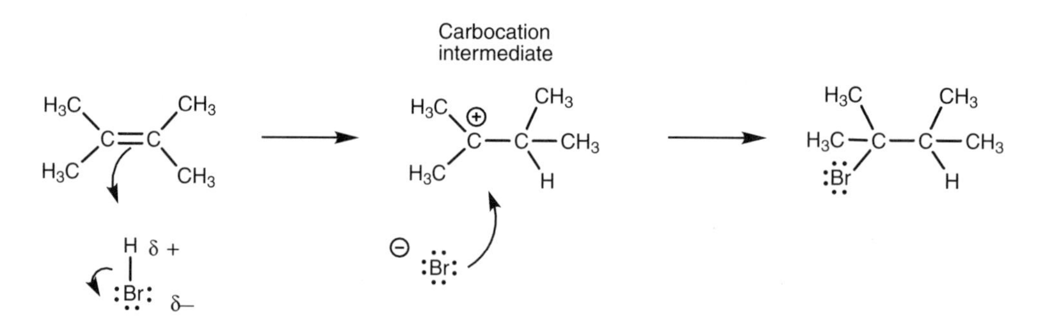

Fig. 4. Mechanism of electrophilic addition of HBr to 2,3-dimethyl-2-butene.

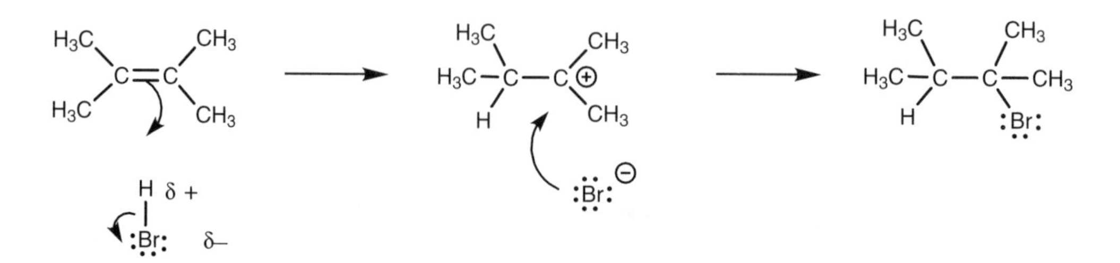

Fig. 5. 'Alternative' mechanism for electrophilic addition of HBr to 2,3-dimethyl-2-butene.

With a symmetrical alkene, the product is the same and so it does not matter which end of the double bond is used for the new bond to hydrogen. The chances are equal of the hydrogen adding to one side or the other.

The electrophilic additions of H–Cl and H–I follow the same mechanism to produce alkyl chlorides and alkyl iodides.

Halogen addition

The reaction of an alkene with a halogen such as bromine or chlorine results in the formation of a vicinal dihalide. The halogen molecule is split and the halogens are added to each end of the double bond (*Fig. 6*). Vicinal dibromides are useful in the purification or protection of alkenes since the bromine atoms can be removed under different reaction conditions to restore the alkene. Vicinal dibromides can also be converted to alkynes (Topic H1).

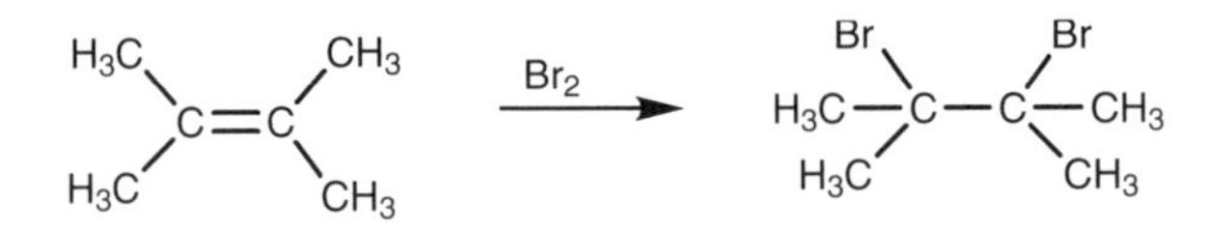

Fig. 6. Electrophilic addition of bromine to 2,3-dimethyl-2-butene.

The same mechanism described above is followed in this reaction. However, the first stage of the mechanism should involve the nucleophilic alkene reacting with an electrophilic centre, and yet there is no obvious electrophilic center in bromine. The bond between the two bromine atoms is a covalent σ bond with both electrons equally shared between the bromine atoms.

If there is no electrophilic center, how can a molecule like bromine react with a nucleophilic alkene? The answer lies in the fact that the bromine molecule approaches end-on to the alkene double bond and an electrophilic center is induced (*Fig. 7*). Since the alkene double bond is electron rich it repels the electrons in the bromine molecule and this results in a polarization of the Br–Br bond such that the nearer bromine becomes electron deficient (electrophilic). Now that an electrophilic center has been generated, the mechanism is the same as before.

There is more to this mechanism than meets the eye. The carbocation intermediate can be stabilized by neighboring alkyl groups through inductive and hyperconjugation effects (see Topic H5). However, it can also be stabilized by sharing the positive charge with the bromine atom and a second carbon atom (*Fig. 8*).

The positively charged carbon is an electrophilic center. The bromine is a weak nucleophilic center. A neutral halogen does not normally act as a nucleophile (see Topic E4), but in this case the halogen is held close to the carbocation

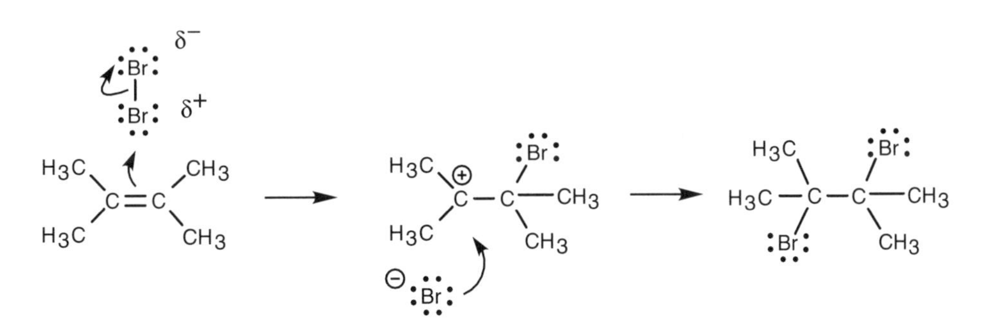

Fig. 7. Mechanism for the electrophilic addition of Br_2 with 2,3-dimethyl-2-butene.

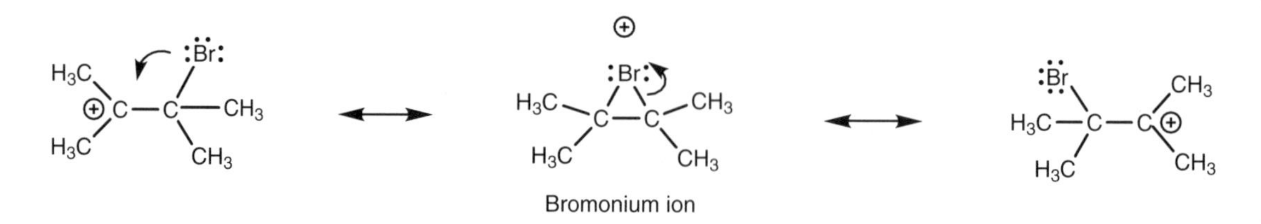

Fig. 8. Formation of the bromonium ion.

making reaction more likely. Once the lone pair of electrons on bromine is used to form a bond to the carbocation, a **bromonium** ion is formed where the bromine gains a positive charge. The mechanism can go in reverse to regenerate the original carbocation. Alternatively, the other carbon–bromine bond can break with both electrons moving onto the bromine. This gives a second carbocation where the other carbon bears the positive charge. Thus, the positive charge is shared between three different atoms and is further stabilized.

Evidence for the existence of the bromonium ion is provided from the observation that bromine adds to cyclic alkenes (e.g. cyclopentene) in an *anti*-stereochemistry (*Fig. 9*). In other words, each bromine adds to opposite faces of the alkene to produce only the *trans* isomer. None of the *cis* isomer is formed. If the intermediate was a carbocation, a mixture of *cis* and *trans* isomers would be expected since the second bromine could add from either side. With a bromonium ion, the second bromine must approach from the opposite side.

The reaction of an alkene with a halogen such as bromine and chlorine normally gives a vicinal dihalide. However, if the reaction is carried out in water as solvent, the product obtained is a halohydrin where the halogen adds to one end of the double bond and a hydroxyl group from water adds to the other (*Fig. 10*).

In this reaction, the first stage of the mechanism proceeds as normal, but then water acts as a nucleophile and 'intercepts' the carbocation intermediate (*Fig. 11*). Since water is the solvent, there are far more molecules of it present compared to the number of bromide ions generated from the first stage of the mechanism.

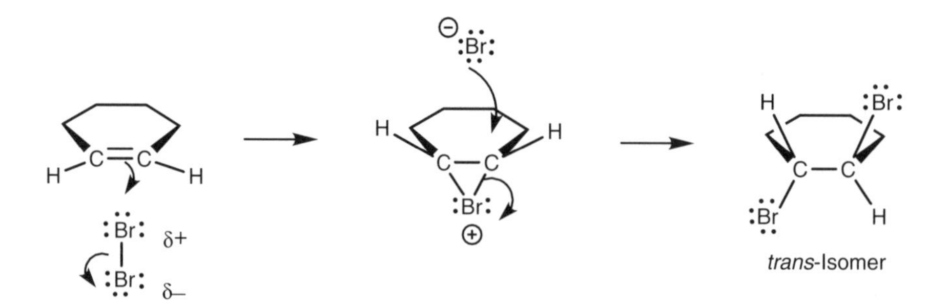

Fig. 9. Anti-stereochemistry of bromine addition to a cyclic alkene.

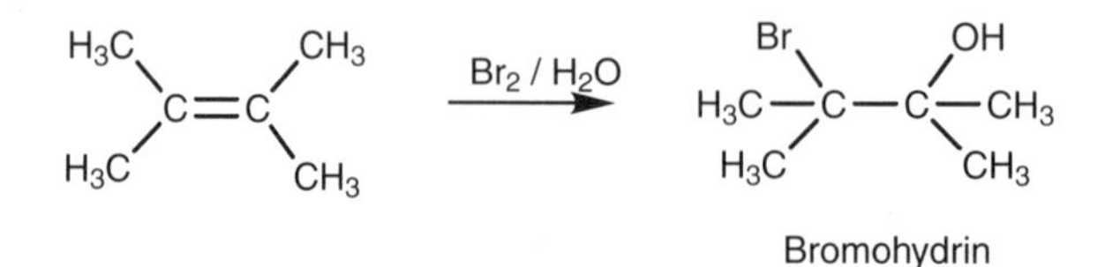

Fig. 10. Formation of a bromohydrin from 2,3-dimethyl-2-butene.

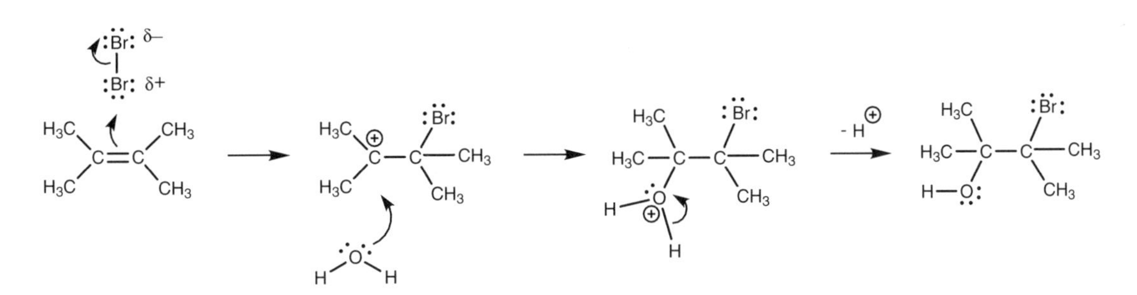

Fig. 11. Mechanism of bromohydrin formation.

Water uses a lone pair of electrons on oxygen to form a bond to the carbocation. As a result, the oxygen effectively 'loses' an electron and gains a positive charge. This charge is lost and the oxygen regains its second lone pair when one of the O–H bonds breaks and both electrons move onto the oxygen.

Alkenes to alcohols

Alkenes can be converted to alcohols by treatment with aqueous acid (sulfuric or phosphoric acid; *Fig. 12*). This electrophilic addition reaction involves the addition of water across the double bond. The hydrogen adds to one carbon while a hydroxyl group adds to the other carbon.

Sometimes the reaction conditions used in this reaction are too harsh since heating is involved and rearrangement reactions can take place. A milder method which gives better results is to treat the alkene with mercuric acetate [$Hg(OAc)_2$] then sodium borohydride (*Fig. 13*). The reaction involves electrophilic addition of the mercury reagent to form an intermediate **mercuronium** ion. This reacts with water to give an organomercury intermediate. Reduction with sodium borohydride replaces the mercury substituent with hydrogen and gives the final product (*Fig. 13*)

Alkenes can also be converted to alcohols by hydroboration (see Topic H7).

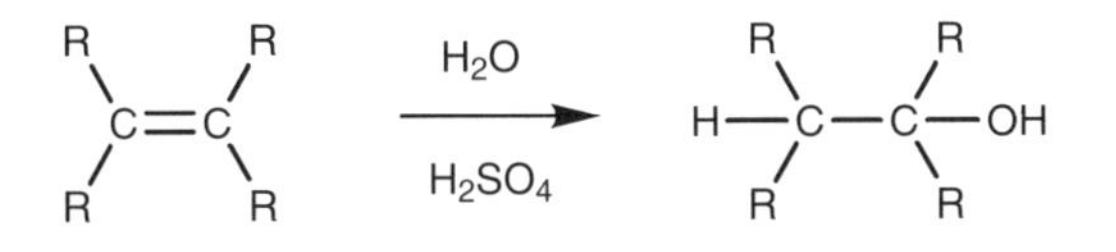

Fig. 12. Synthesis of an alcohol from an alkene.

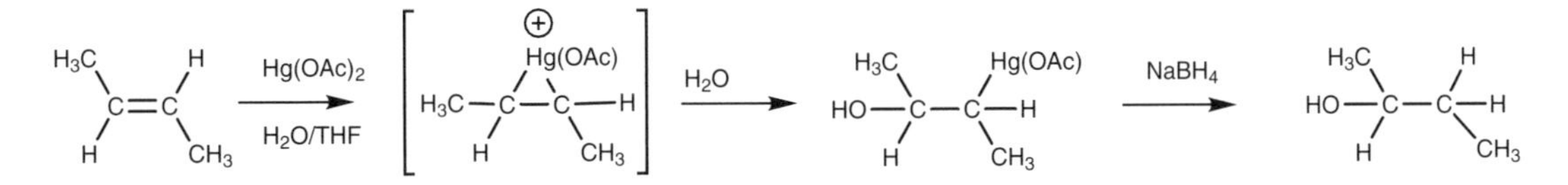

Fig. 13. Synthesis of an alcohol from an alkene using mercuric acetate.

Alkenes to ethers

A similar reaction to the mercuric acetate/sodium borohydride synthesis of alcohols allows the conversion of alkenes to ethers. In this case, mercuric trifluoracetate is used (*Fig. 14*).

Alkenes to arylalkanes

The reaction of an aromatic ring such as benzene with an alkene under acid conditions results in the formation of an arylalkane (*Fig. 15*). As far as the alkene

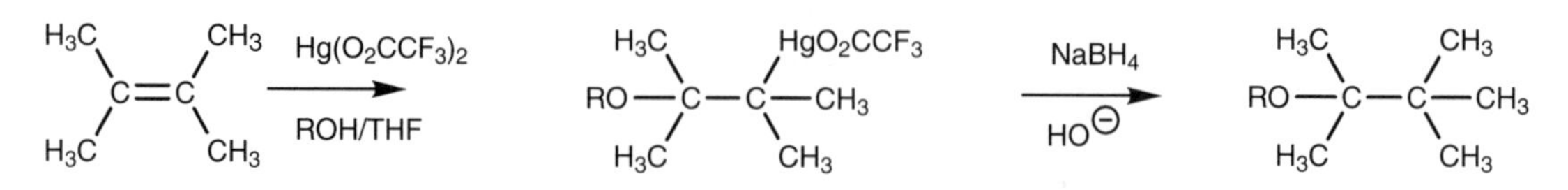

Fig. 14. Synthesis of an ether from an alkene using mercuric trifluoroacetate.

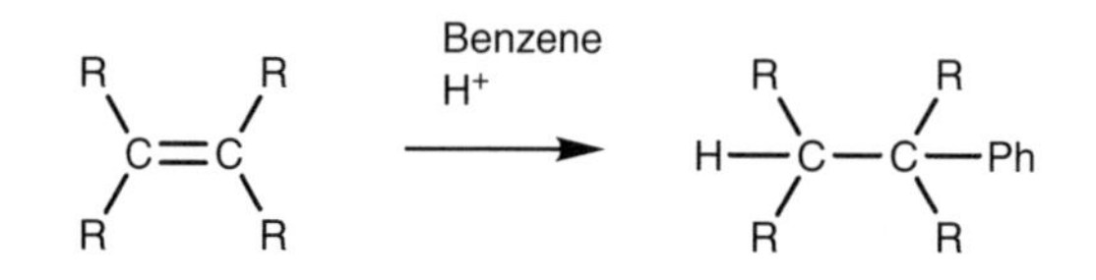

Fig. 15. Synthesis of arylalkanes from alkenes.

is concerned this is another example of electrophilic addition involving the addition of a proton to one end of the double bond and the addition of the aromatic ring to the other. As far as the aromatic ring is concerned this is an example of an electrophilic substitution reaction called the **Friedel–Crafts alkylation** (Topic I3).

H4 ELECTROPHILIC ADDITION TO UNSYMMETRICAL ALKENES

Key Notes

Addition of hydrogen halides

The addition of a hydrogen halide to an unsymmetrical alkene can result in two different products. These products are not formed in equal amounts. Markovnikov's rule states that 'in the addition of HX to an alkene, the hydrogen atom adds to the carbon atom that already has the greater number of hydrogen atoms'. This produces the more substituted alkyl halide.

Carbocation stabilities

The favored product arising from addition of a hydrogen halide to an unsymmetrical alkene will be formed from the more stable of the two possible carbocations. The more stable carbocation will have more alkyl groups attached to the positive center.

Addition of halogens

Different products are not possible from the reaction of a halogen with an unsymmetrical alkene unless water is used as a solvent, in which case a hydroxyl group ends up on the more substituted carbon. This demonstrates that the bromonium ion does not share the positive charge equally amongst the bromine and the two carbons.

Addition of water

The more substituted alcohol is the preferred product from the acidic hydrolysis of an alkene as well as from the organomercuric synthesis of alcohols.

Related topics

Electrophilic addition to symmetrical alkenes (H3)

Carbocation stabilization (H5)

Addition of hydrogen halides

The reaction of a symmetrical alkene with hydrogen bromide produces the same product regardless of whether the hydrogen of HBr is added to one end of the double bond or the other. However, this is not the case with unsymmetrical alkenes (*Fig. 1*). In this case, two different products are possible. These are not

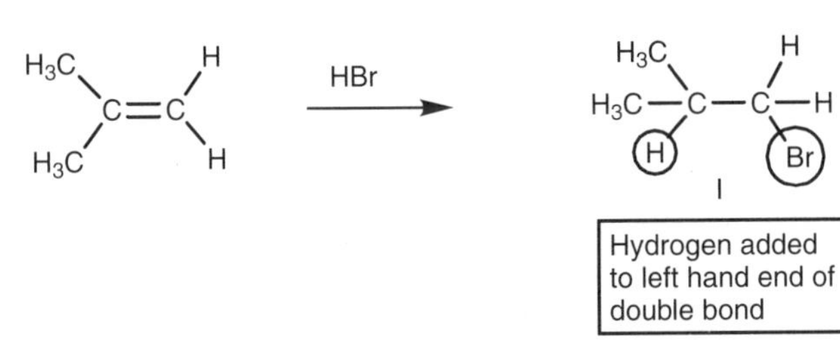

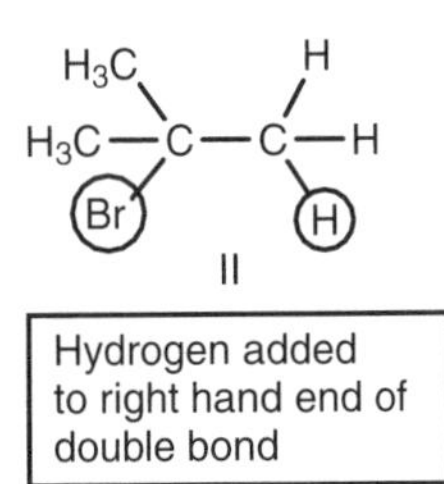

Fig. 1. Electrophilic addition of HBr to an unsymmetrical alkene.

formed to an equal extent and the more substituted alkyl halide (II) is preferred. The reaction proceeds in a **Markovnikov** sense with hydrogen ending up on the least substituted position and the halogen ending up on the most substituted position. Markovnikov's rule states that 'in the addition of HX to an alkene, the hydrogen atom adds to the carbon atom that already has the greater number of hydrogen atoms'. This produces the more substituted alkyl halide.

Carbocation stabilities

This reaction can be rationalized by proposing that the carbocation intermediate leading to product II is more stable than the carbocation intermediate leading to product I (*Fig. 2*). It is possible to predict the more stable carbocation by counting the number of alkyl groups attached to the positive center. The more stable carbocation on the right has three alkyl substituents attached to the positively charged carbon whereas the less stable carbocation on the left only has one such alkyl substituent. The reasons for this difference in stability is explained in Topic H5, but the result is summarized by Markovnikov's rule.

However, Markovnikov's rule does not always hold true. For example, the reaction of $CF_3CH{=}CH_2$ with HBr gives $CF_3CH_2CH_2Br$ rather than $CF_3CHBrCH_3$. Here, the presence of electron-withdrawing fluorine substituents has a destabilizing influence on the two possible intermediate carbocations (*Fig. 3*). The destabilizing effect will be greater for the more substituted carbocation since the carbocation is closer to the fluorine substituents and so the favoured carbocation is the least substituted one in this case.

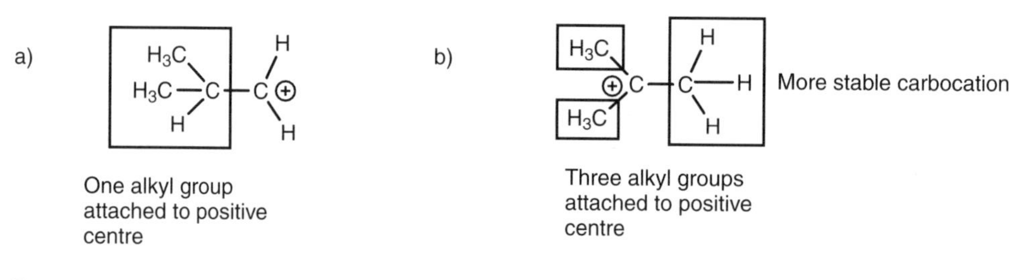

Fig. 2. (a) Carbocation leading to product I; (b) carbocation leading to product II.

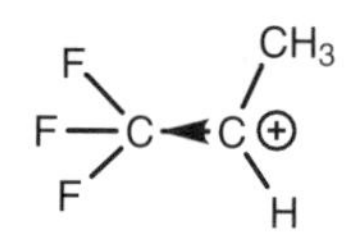

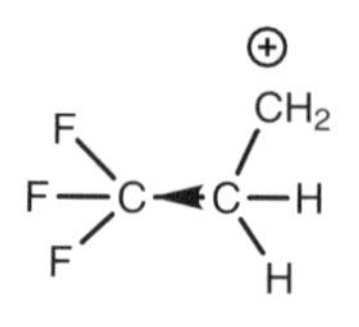

Fig. 3. Comparison of carbocations.

Addition of halogens

There is no possibility of different products when a halogen such as bromine or chlorine is added to an unsymmetrical alkene. However, this is not the case if water is used as a solvent. In such cases, the halogen is attached to the least substituted carbon and the hydroxyl group is attached to the more substituted carbon (*Fig. 4*). This result can be explained by proposing that the bromonium ion is not symmetrical and that although the positive charge is shared between the bromine

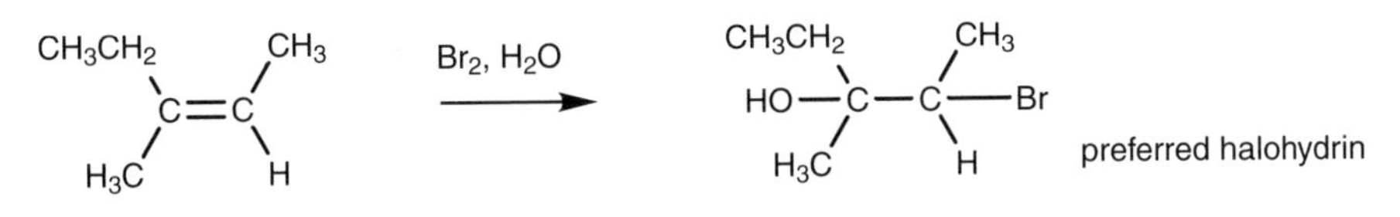

Fig. 4. Reaction of 3-methyl-2-pentene with bromine and water.

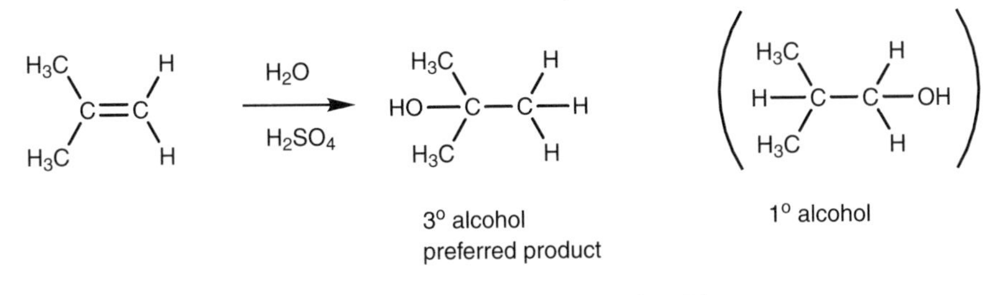

Fig. 5. Reaction of 2-methyl-1-propene with aqueous sulfuric acid.

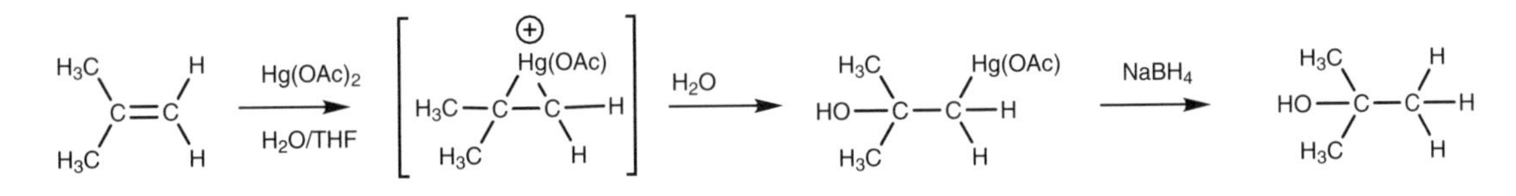

Fig. 6. Organomercuric synthesis of alcohols.

and the two carbon atoms, the positive charge is greater on the more substituted carbon compared with the less substituted carbon.

Addition of water

With unsymmetrical alkenes, the more substituted alcohol is the preferred product (*Fig. 5*).

The same holds true for the organomercuric synthesis of alcohols (*Fig. 6*).

H5 Carbocation stabilization

Key Notes

Stabilization — Carbocations are stabilized by induction, hyperconjugation, or delocalization.

Inductive effects — Alkyl groups have an electron-donating effect on any neighboring positive charge. The more alkyl groups attached, the greater the stabilizing effect.

Hyperconjugation — In hyperconjugation, the vacant $2p$ orbital of the carbocation can interact with the σ orbitals of neighboring C–H bonds. As a result, the σ electrons of the C–H bond can spend a small amount of time entering the space occupied by the $2p$ orbital such that the latter orbital is not completely empty. This interaction serves to spread or delocalize the positive charge to neighboring σ bonds and thus stabilize it. The more substituents present, the more chances there are for hyperconjugation.

Related topics — Electrophilic substitution to symmetrical alkenes (H3); Electrophilic substitution to unsymmetrical alkenes (H4)

Stabilization

Positively charged species such as carbocations are inherently reactive and unstable. The more unstable they are, the less easily they are formed and the less likely the overall reaction. Any factor which helps to stabilize the positive charge (and by inference the carbocation) will make the reaction more likely. There are three ways in which a positive charge can be stabilized: inductive effects, hyperconjugation, and delocalization. We have already seen the effects of delocalization in stabilizing the bromonium ion (Topic H3). We will now look at the effects of induction and hyperconjugation.

Inductive effects

Alkyl groups can donate electrons towards a neighboring positive center and this helps to stabilize the ion since some of the positive charge is partially dispersed over the alkyl group (*Fig. 1*). The more alkyl groups which are attached, the greater the electron donating power and the more stable the carbocation.

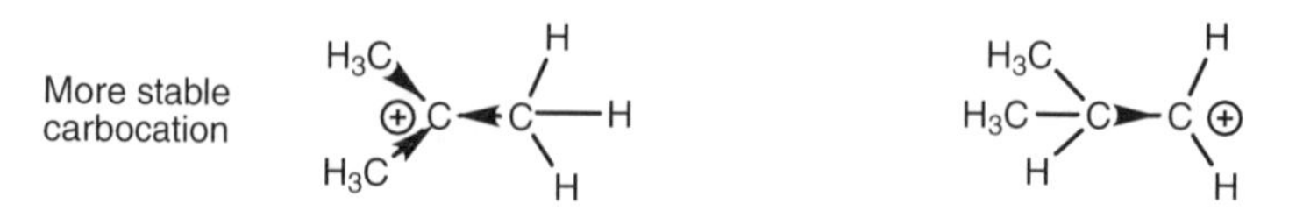

Fig. 1. Comparison of possible carbocations.

Hyperconjugation

Both carbons of an alkene are sp^2 hybridized. However, this is altered on formation of the carbocation (*Fig. 2*). When an alkene reacts with an electrophile such as a proton, both electrons in the π bond are used to form a new σ bond to

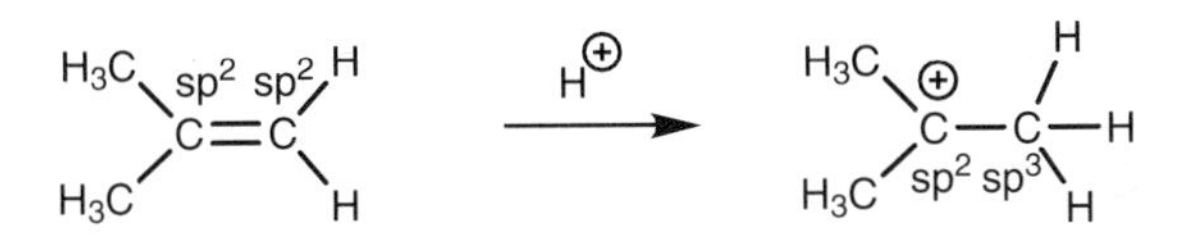

Fig. 2. *Hybridization of alkene and carbocation.*

the electrophile. As a result, the carbon which gains the electrophile becomes an sp^3 center. The other carbon containing the positive charge remains as an sp^2 center. This means that it has three sp^2 hybridized orbitals (used for the three σ bonds still present) and one vacant 2*p* orbital which is not involved in bonding. Hyperconjugation involves the overlap of the vacant 2*p* orbital with a neighboring C–H σ-bond orbital (*Fig. 3*).

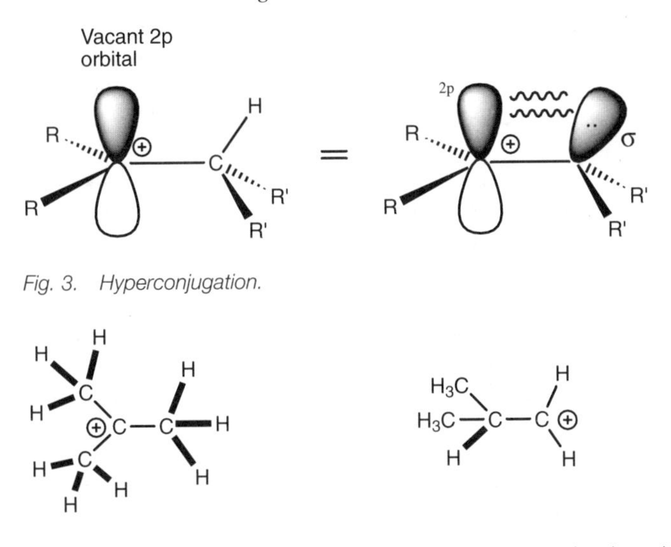

Fig. 3. *Hyperconjugation.*

Fig. 4. *(a) More substituted carbocation; (b) less substituted carbocation.*

This interaction means that the 2*p* orbital is not completely vacant since the σ electrons of the C–H bond can spend a small amount of time entering the space occupied by the 2*p* orbital. This means that the C–H bond becomes slightly electron deficient. As a result, the positive charge is delocalized and hence stabilized. The more alkyl groups attached to the carbocation, the more possibilities there are for hyperconjugation and the more stable the carbocation. For example, the more substituted carbocation (*Fig. 4a*) can be stabilized by hyperconjugation to nine C–H bonds, whereas the less substituted carbocation (*Fig. 4b*) can only be stabilized by hyperconjugation to one C–H bond.

H6 Reduction and oxidation of alkenes

Key Notes

Alkenes to alkanes

Alkenes can be converted to alkanes by reduction with hydrogen gas. A metal catalyst is necessary in order to lower the free energy of activation.

Alkenes to aldehydes and ketones

Treating an alkene with ozone then with zinc and water results in cleavage of the alkene across the double bond to give two carbonyl compounds (ketones and/or aldehydes). The reaction is known as an ozonolysis and is an example of an oxidation reaction.

Alkenes to carboxylic acids and ketones

Alkenes can be cleaved by oxidation with hot alkaline potassium permanganate. The products obtained are carboxylic acids and/or ketones depending on the substituents present on the alkene.

Alkenes to 1,2-diols

Alkenes can be converted to 1,2-diols (or glycols) by reaction with osmium tetroxide or by reaction with cold alkaline potassium permanganate. In both cases, the two hydroxyl groups are added to the same face of the alkene – *syn*-hydroxylation. Osmium tetroxide gives better yields, but is more toxic.

Alkenes to epoxides

Treatment of alkenes with a peroxyacid (RCO_3H) results in the formation of an epoxide. The reaction is a one-step process without intermediates. Epoxides can also be obtained in a two-step process via a halohydrin.

Related topic

Reduction of alkynes (H9)

Alkenes to alkanes

Alkenes are converted to alkanes by treatment with hydrogen over a finely divided metal catalyst such as palladium, nickel, or platinum (*Fig. 1*). This is an **addition** reaction since it involves the addition of hydrogen atoms to each end of the double bond. It is also called a **catalytic hydrogenation** or a **reduction** reaction.

The catalyst is crucial since the reaction will not take place at room temperature in its absence. This is because hydrogenation has a high **free energy of activation** (ΔG_1*) (*Fig. 2*). The role of the catalyst is to bind the alkene and the hydrogen to a

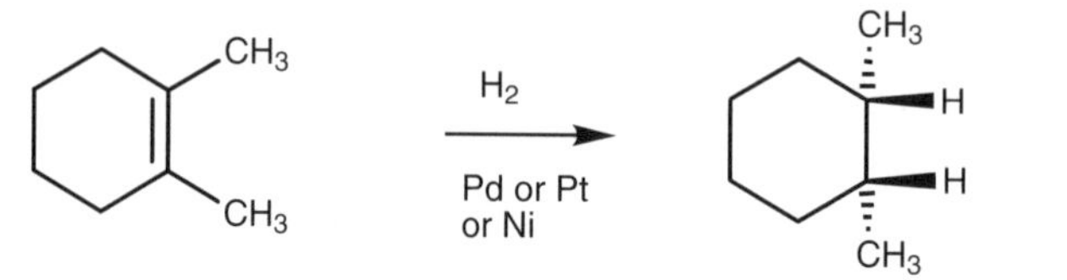

Fig. 1. Hydrogenation of an alkene.

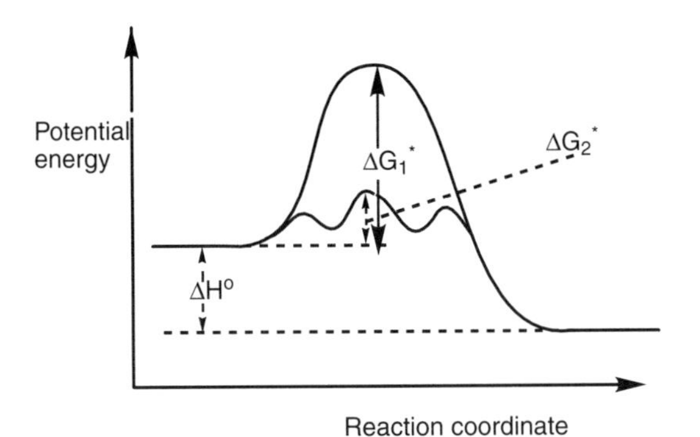

Fig. 2. Graph of potential energy versus reaction coordinate for an uncatalyzed and a catalyzed hydrogenation reaction of an alkene.

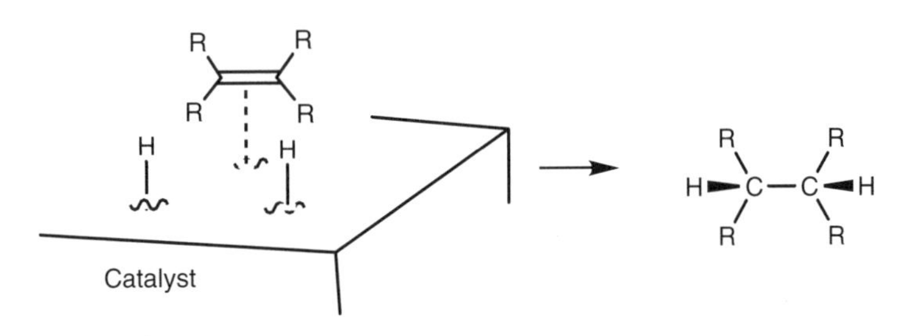

Fig. 3. Binding of alkene and hydrogen to catalytic surface.

common surface such that they can react more easily. This results in a much lower energy of activation (ΔG_2*) allowing the reaction to proceed under much milder conditions. The catalyst itself is unchanged after the reaction and can be used in small quantity.

Both the hydrogen and the alkene are bound to the catalyst surface before the hydrogen atoms are transferred, which means that both hydrogens are added to the same side of the double bond (see *Fig. 3*) – ***syn*-addition**. Note that the hydrogen molecule is split once it has been added to the catalyst.

Alkenes to aldehydes and ketones

Treating an alkene with ozone (*Fig. 4*) results in **oxidation** of the alkene and formation of an initial **ozonide** which then rearranges to an isomeric ozonide. This second ozonide is unstable and potentially explosive and so it is not usually isolated. Instead, it is reduced with zinc and water resulting in the formation of two separate molecules.

The alkene is split across the double bond to give two carbonyl compounds. These will be ketones or aldehydes depending on the substituents present. For example, 3-methyl-2-pentene gives an aldehyde and a ketone (*Fig. 5*).

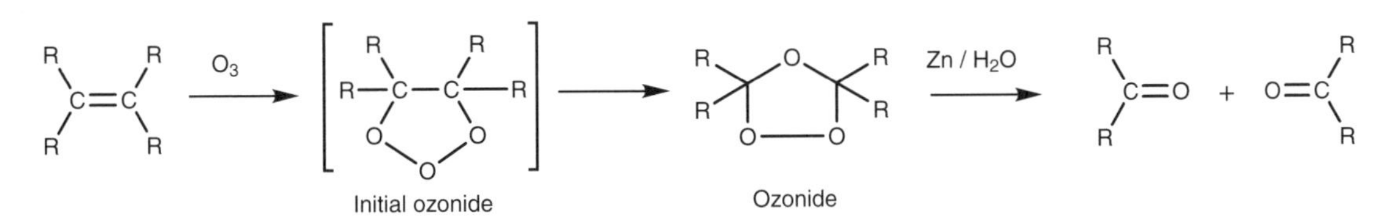

Fig. 4. Ozonolysis of an alkene.

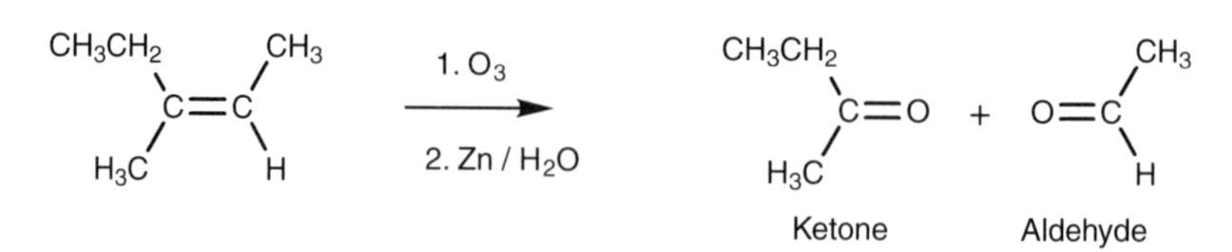

Fig. 5. Ozonolysis of 3-methyl-2-pentene.

Alkenes to carboxylic acids and ketones

Alkenes can be oxidatively cleaved with hot permanganate solution to give carboxylic acids and/or ketones (*Fig. 6*). The products obtained depend on the substituents present on the alkene.

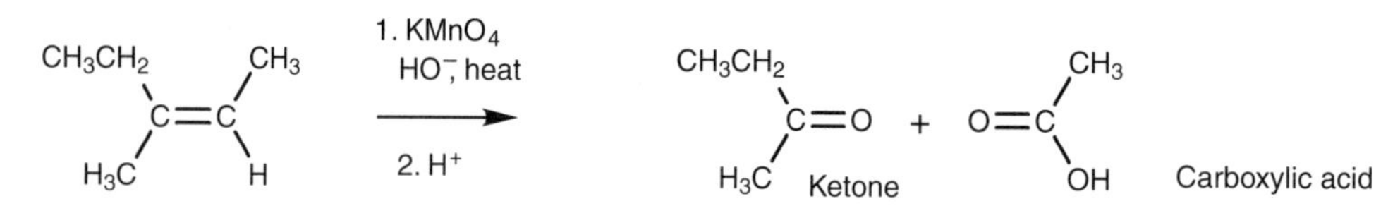

Fig. 6. Oxidative cleavage of 3-methyl-2-pentene.

Alkenes to 1,2-diols

The reaction of alkenes with osmium tetroxide (OsO_4) is another example of an oxidation reaction (*Fig. 7*). However, in this case the alkene is not split. Instead, a 1,2-diol is obtained – also known as a glycol. The reaction involves the formation of a cyclic intermediate where the osmium reagent is attached to one face of the alkene. On treatment with sodium bisulfite, the intermediate is cleaved such that the two oxygen atoms linking the osmium remain attached. This results in both alcohols being added to the same side of the double bond – ***syn*-hydroxylation**.

The same reaction can also be carried out using cold alkaline potassium permanganate ($KMnO_4$) followed by treatment with aqueous base (*Fig. 8*). It is important to keep the reaction cold since potassium permanganate can cleave the diol by further oxidation (*Fig. 6*).

The reaction works better with osmium tetroxide. However, this is a highly toxic and expensive reagent and has to be handled with care.

Anti-hydroxylation of the double bond can be achieved by forming an epoxide, then carrying out an acid-catalyzed hydrolysis (Topic N3).

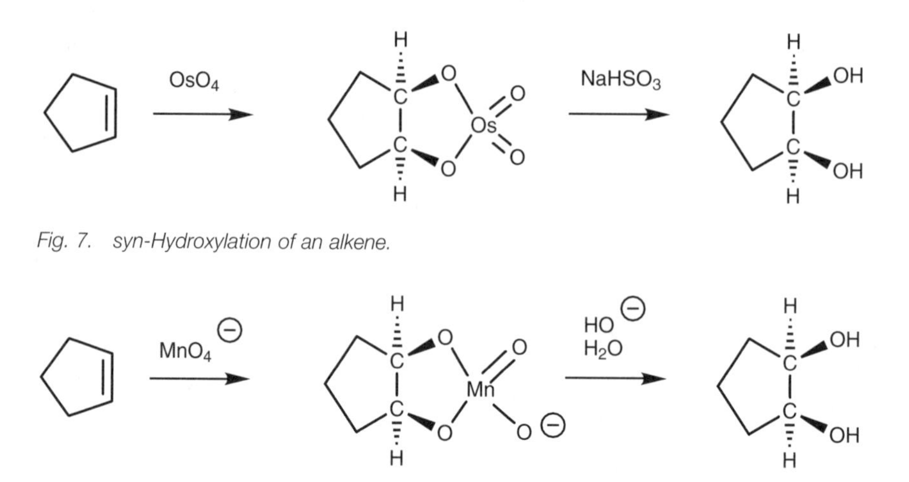

Fig. 7. syn-Hydroxylation of an alkene.

Fig. 8. syn-Hydroxylation with $KMnO_4$.

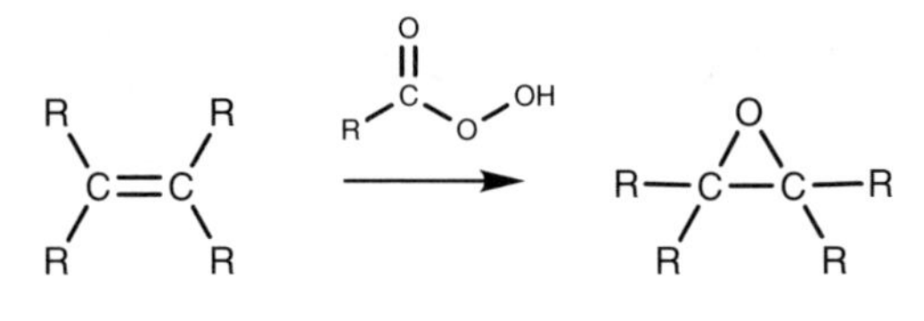

Fig. 9. Epoxidation of an alkene.

Alkenes to epoxides

Treatment of an alkene with a **peroxyacid** (RCO_3H) results in the formation of an epoxide (*Fig. 9*). *m*-Chloroperoxybenzoic acid is one of the most frequently used peroxyacids for this reaction. The reaction is unusual in that there is no carbocation intermediate, and involves a one-step process without intermediates.

H7 Hydroboration of alkenes

Key Notes

Reaction

Alkenes can be converted to alcohols by treatment with diborane followed by hydrogen peroxide. With unsymmetrical alkenes, the least substituted alcohol is obtained in contrast to the electrophilic addition of water where the most substituted alcohol is obtained.

Mechanism

The mechanism involves an initial π complex between the alkene and BH_3. A four-centered transition state is then formed where the π bond of the alkene and one of the B–H bonds is partially broken and the bonds linking H and B to the alkene are partially formed. There is an imbalance of charge in the transition state, resulting in one of the carbon atoms being slightly positive. The reaction proceeds such that the most substituted carbon bears the partial charge and this results in the boron adding to the least substituted carbon. Oxidation with hydrogen peroxide involves a hydroperoxide molecule bonding to boron, followed by migration of an alkyl group from boron to oxygen. This is repeated twice more to form a trialkyl borate which is hydrolyzed to give the final alcohol.

Related topics

Electrophilic addition to symmetrical alkenes (H3)

Electrophilic addition to unsymmetrical alkenes (H4)

Reaction

Alcohols can be generated from alkenes by reaction with diborane (B_2H_6 or BH_3), followed by treatment with hydrogen peroxide (*Fig. 1*). The first part of the reaction involves the splitting of a B–H bond in BH_3 with the hydrogen joining one end of the alkene and the boron joining the other. Each of the B–H bonds is split in this way such that each BH_3 molecule reacts with three alkenes to give an organoborane intermediate where boron is linked to three alkyl groups. This can then be oxidized with alkaline hydrogen peroxide to produce the alcohol.

With unsymmetrical alkenes, the least substituted alcohol is obtained (anti-Markovnikov; *Fig. 2*) and so the organoborane reaction is complementary to the electrophilic addition reaction with aqueous acid (Topic H3). Steric factors appear to play a role in controlling this preference with the boron atom preferring to approach the least sterically hindered site. Electronic factors also play a role as described in the mechanism below.

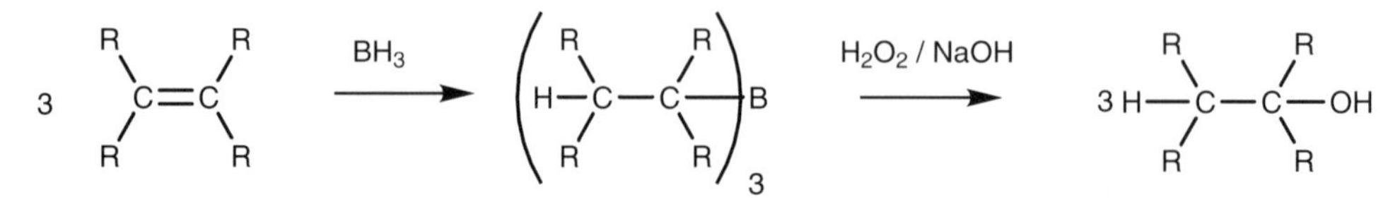

Fig. 1. Hydroboration of an alkene.

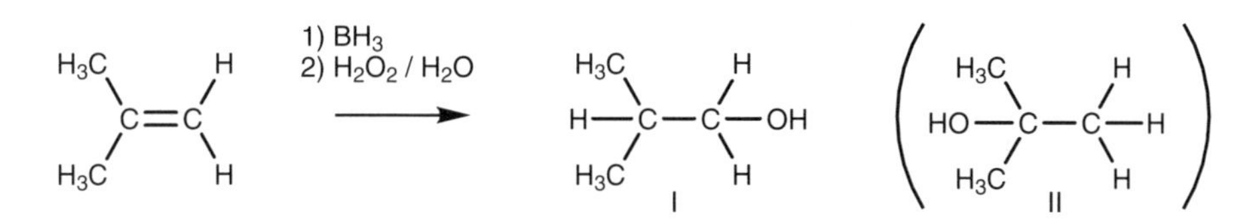

Fig. 2. Hydroboration of 2-methylpropene to give a primary alcohol (I). The tertiary alcohol (II) is not obtained.

Mechanism

The mechanism (*Fig. 3*) involves the alkene π bond interacting with the empty *p* orbital of boron to form a π complex. One of BH_3's hydrogen atoms is then transferred to one end of the alkene as boron itself forms a σ bond to the other end. This takes place through a four-centered transition state where the alkene's π bond and the B–H bond are partially broken, and the eventual C–H and C–B bonds are partially formed. There is an imbalance of electrons in the transition state which results in the boron being slightly negative and one of the alkene carbons being slightly positive. The carbon best able to handle this will be the most substituted carbon and so the boron will end up on the least substituted carbon. (Note that boron has six valence electrons and is electrophilic. Therefore, the addition of boron to the least substituted position actually follows Markovnikov's rule.)

Since subsequent oxidation with hydrogen peroxide replaces the boron with a hydroxyl group, the eventual alcohol will be on the least substituted carbon. Furthermore, the addition of the boron and hydrogen atoms take place such that they are on the same side of the alkene. This is called ***syn*-addition**. The mechanism of oxidation (*Fig. 4*) involves addition of the hydroperoxide to the electron deficient

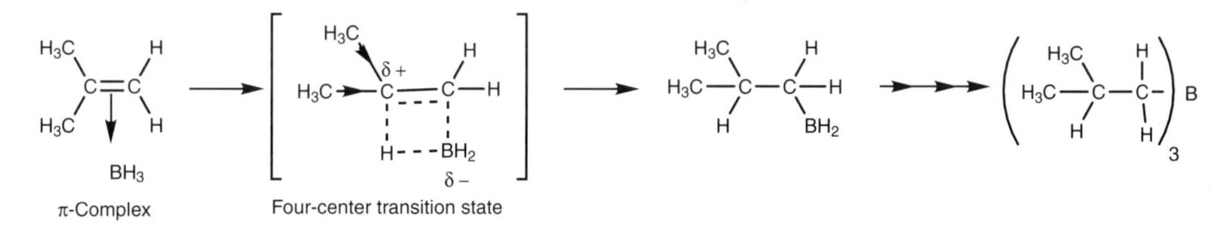

Fig. 3. Mechanism of hydroboration.

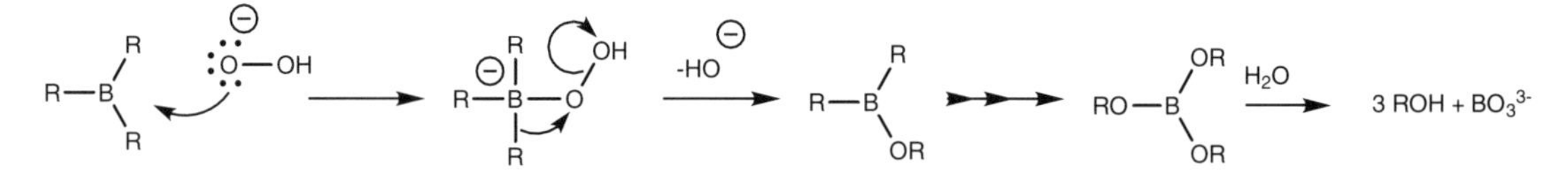

Fig. 4. Mechanism of oxidation with hydroperoxide.

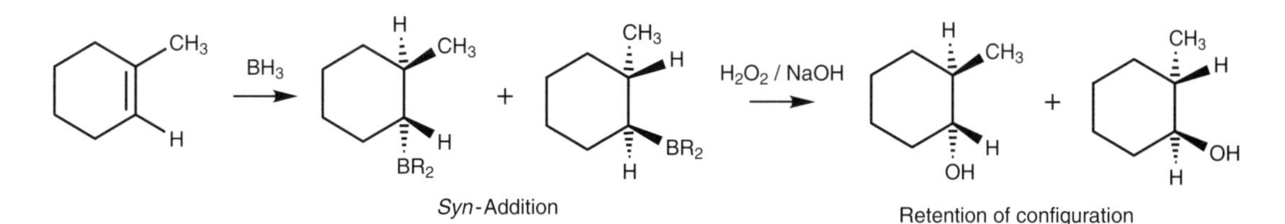

Fig. 5. Stereochemical aspects of hydroboration.

boron to form an unstable intermediate which then rearranges such that an alkyl group migrates from the boron atom to the neighboring oxygen and expels a hydroxide ion. This process is then repeated for the remaining two hydrogens on boron and the final trialkyl borate $B(OR)_3$ can then be hydrolyzed with water to give three molecules of alcohol plus a borate ion.

The mechanism of oxidation takes place with retention of stereochemistry at the alcohol's carbon atom and so the overall reaction is stereospecific (*Fig. 5*). Note that the reaction is stereospecific such that the alcohol group is *trans* to the methyl group in the product. However, it is not enantiospecific and both enantiomers are obtained in equal amounts (a racemate).

H8 Electrophilic additions to alkynes

Key Notes

Additions to symmetrical alkynes

Symmetrical alkynes undergo electrophilic addition with halogens or hydrogen halides by the same mechanism described for alkenes. The reaction can go once or twice depending on the amount of reagent used. Alkynes are less reactive than alkenes since the intermediate formed (a vinylic carbocation) is more unstable than the corresponding carbocation formed from electrophilic addition to an alkene. Reaction of an alkyne with aqueous acid and mercuric acid produces an intermediate enol which rearranges by keto–enol tautomerism to give a ketone.

Additions to terminal alkynes

Addition of hydrogen halide to a terminal alkyne results in the hydrogen(s) adding to the terminal carbon and the halogen(s) adding to the more substituted carbon. This is another example of Markovnikov's rule. Similarly, reaction of a terminal alkyne with aqueous acid and mercuric sulfate leads to a ketone rather than an aldehyde following addition of one molecule of water followed by a keto–enol tautomerism.

Related topics

Configurational isomers – alkenes and cycloalkanes (D2)
Electrophilic addition to symmetrical alkenes (H3)
Electrophilic addition to unsymmetrical alkenes (H4)
Carbocation stabilization (H5)

Additions to symmetrical alkynes

Alkynes undergo electrophilic addition reactions with the same reagents as alkenes (e.g. halogens and hydrogen halides). Since there are two π bonds in alkynes, it is possible for the reaction to go once or twice depending on the amount of reagent added. For example, reaction of 2-butyne with one equivalent of bromine gives an (*E*)-dibromoalkene (*Fig. 1a*). With two equivalents of bromine, the initially formed (*E*)-dibromoalkene reacts further to give a tetrabromoalkane (*Fig. 1b*).

Treatment of an alkyne with one equivalent of HBr gives a bromoalkene (*Fig. 2a*). If two equivalents of hydrogen bromide are present the reaction can go twice to give a **geminal** dibromoalkane where both bromine atoms are added to the same carbon (*Fig. 2b*).

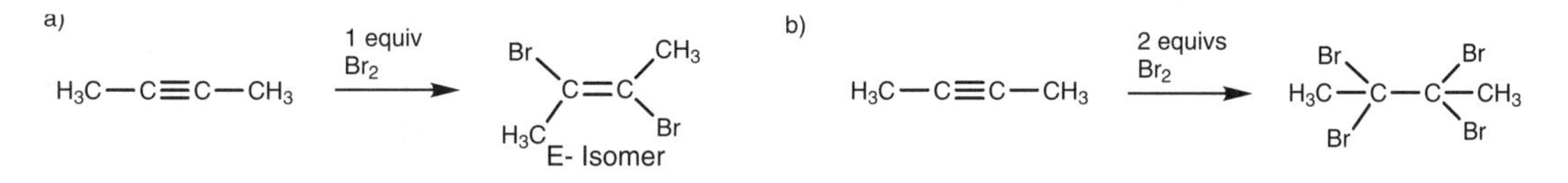

Fig. 1. Reaction of 2-butyne with (a) 1 equivalent of bromine; (b) 2 equivalents of bromine.

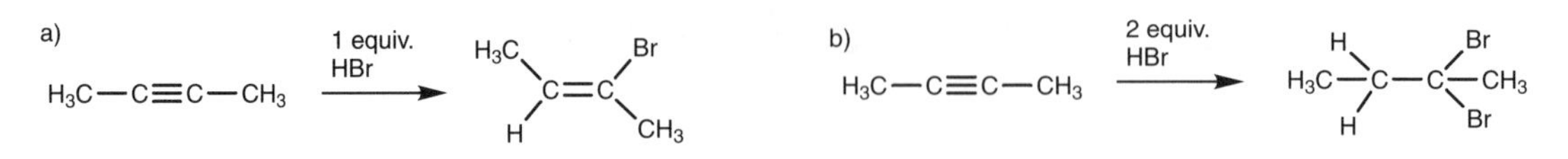

Fig. 2. Reaction of 2-butyne with (a) 1 equivalent of HBr; (b) 2 equivalents of HBr.

The above addition reactions are similar to the addition reactions of alkenes (Topic H3). However, the reaction is much slower for an alkyne, since alkynes are less reactive. One might expect alkynes to be more nucleophilic, since they are more electron rich in the vicinity of the multiple bond, that is, six electrons in a triple bond as compared to four in a double bond. However, electrophilic addition to an alkyne involves the formation of a **vinylic** carbocation (*Fig. 3*). This carbocation is much less stable than the carbocation intermediate resulting from electrophilic addition to an alkene.

Due to this low reactivity, alkynes react slowly with aqueous acid and mercuric sulfate has to be added as a catalyst. The product which might be expected from this reaction would be a diol (*Fig. 4*).

In fact, a diol is not formed. The intermediate (an enol) undergoes acid-catalyzed rearrangement to give a ketone instead (*Fig. 5*). This process is known as a **keto–enol tautomerism** (Topic J2).

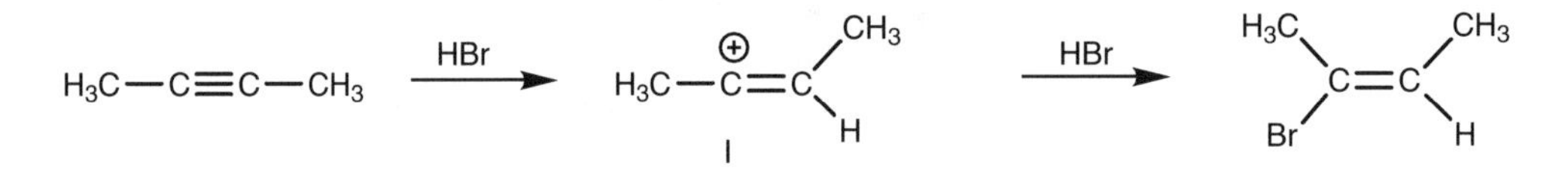

Fig. 3. Electrophilic addition to an alkyne via a vinylic carbocation (I).

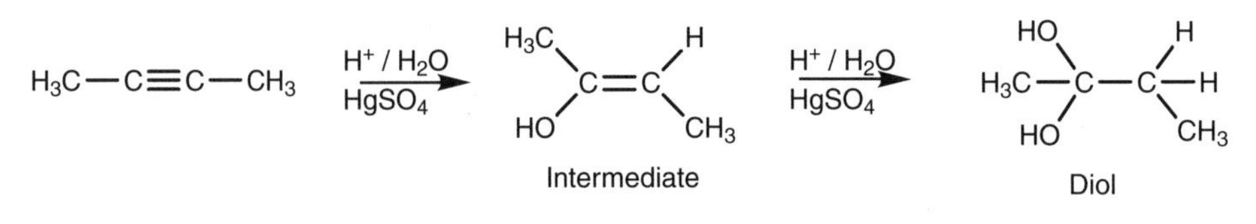

Fig. 4. Reaction of 2-butyne with aqueous acid and mercuric sulfate.

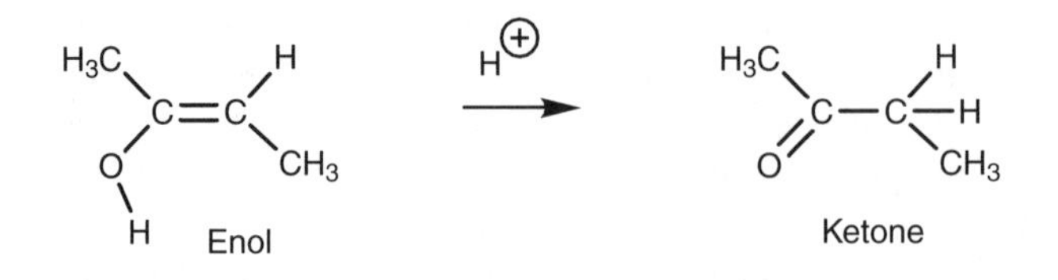

Fig. 5. Keto–enol tautomerism.

Tautomerism is the term used to describe the rapid interconversion of two different isomeric forms (**tautomers**) – in this case the keto and enol tautomers of a ketone. The keto tautomer is by far the dominant species for a ketone and the enol tautomer is usually present in only very small amounts (typically 0.0001%). Therefore, as soon as the enol is formed in the above reaction, it rapidly tautomerizes to the keto form and further electrophilic addition does not take place.

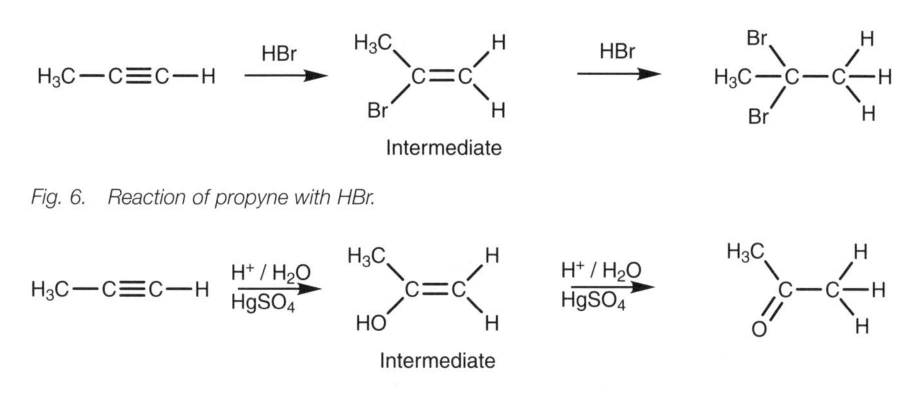

Fig. 6. Reaction of propyne with HBr.

Fig. 7. Reaction of propyne with aqueous acid and mercuric sulfate.

Additions to terminal alkynes

If a terminal alkyne is treated with an excess of hydrogen halide the halogens both end up on the more substituted carbon (*Fig. 6*).

This is another example of Markovnikov's rule which states that the additional hydrogens end up on the carbon which already has the most hydrogens (see Topics H4 and H5). The same rule applies for the reaction with acid and mercuric sulfate, which means that a ketone is formed after keto–enol tautomerism instead of an aldehyde (*Fig. 7*).

H9 Reduction of alkynes

Key Notes

Hydrogenation

Alkynes are reduced to alkanes with hydrogen gas over a platinum catalyst. Two molecules of hydrogen are added and the reaction goes through an alkene intermediate. The reaction can be stopped at the alkene stage if a less active or 'poisoned' catalyst is used. The stereochemistry of the alkene from such a reaction is (Z) since both hydrogen atoms are added to the same side of the alkyne.

Dissolving metal reduction

Alkynes can also be reduced to (*E*)-alkenes using sodium or lithium in liquid ammonia.

Related topics

Configurational isomers – alkenes and cycloalkanes (D2)
Mechanisms (F2)
Reduction and oxidation of alkenes (H6)

Hydrogenation

Alkynes react with hydrogen gas in the presence of a metal catalyst in a process known as hydrogenation – an example of a **reduction** reaction. With a fully active catalyst such as platinum metal, two molecules of hydrogen are added to produce an alkane (*Fig. 1*).

The reaction involves the addition of one molecule of hydrogen to form an alkene intermediate which then reacts with a second molecule of hydrogen to form the alkane. With less active catalysts, it is possible to stop the reaction at the alkene stage. In particular, (Z)-alkenes can be synthesized from alkynes by reaction with hydrogen gas and Lindlar's catalyst (*Fig. 2*). This catalyst consists of metallic palladium deposited on calcium carbonate which is then treated with lead acetate and quinoline. The latter treatment 'poisons' the catalyst such that the alkyne reacts with hydrogen to give an alkene, but does not react further. Since the starting materials are absorbed onto the catalytic surface, both hydrogens are added to the same side of the molecule to produce the (Z) isomer.

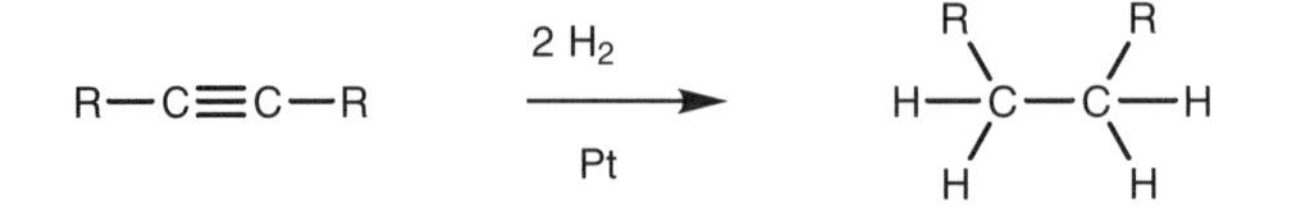

Fig. 1. Reduction of an alkyne to an alkane.

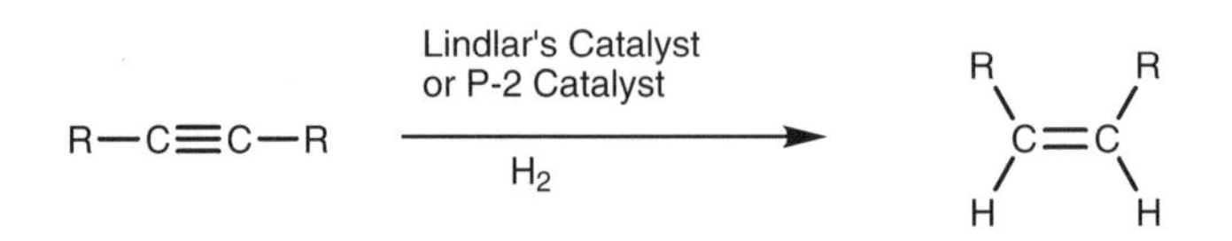

Fig. 2. Reduction of an alkyne to a (Z)-alkene.

An alternative catalyst which achieves the same result is nickel boride (Ni_2B) – the P-2 catalyst.

Dissolving metal reduction

Reduction of an alkyne to an (*E*)-alkene can be achieved if the alkyne is treated with lithium or sodium metal in ammonia at low temperatures (*Fig. 3*). This is known as a dissolving metal reduction.

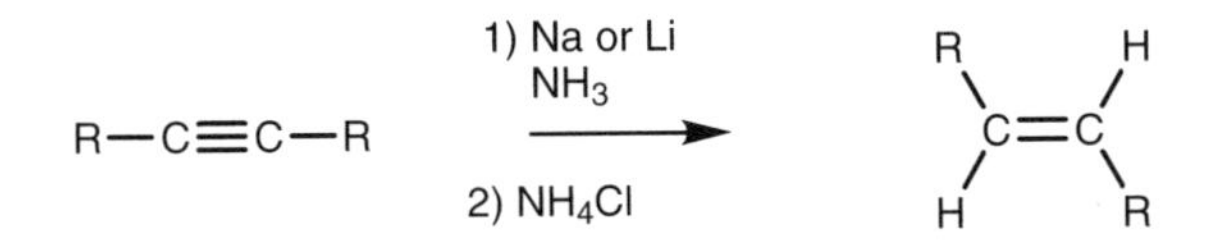

Fig. 3. Reduction of an alkyne to an E-alkene.

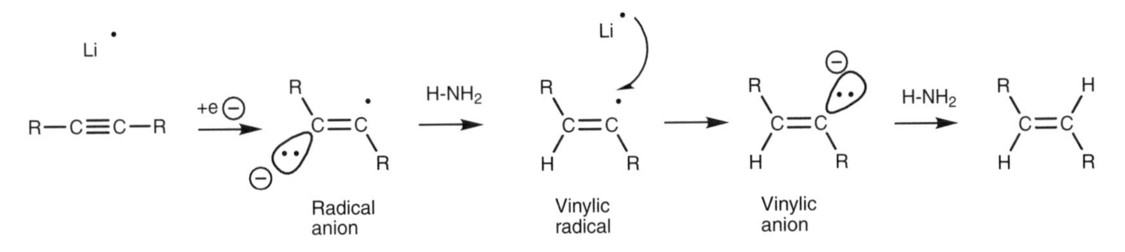

Fig. 4. Mechanism for the dissolving metal reduction of an alkyne.

In this reaction, the alkali metal donates its valence electron to the alkyne to produce a radical anion (*Fig. 4*). This in turn removes a proton from ammonia to produce a vinylic radical which receives an electron from a second alkali metal to produce a *trans*-vinylic anion. This anion then removes a proton from a second molecule of ammonia and produces the *trans*- or (*E*)-alkene. Note that half curly arrows are used in the mechanism since this is a radical reaction involving the movement of single electrons (Topic F2).

H10 ALKYLATION OF TERMINAL ALKYNES

Key Notes

Terminal alkynes A terminal alkyne is defined as an alkyne having a hydrogen substituent. The hydrogen of a terminal alkyne is weakly acidic and can be removed with a strong base such as sodium amide to produce an alkynide ion.

Alkylation The alkynide ion can be treated with a primary alkyl halide to produce a disubstituted alkyne.

Related topics Brønsted–Lowry acids and bases (G1); Nucleophilic substitution (L2); Elimination (L4)

Terminal alkynes

A terminal alkyne is defined as an alkyne having a hydrogen substituent (*Fig. 1*). This hydrogen substituent is acidic and can be removed with strong base (e.g. sodium amide $NaNH_2$) to produce an **alkynide** (*Fig. 2*). This is an example of an acid–base reaction (Topic G1).

Alkylation

Once the alkynide has been formed, it can be treated with an alkyl halide to produce more complex alkynes (*Fig. 3*). This reaction is known as an alkylation as far as the alkynide is concerned, and is an example of nucleophilic substitution as far as the alkyl halide is concerned (Topic L2).

Fig. 1. Terminal alkyne.

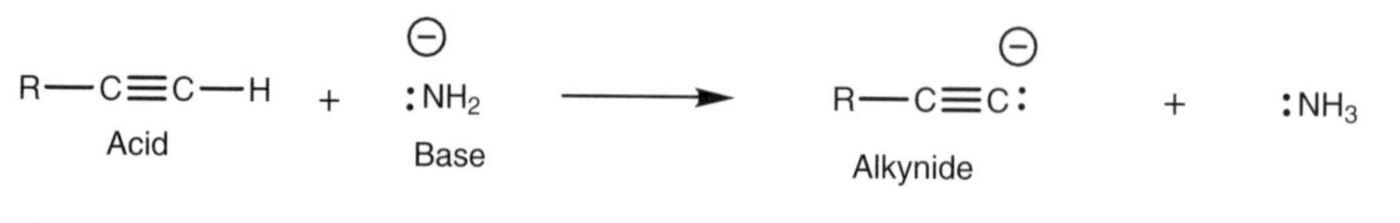

Fig. 2. Reaction of a terminal alkyne with a strong base.

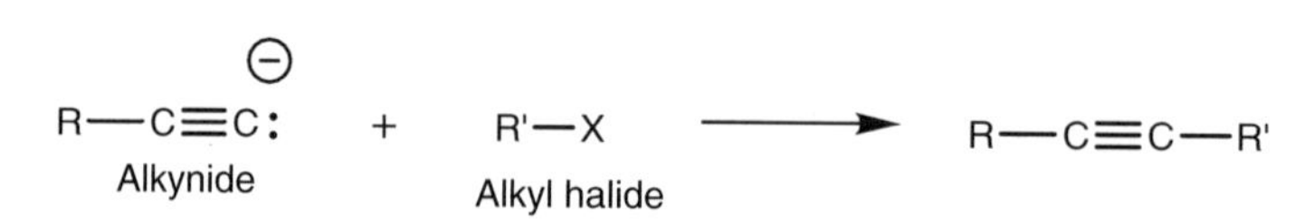

Fig. 3. Reaction of an alkynide ion with an alkyl halide.

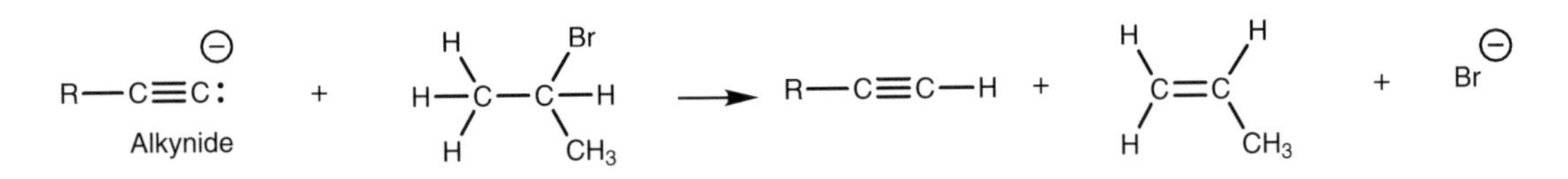

Fig. 4. Elimination of HBr.

This reaction works best with primary alkyl halides. When secondary or tertiary alkyl halides are used, the alkynide reacts like a base and this results in elimination of hydrogen halide from the alkyl halide to produce an alkene (Topic L4; *Fig. 4*).

H11 CONJUGATED DIENES

Key Notes

Structure

A conjugated diene consists of two alkene units separated by a single bond.

Bonding

The 'single' bond connecting the two alkene units of a conjugated diene has partial double-bond character. There are two possible explanations. Firstly, the use of sp^2 hybridized orbitals to form the σ bond results in a shorter bond. Secondly, the π systems of the two alkene units can overlap to give partial double-bond character.

Electrophilic addition

Electrophilic addition to a diene produces a mixture of two products arising from 1,2-addition and 1,4-addition. The reaction proceeds through an allylic carbocation which is stabilized by delocalization and which accounts for the two possible products obtained.

Diels–Alder cycloaddition

Heating a conjugated diene with an electron-deficient alkene results in the formation of a cyclohexene ring. The mechanism is concerted and involves no intermediates.

Related topics

sp^2 Hybridization (A4)
Acid strength (G2)
Enolates (G5)
Electrophilic addition to symmetrical alkenes (H3)
Electrophilic addition to unsymmetrical alkenes (H4)

Structure

A **conjugated diene** consists of two alkene units separated by a single bond (*Fig. 1a*). Dienes which are separated by more than one single bond are called nonconjugated dienes (*Fig. 1b*).

Bonding

A conjugated diene does not behave like two isolated alkenes. For example, the length of the 'single' bond connecting the two alkene units is slightly shorter than one would expect for a typical single bond (1.48 Å vs. 1.54 Å). This demonstrates that there is a certain amount of double-bond character present in this bond. Two explanations can be used to account for this. First of all, the bond in question links

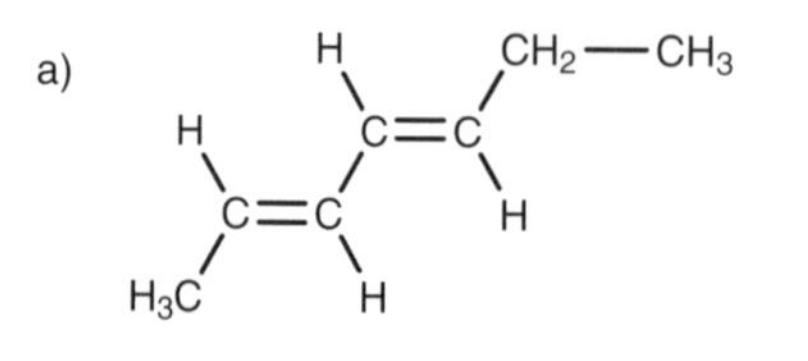

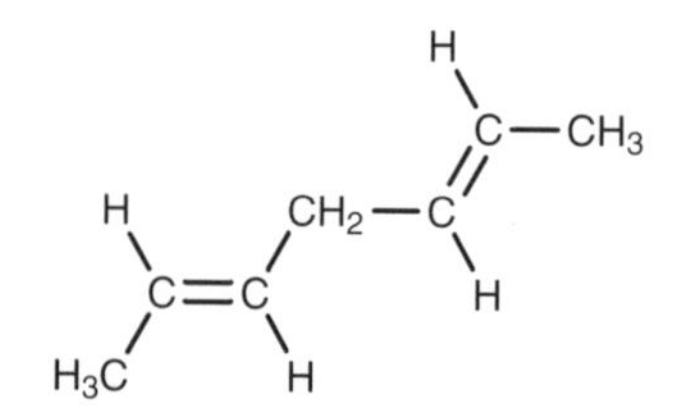

Fig. 1. (a) Conjugated diene; (b) nonconjugated diene.

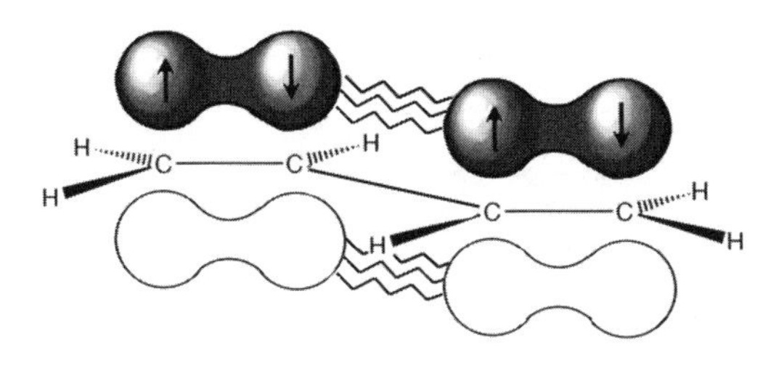

Fig. 2. π-Orbital overlap.

two sp^2 hybridized carbons rather than two sp^3 hybridized carbons. Therefore, an sp^2 hybridized orbital from each carbon is used for the single bond. Since this hybridized orbital has more *s*-character than an sp^3 hybridized orbital, the bond is expected to be shorter. An alternative explanation is that the π orbitals of the two alkene systems can overlap to produce the partial double-bond character (*Fig. 2*).

Electrophilic addition

The reactions of a conjugated diene reflect the fact that a conjugated diene should be viewed as a functional group in its own right, rather than as two separate alkenes. Electrophilic addition to a conjugated diene results in a mixture of two possible products arising from **1,2-addition** and **1,4 addition** (*Fig. 3*).

In 1,2-addition, new atoms have been added to each end of one of the alkene units. This is the normal electrophilic addition of an alkene with which we are familiar (Topics H3 and H4). In 1,4-addition, new atoms have been added to each end of the entire diene system. Furthermore, the double bond remaining has shifted position (isomerized) to the 2,3 position.

The mechanism of 1,4-addition starts off in the same way as a normal electrophilic addition. We shall consider the reaction of a conjugated diene with hydrogen bromide as an example (*Fig. 4*). One of the alkene units of the diene uses its π electrons to form a bond to the electrophilic hydrogen of HBr. The H–Br bond breaks at the same time to produce a bromide ion. The intermediate carbocation produced has a double bond next to the carbocation center and is called an **allylic carbocation**.

This system is now set up for resonance (cf. Topics G2 and G5) involving the remaining alkene and the carbocation center, resulting in delocalization of the

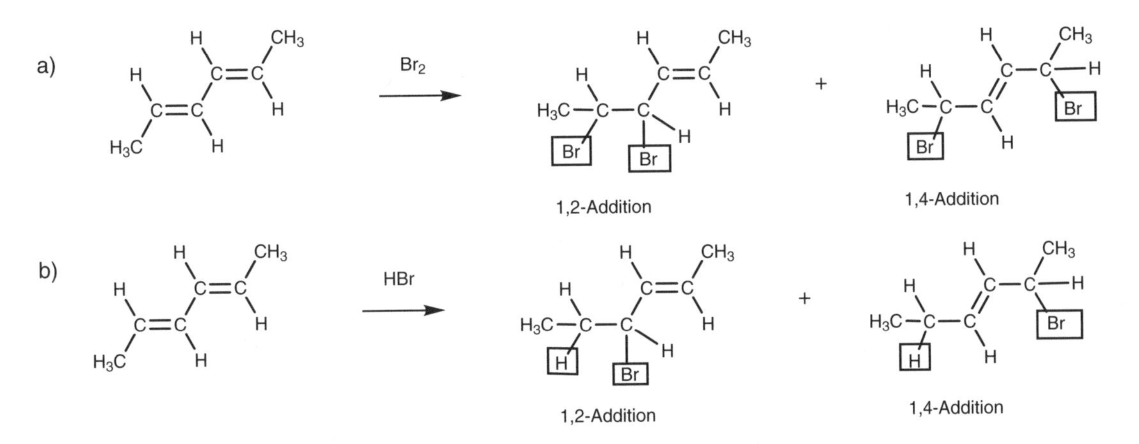

Fig. 3. Electrophilic addition to a conjugated diene of (a) bromine and (b) HBr.

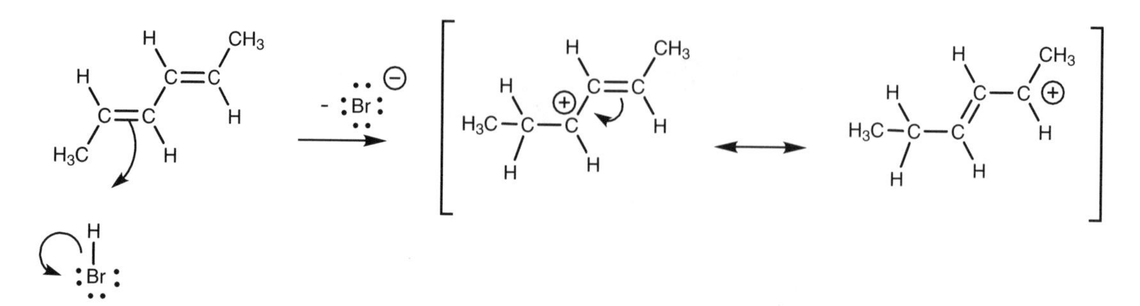

Fig. 4. Mechanism of 1,4-addition – first step.

positive charge between positions 2 and 4. Due to this delocalization, the carbocation is stabilized and this in turn explains two features of this reaction. First of all, the formation of two different products is now possible since the second stage of the mechanism involves the bromide anion attacking either at position 2 or at position 4 (*Fig. 5*). Secondly, it explains why the alternative 1,2-addition product is not formed (*Fig. 6*). The intermediate carbocation required for this 1,2-addition cannot be stabilized by resonance. Therefore, the reaction proceeds through the allylic carbocation instead.

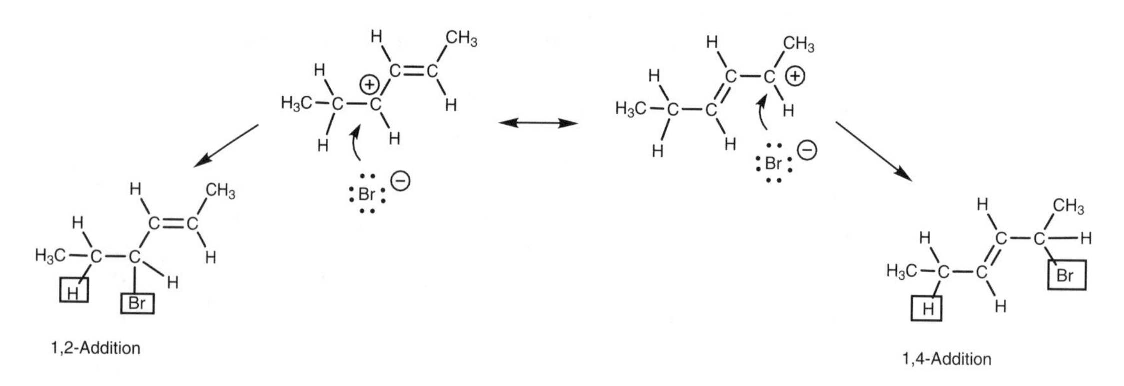

Fig. 5. Mechanism of 1,2- and 1,4-addition – second step.

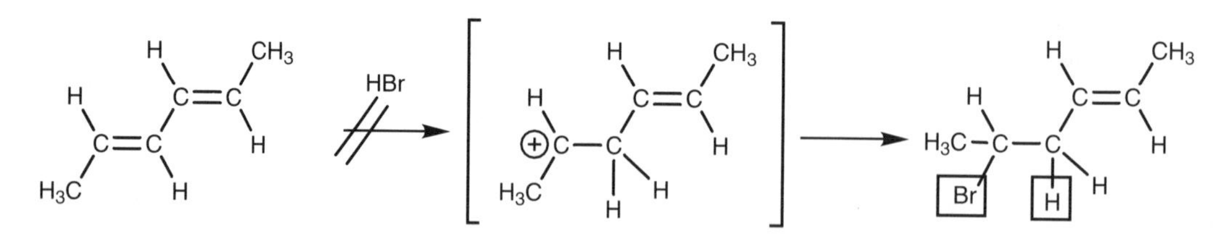

Fig. 6. Unfavored reaction mechanism.

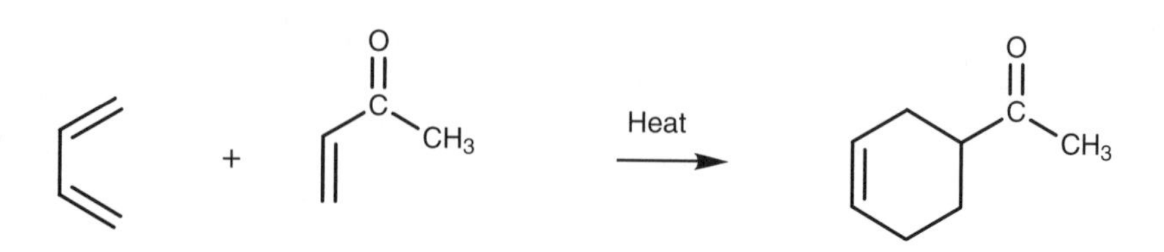

Fig. 7. Diels–Alder cycloaddition.

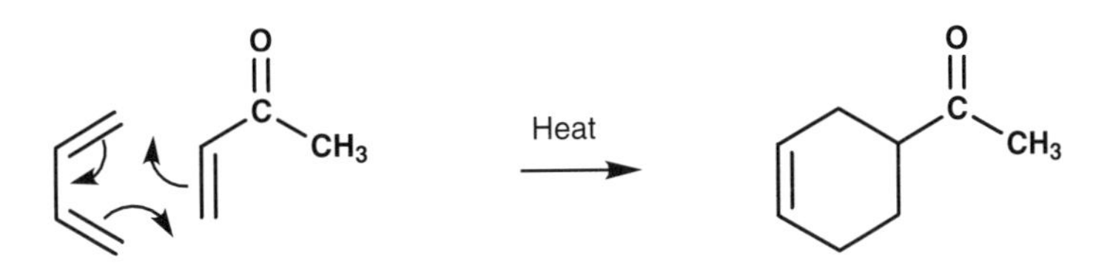

Fig. 8. Mechanism of Diels–Alder cycloaddition.

Diels–Alder cycloaddition

The Diels–Alder cycloaddition reaction is an important method by which six-membered rings can be synthesized. The reaction involves a conjugated diene and an alkene (*Fig. 7*). The alkene is referred to as a **dienophile** (diene-lover) and usually has an electron-withdrawing group linked to it (e.g. a carbonyl group or a nitrile). The mechanism is concerted with new bonds being formed at the same time as old bonds are being broken (*Fig. 8*). No intermediates are involved.

I1 AROMATICITY

Key Notes

Definition

Aromatic compounds such as benzene are more stable than suggested from their structure. They undergo reactions which retain the aromatic ring system and behave differently from alkenes or polyenes.

Hückel rule

Aromatic compounds are cyclic and planar with sp^2 hybridized atoms. They also obey the Hückel rule and have $4n + 2$ π electrons where $n = 1, 2, 3, ...$ Aromatic systems can be monocyclic or polycyclic, neutral, or charged.

Related topic

sp^2 Hybridization (A4)

Definition

The term **aromatic** was originally applied to benzene-like structures because of the distinctive aroma of these compounds, but the term now means something different in modern chemistry. Aromatic compounds undergo distinctive reactions which set them apart from other functional groups. They are highly unsaturated compounds, but unlike alkenes and alkynes, they are relatively unreactive and will tend to undergo reactions which involve a retention of their unsaturation. We have already discussed the reasons for the stability of benzene in Topic A4. Benzene is a six-membered ring structure with three formal double bonds (*Fig. 1a*). However, the six π electrons involved are not localized between any two carbon atoms. Instead, they are delocalized around the ring which results in an increased stability. This is why benzene is often written with a circle in the center of the ring to signify the delocalization of the six π electrons (*Fig. 1b*). Reactions which disrupt this delocalization are not favored since it means a loss of stability, so benzene undergoes reactions where the aromatic ring system is retained. All six carbon atoms in benzene are sp^2 hybridized, and the molecule itself is cyclic and planar – the planarity being necessary if the $2p$ atomic orbitals on each carbon atom are to overlap and result in delocalization.

Fig. 1. Representations of benzene.

Hückel rule

An aromatic molecule must be cyclic and planar with sp^2 hybridized atoms (i.e. conjugated), but it must also obey what is known as the **Hückel rule**. This rule states that the ring system must have $4n + 2$ π electrons where $n = 1, 2, 3$, *etc.* Therefore, ring systems which have 6, 10, 14, ... π electrons are aromatic. Benzene fits the Hückel rule since it has six π electrons. Cyclooctatetraene has eight π electrons and does not obey the Hückel rule. Although all the carbon atoms in the ring are sp^2 hybridized, cyclooctatetraene reacts like a conjugated alkene. It is not planar, the π electrons are not delocalized and the molecule consists of alternating single and double bonds (*Fig. 2a*). However, the 18-membered cyclic system (*Fig. 2b*) does fit the Hückel rule (n=4) and is a planar molecule with aromatic properties and a delocalized π system.

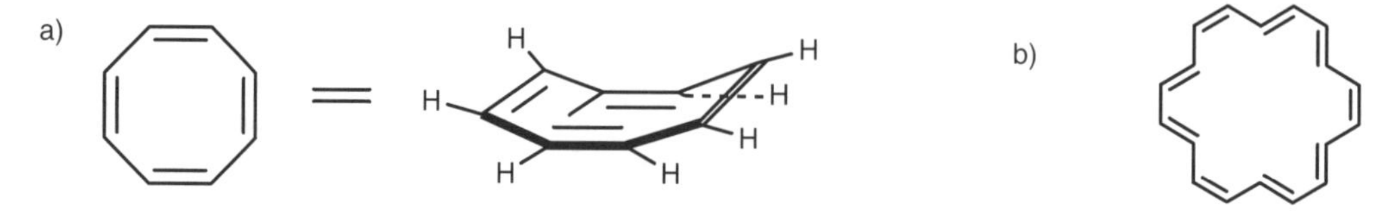

Fig. 2. (a) Cyclooctatetraene; (b) 18-membered aromatic ring.

It is also possible to get aromatic ions. The cyclopentadienyl anion and the cycloheptatrienyl cation are both aromatic (*Fig. 3*). Both are cyclic and planar, containing six π electrons, and all the atoms in the ring are sp^2 hybridized.

Bicyclic and polycyclic systems can also be aromatic (*Fig. 4*).

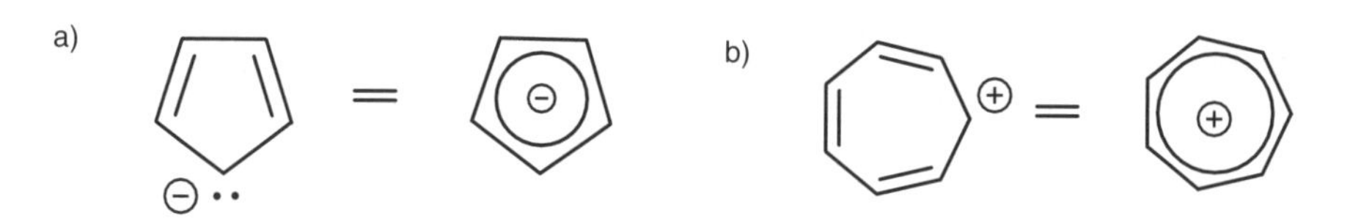

Fig. 3. (a) Cyclopentadienyl anion; (b) cycloheptatrienyl cation.

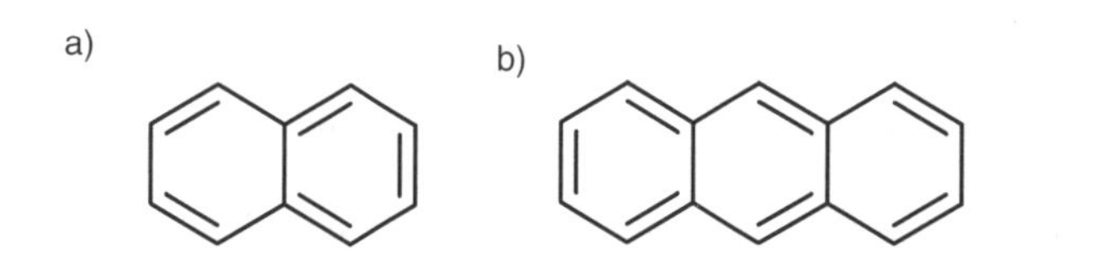

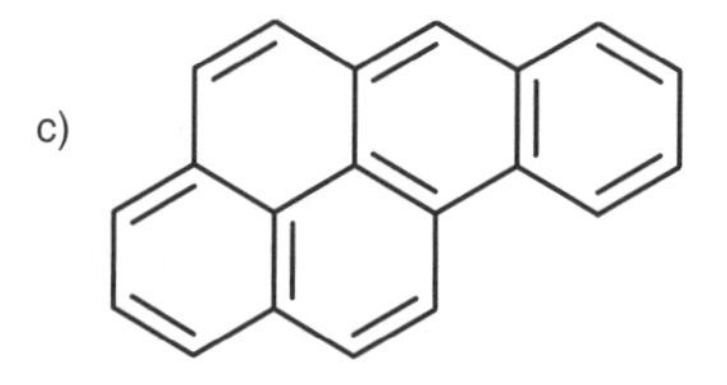

Fig. 4. (a) Naphthalene; (b) anthracene; (c) benzo[a]pyrene.

I2 Preparation and properties of aromatic compounds

Key Notes

Preparation — Simple aromatic structures such as benzene, toluene, or naphthalene are isolated from natural sources and converted to more complex aromatic structures.

Properties — Many aromatic compounds have a characteristic aroma and burn with a smoky flame. They are nonpolar, hydrophobic molecules which dissolve in organic solvents rather than water. Aromatic molecules can interact by van der Waals interactions or with a cation through an induced dipole interaction. Aromatic compounds undergo reactions where the aromatic ring is retained. Electrophilic substitution is the most common type of reaction. However, reduction is also possible.

Related topics — Electrophilic substitutions of benzene (I3); Oxidation and reduction (I7)

Preparation

It is not practical to synthesize aromatic structures in the laboratory from scratch and most aromatic compounds are prepared from benzene or other simple aromatic compounds (e.g. toluene and naphthalene). These in turn are isolated from natural sources such as coal or petroleum.

Properties

Many aromatic compounds have a characteristic aroma and will burn with a smoky flame. They are hydrophobic, nonpolar molecules which will dissolve in organic solvents and are poorly soluble in water. Aromatic molecules can interact with each other through intermolecular bonding by van der Waals interactions (*Fig. 1a*). However, induced dipole interactions are also possible with alkyl ammonium ions or metal ions where the positive charge of the cation induces a dipole in the aromatic ring such that the face of the ring is slightly negative and the edges are slightly positive (*Fig. 1b*). This results in the cation being sandwiched between two aromatic rings.

a)

van der Waals interactions

b)

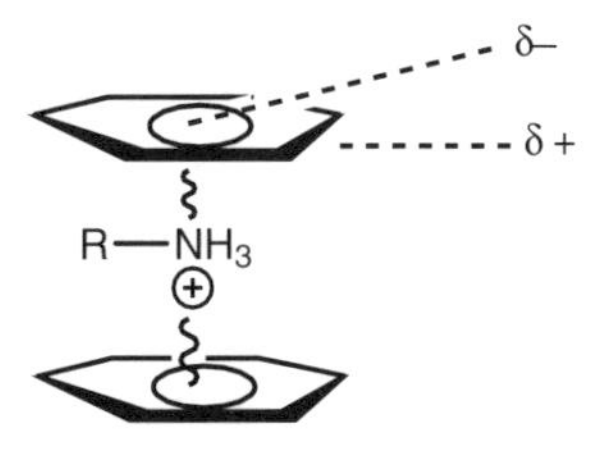

Induced dipole interaction

Fig. 1. Intermolecular bonding involving aromatic rings.

Aromatic compounds are unusually stable and do not react in the same way as alkenes. They prefer to undergo reactions where the stable aromatic ring is retained. The most common type of reaction for aromatic rings is **electrophilic substitution**, but **reduction** is also possible.

13 ELECTROPHILIC SUBSTITUTIONS OF BENZENE

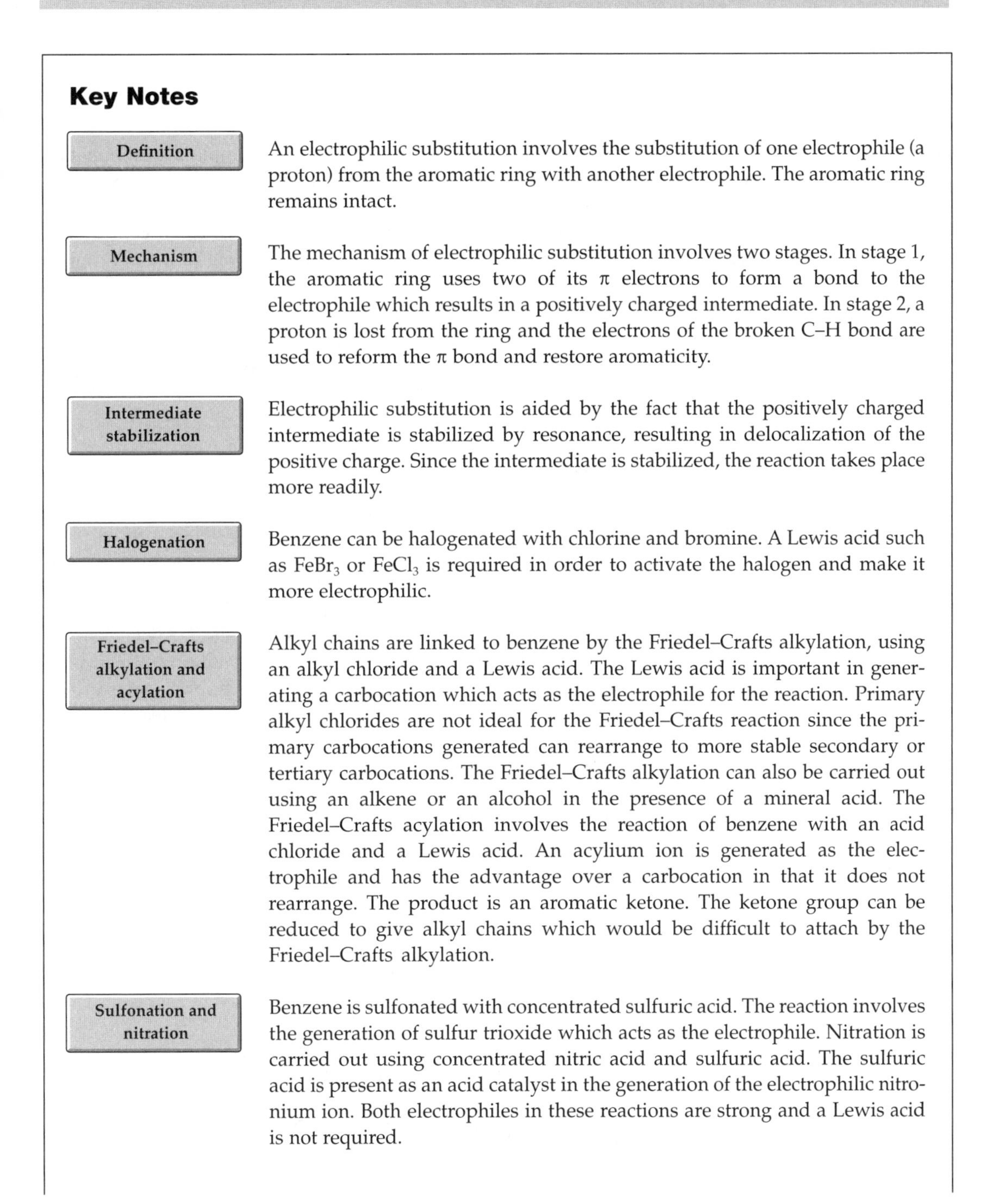

Key Notes

Definition

An electrophilic substitution involves the substitution of one electrophile (a proton) from the aromatic ring with another electrophile. The aromatic ring remains intact.

Mechanism

The mechanism of electrophilic substitution involves two stages. In stage 1, the aromatic ring uses two of its π electrons to form a bond to the electrophile which results in a positively charged intermediate. In stage 2, a proton is lost from the ring and the electrons of the broken C–H bond are used to reform the π bond and restore aromaticity.

Intermediate stabilization

Electrophilic substitution is aided by the fact that the positively charged intermediate is stabilized by resonance, resulting in delocalization of the positive charge. Since the intermediate is stabilized, the reaction takes place more readily.

Halogenation

Benzene can be halogenated with chlorine and bromine. A Lewis acid such as $FeBr_3$ or $FeCl_3$ is required in order to activate the halogen and make it more electrophilic.

Friedel–Crafts alkylation and acylation

Alkyl chains are linked to benzene by the Friedel–Crafts alkylation, using an alkyl chloride and a Lewis acid. The Lewis acid is important in generating a carbocation which acts as the electrophile for the reaction. Primary alkyl chlorides are not ideal for the Friedel–Crafts reaction since the primary carbocations generated can rearrange to more stable secondary or tertiary carbocations. The Friedel–Crafts alkylation can also be carried out using an alkene or an alcohol in the presence of a mineral acid. The Friedel–Crafts acylation involves the reaction of benzene with an acid chloride and a Lewis acid. An acylium ion is generated as the electrophile and has the advantage over a carbocation in that it does not rearrange. The product is an aromatic ketone. The ketone group can be reduced to give alkyl chains which would be difficult to attach by the Friedel–Crafts alkylation.

Sulfonation and nitration

Benzene is sulfonated with concentrated sulfuric acid. The reaction involves the generation of sulfur trioxide which acts as the electrophile. Nitration is carried out using concentrated nitric acid and sulfuric acid. The sulfuric acid is present as an acid catalyst in the generation of the electrophilic nitronium ion. Both electrophiles in these reactions are strong and a Lewis acid is not required.

Related topics

Acid strength (G2)
Lewis acids and bases (G4)
Enolates (G5)
Electrophilic addition to symmetrical alkenes (H3)
Carbocation stabilization (H5)
Conjugated dienes (H11)
Electrophilic substitution of mono-substituted aromatic rings (I5)

Definition

Aromatic rings undergo electrophilic substitution, for example the bromination of benzene (*Fig. 1*). The reaction involves an electrophile (Br^+) replacing another electrophile (H^+) with the aromatic ring remaining intact. Therefore, one electrophile replaces another and the reaction is known as an electrophilic substitution. (At this stage we shall ignore how the bromine cation is formed.)

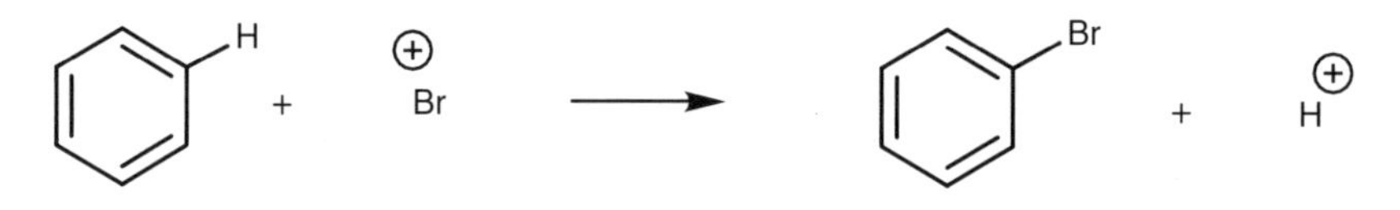

Fig. 1. Electrophilic substitution of benzene.

Mechanism

In the mechanism (*Fig. 2*) the aromatic ring acts as a nucleophile and provides two of its π electrons to form a bond to Br^+. The aromatic ring has now lost one of its formal double bonds resulting in a positively charged carbon atom. This first step in the mechanism is the same as the one described for the electrophilic addition to alkenes, and so the positively charged intermediate here is equivalent to the carbocation intermediate in electrophilic addition. However in step 2, the mechanisms of electrophilic addition and electrophilic substitution differ. Whereas the carbocation intermediate from an alkene reacts with a nucleophile to give an addition product, the intermediate from the aromatic ring loses a proton. The C–H σ bond breaks and the two electrons move into the ring to reform the π bond, thus regenerating the aromatic ring and neutralizing the positive charge on the carbon. This is the mechanism undergone in all electrophilic substitutions. The only difference is the nature of the electrophile (*Fig. 3*).

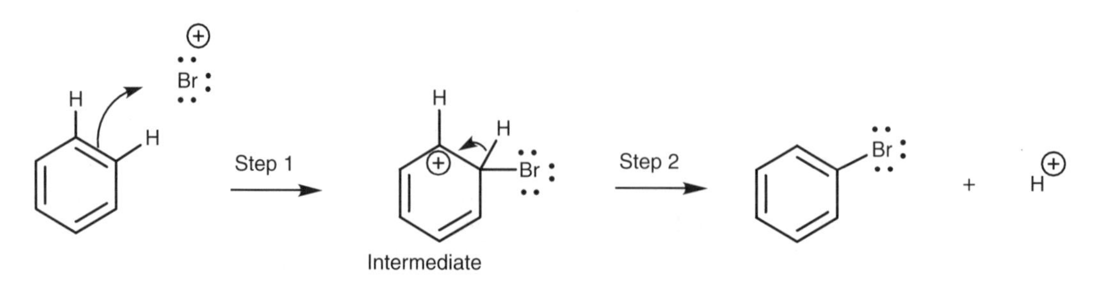

Fig. 2. Mechanism of electrophilic substitution.

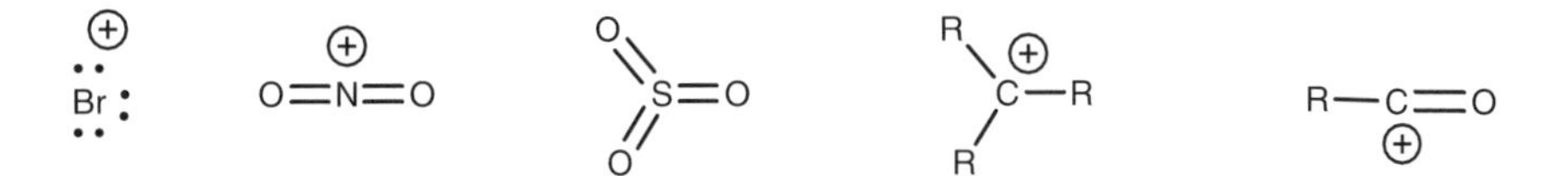

Fig. 3. Examples of electrophiles used in electrophilic substitution.

Intermediate stabilization

The rate-determining step in electrophilic substitution is the formation of the positively charged intermediate, and so the rate of the reaction is determined by the energy level of the transition state leading to that intermediate. The transition state resembles the intermediate in character and so any factor stabilizing the intermediate also stabilizes the transition state and lowers the activation energy required for the reaction. Therefore, electrophilic substitution is more likely to take place if the positively charged intermediate can be stabilized. Stabilization is possible if the positive charge can be spread amongst different atoms – a process called **delocalization**. The process by which this can take place is known as **resonance** (*Fig. 4*) – see also Topics H11, G2, and G5.

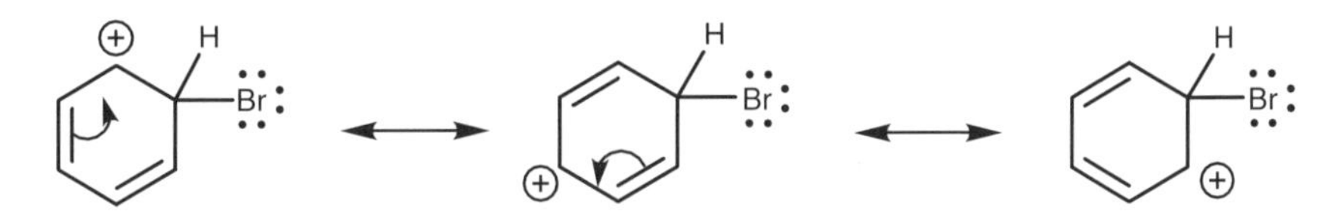

Fig. 4. Resonance stabilization of the charged intermediate.

The resonance process involves two π electrons shifting their position round the ring to provide the 'top' carbon with a fourth bond and thus neutralize its positive charge. In the process, another carbon in the ring is left short of bonds and gains the positive charge. This process can be repeated such that the positive charge is spread to a third carbon. The structures drawn in *Fig. 4* are known as resonance structures.

Halogenation

The stable aromatic ring means that aromatic compounds are less reactive than alkenes to electrophiles. For example, an alkene will react with Br_2 whereas an aromatic ring will not. Therefore, we have to activate the aromatic ring (i.e. make it a better nucleophile) or activate the Br_2 (i.e. make it a better electrophile) if we want a reaction to occur. In Topic I5, we will explain how electron-donating substituents on an aromatic ring increase the nucleophilicity of the aromatic ring. Here, we shall see how a Br_2 molecule can be activated to make it a better electrophile. This can be done by adding a Lewis acid such as $FeCl_3$, $FeBr_3$, or $AlCl_3$ (Topic G4) to the reaction medium. These compounds all contain a central atom (iron or aluminum) which is strongly electrophilic and does not have a full valence shell of electrons. As a result, the central atom can accept a lone pair of electrons, even from a weakly nucleophilic atom such as a halogen. In the example shown (*Fig. 5*) bromine uses a lone pair of electrons to form a bond to the Fe atom in $FeBr_3$ and becomes positively charged. Bromine is now activated to behave as an electrophile and will react more easily with a nucleophile (the aromatic ring) by the normal mechanism for electrophilic substitution.

An aromatic ring can be chlorinated in a similar fashion, using Cl_2 in the presence of $FeCl_3$.

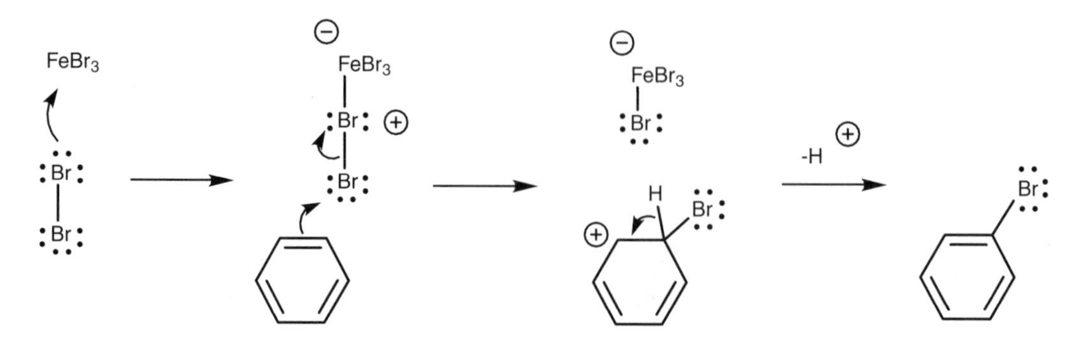

Fig. 5. Mechanism by which a Lewis acid activates bromine towards electrophilic substitution.

Friedel–Crafts alkylation and acylation

Friedel–Crafts alkylation and acylation (*Fig. 6*) are two other examples of electrophilic substitution requiring the presence of a Lewis acid, and are particularly important because they allow the construction of larger organic molecules by adding alkyl (R) or acyl (RCO) side chains to an aromatic ring.

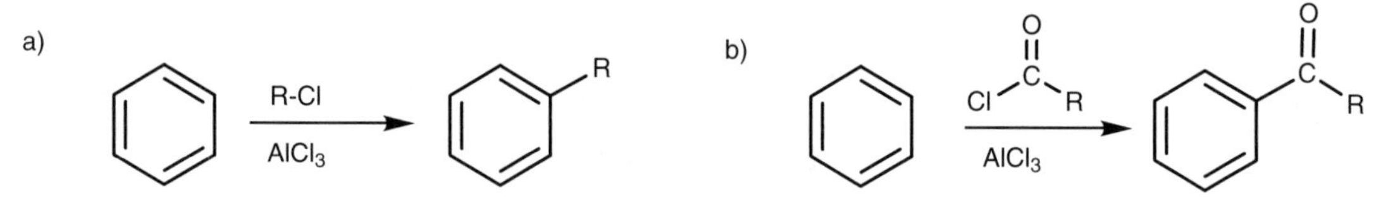

Fig. 6. (a) Friedel–Crafts alkylation; (b) Freidel–Crafts acylation.

An example of Friedel–Crafts alkylation is the reaction of benzene with 2-chloropropane (*Fig. 7*). The Lewis acid ($AlCl_3$) promotes the formation of the carbocation required for the reaction and does so by accepting a lone pair of electrons from chlorine to form an unstable intermediate which fragments to give a carbocation and $AlCl_4^-$ (*Fig. 8*). Once the carbocation is formed it reacts as an electrophile with the aromatic ring by the electrophilic substitution mechanism already described (*Fig. 9*).

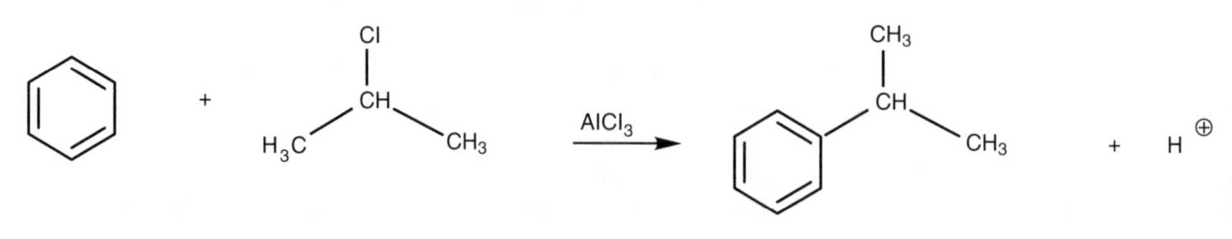

Fig. 7. Freidel–Crafts reaction of benzene with 2-chloropropane.

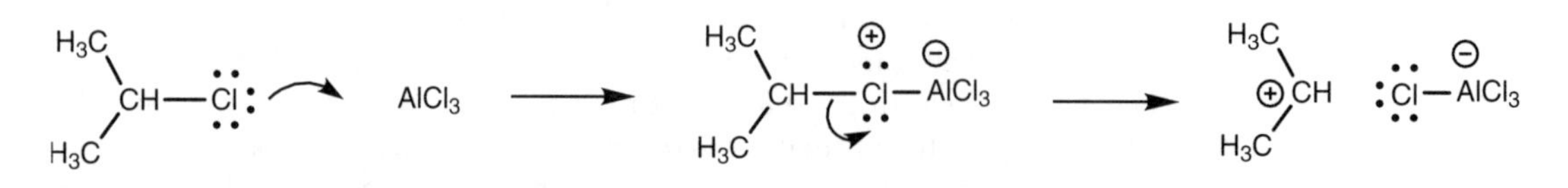

Fig. 8. Mechanism of carbocation formation.

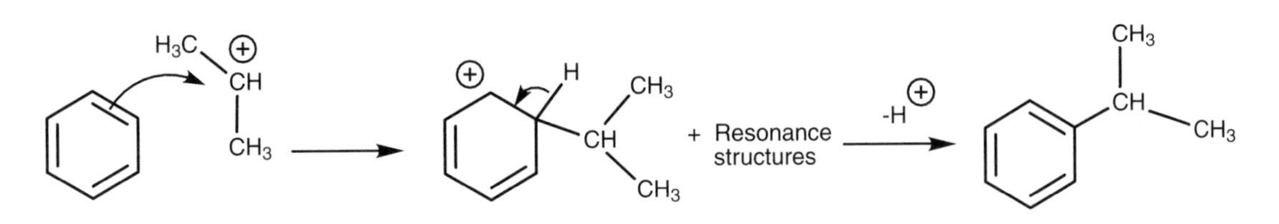

Fig. 9. Mechanism for the Friedel–Crafts alkylation.

There are limitations to the Friedel–Crafts alkylation. For example, the reaction of 1-chlorobutane with benzene gives two products with only 34% of the desired product (*Fig. 10*). This is due to the fact that the primary carbocation which is generated can rearrange to a more stable secondary carbocation where a hydrogen (and the two sigma electrons making up the C–H bond) 'shift' across to the neighboring carbon atom (*Fig. 11*). This is known as a **hydride shift** and it takes place because the secondary carbocation is more stable than the primary carbocation (see Topic H5). Such rearrangements limit the type of alkylations which can be carried out by the Friedel–Crafts reaction.

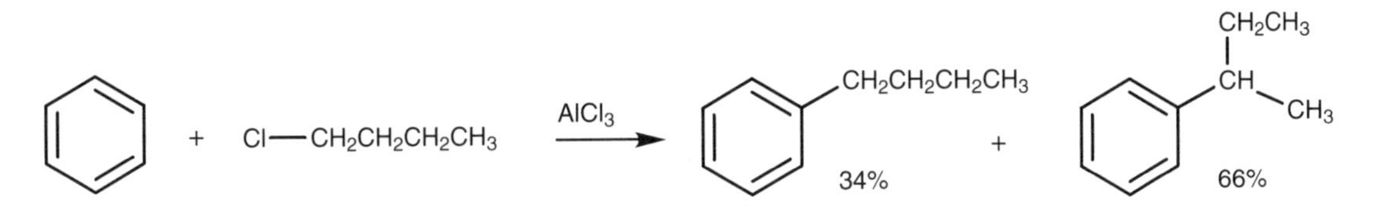

Fig. 10. Friedel–Crafts reaction of 1-chlorobutane with benzene.

Fig. 11. Hydride shift.

Bearing this in mind, how is it possible to make structures like 1-butylbenzene in good yield? The answer to this problem lies in the **Friedel–Crafts acylation** (*Fig. 12*). By reacting benzene with butanoyl chloride instead of 1-chlorobutane, the necessary 4-C skeleton is linked to the aromatic ring and no rearrangement takes place. The carbonyl group can then be removed by reducing it with hydrogen over a palladium catalyst to give the desired product.

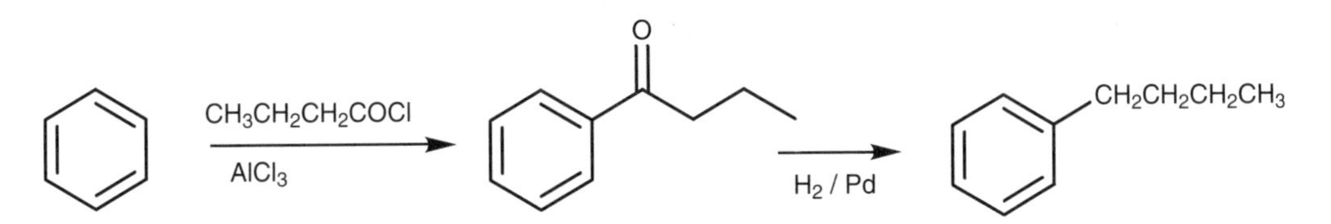

Fig. 12. Synthesis of 1-butylbenzene by Friedel–Crafts acylation and reduction.

The mechanism of the Friedel–Crafts acylation is the same as the Friedel–Crafts alkylation involving an **acylium** ion instead of a carbocation. As with the Friedel–Crafts alkylation, a Lewis acid is required to generate the acylium ion $(R–C=O)^+$, but unlike a carbocation the acylium ion does not rearrange since there is resonance stabilization from the oxygen (*Fig. 13*).

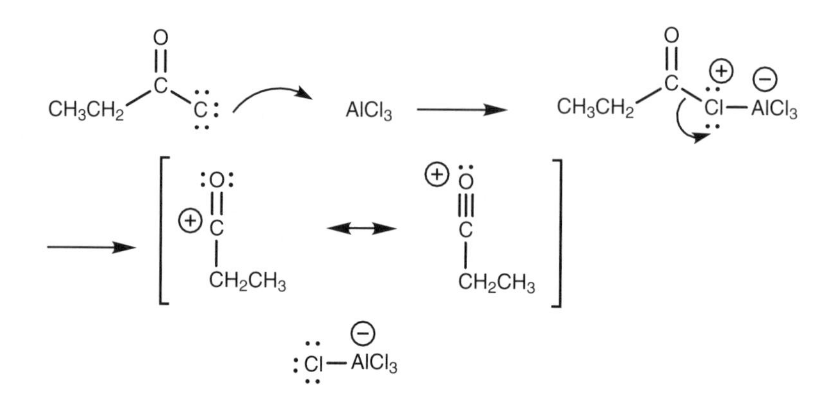

Fig. 13. Generation of the acylium ion.

Friedel–Crafts alkylations can also be carried out using alkenes instead of alkyl halides. A Lewis acid is not required, but a mineral acid is. Treatment of the alkene with the acid leads to a carbocation which can then react with an aromatic ring by the same electrophilic substitution mechanism already described (*Fig. 14*). As far as the alkene is concerned, this is another example of electrophilic addition where a proton is attached to one end of the double bond and a phenyl group is added to the other (Topic H3).

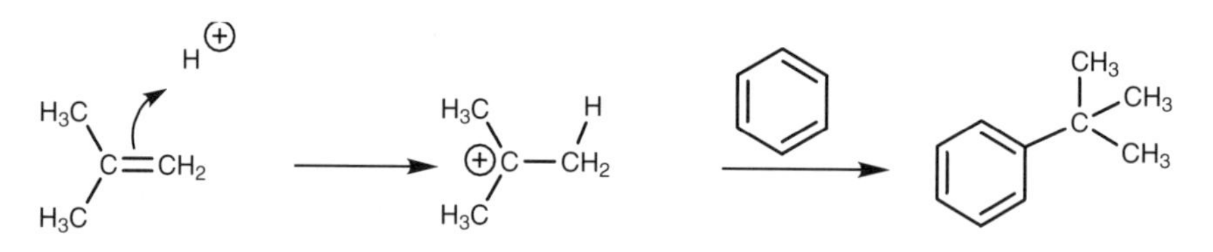

Fig. 14. Friedel–Crafts alkylation of benzene with an alkene.

Friedel–Crafts reactions can also be carried out with alcohols in the presence of mineral acid. The acid leads to the elimination of water from the alcohol resulting in the formation of an alkene which can then be converted to a carbocation as already described (*Fig. 15*). The conversion of alcohols to alkenes under acid conditions will be covered in Topic M4.

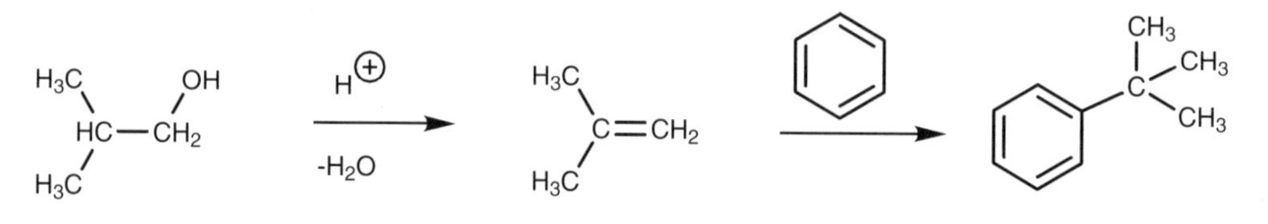

Fig. 15. Freidel–Crafts alkylation of benzene with an alcohol.

Sulfonation and nitration

Sulfonation and nitration are electrophilic substitutions which involve strong electrophiles and do not need the presence of a Lewis acid (*Fig. 16*).

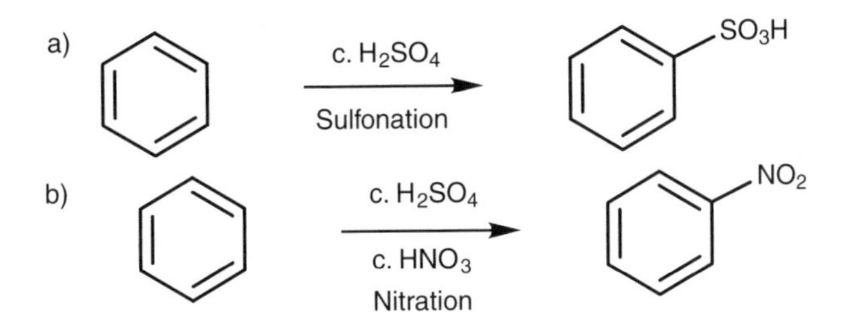

Fig. 16. (a) Sulfonation of benzene; (b) nitration of benzene.

In sulfonation, the electrophile is sulfur trioxide (SO_3) which is generated under the acidic reaction conditions (*Fig. 17*). Protonation of an OH group generates a protonated intermediate (I). Since the oxygen gains a positive charge it becomes a good leaving group and water is lost from the intermediate to give sulfur trioxide. Although sulfur trioxide has no positive charge, it is a strong electrophile. This is because the sulfur atom is bonded to three electronegative oxygen atoms which are all 'pulling' electrons from the sulfur, and making it electron deficient (i.e. electrophilic). During electrophilic substitution (*Fig. 18*), the aromatic ring forms a bond to sulfur and one of the π bonds between sulfur and oxygen is broken. Both electrons move to the more electronegative oxygen to form a third lone pair and produce a negative charge on that oxygen. This will finally be neutralized when the third lone pair of electrons is used to form a bond to a proton.

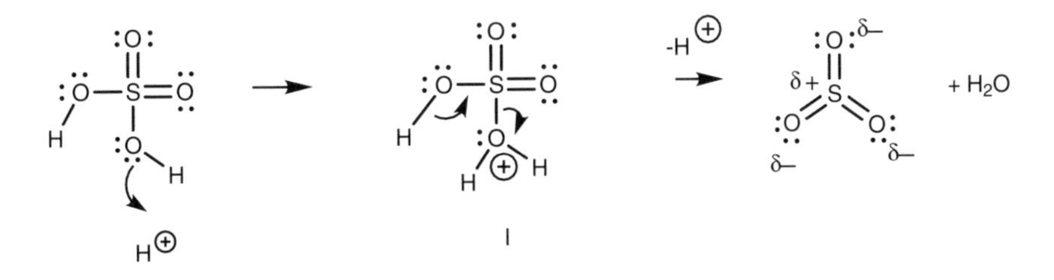

Fig. 17. Generation of sulfur trioxide.

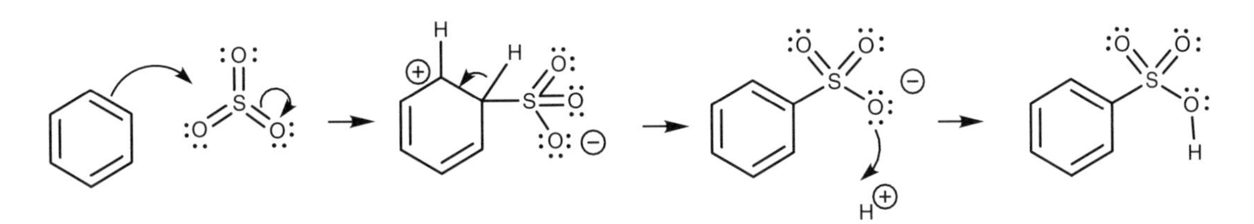

Fig. 18. Sulfonation of benzene.

In nitration, sulfuric acid serves as an acid catalyst for the formation of a nitronium ion (NO_2^+) which is generated from nitric acid by a very similar mechanism to that used in the generation of sulfur trioxide from sulfuric acid (*Fig. 19*).

The mechanism for the nitration of benzene is very similar to sulfonation (*Fig. 20*). As the aromatic ring forms a bond to the electrophilic nitrogen atom, a π bond between N and O breaks and both electrons move onto the oxygen atom. Unlike sulfonation, this oxygen keeps its negative charge and does not pick up a proton. This is because it acts as a counterion to the neighboring positive charge on nitrogen.

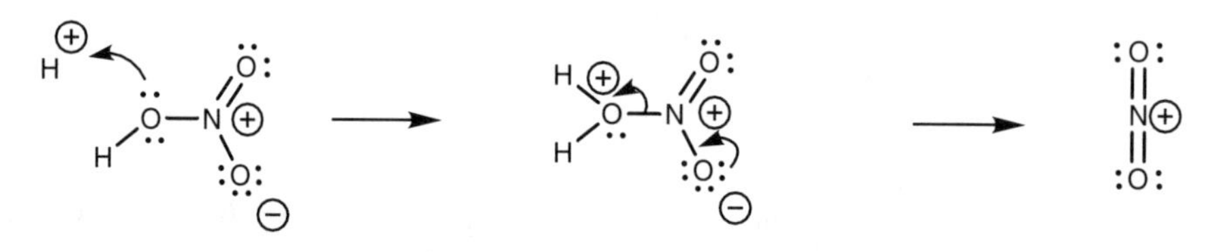

Fig. 19. Generation of the nitronium ion.

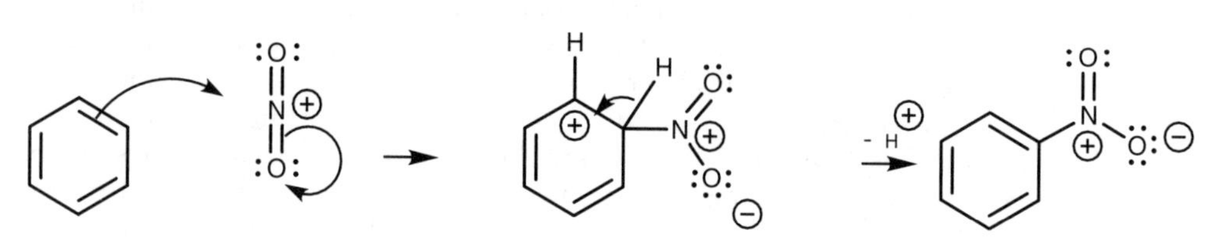

Fig. 20. Nitration of benzene.

14 Synthesis of mono-substituted benzenes

Key Notes

Functional group transformations

Many functional groups cannot be added directly to an aromatic ring by electrophilic substitution but can be obtained by converting functional groups already added. An amino group can be obtained by reduction of a nitro group then converted to a large range of other functional groups. Alkyl groups can be oxidized to a carboxylic acid group which can in turn be converted to other functional groups.

Synthetic planning

When planning the synthesis of an aromatic compound, it is best to work backwards from the product in simple stages (retrosynthesis). If the substituent present cannot be added directly to the aromatic ring, it is best to consider what functional group could be transformed to give the desired substituent.

Related topics

Electrophilic substitutions of benzene (I3)
Electrophilic substitutions of mono-substituted benzenes (I5)
Reactions of amines (O3)

Functional group transformations

Some substituents cannot be introduced directly onto an aromatic ring by electrophilic substitution. These include the following groups: $-NH_2$, $-NHR$, NR_2, $NHCOCH_3$, CO_2H, CN, OH. Although these groups cannot be added directly onto the aromatic ring they can be obtained by transforming a functional group which **can** be applied directly by electrophilic substitution. Three of the most important transformations are shown (*Fig. 1*).

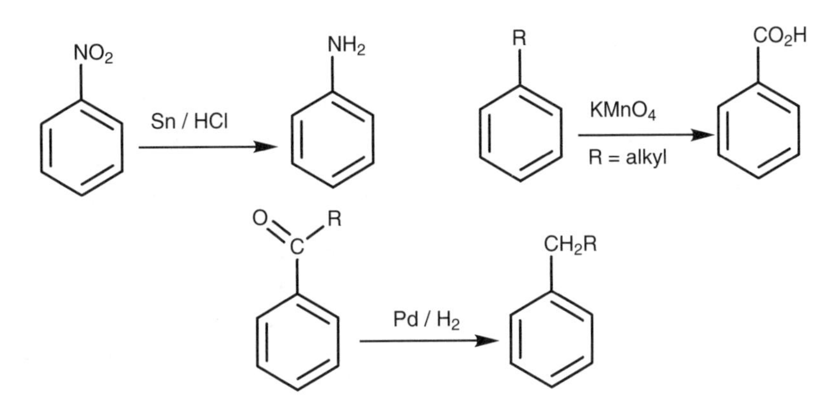

Fig. 1. Functional group transformations of importance in aromatic chemistry.

Nitro, alkyl, and acyl groups can readily be added by electrophilic substitution and can then be converted to amino, carboxylic acid, and alkyl groups respectively. Once the amino and carboxylic acid groups have been obtained, they can be further converted to a large range of other functional groups such as secondary and tertiary amines, amides, diazonium salts, halides, nitriles, esters, phenols, alcohols, and ethers (Topic O3 and Section K).

Synthetic planning

A knowledge of the electrophilic substitutions and functional group transformations which are possible is essential in planning the synthesis of an aromatic compound. When designing such a synthesis, it is best to work backwards from the product and to ask what it could have been synthesized from – a process called **retrosynthesis**. We can illustrate this by designing a synthesis of an aromatic ester (*Fig. 2*). An ester functional group cannot be attached directly by electrophilic substitution, so the synthesis must involve several steps. The usual way to make an ester is from an acid chloride which is synthesized in turn from a carboxylic acid. Alternatively, the ester can be made directly from the carboxylic acid by treating it with an alcohol and an acid catalyst (Topics K2 and K5). Either way, benzoic acid is required to synthesize the ester. Carboxylic acids cannot be added directly to aromatic rings either, so we have to look for a different functional group which can be added directly, then transformed to a carboxylic acid. A carboxylic acid group can be obtained from the oxidation of a methyl group (Topic I7). Methyl groups can be added directly by Friedel–Crafts alkylation. Therefore a possible synthetic route would be as shown in *Fig. 2*.

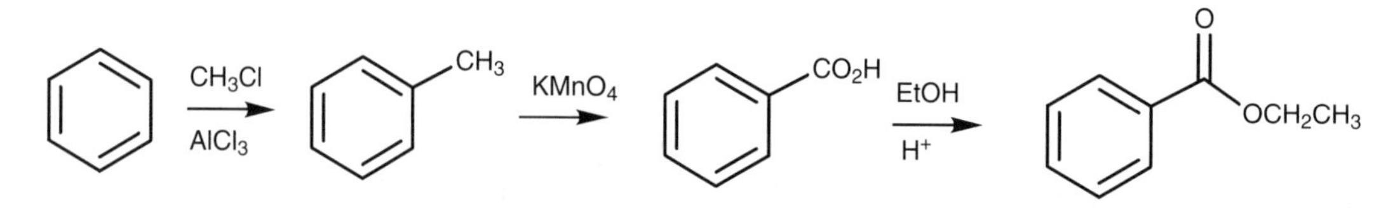

Fig. 2. Possible synthesis of an aromatic ester.

One possible problem with this route is the possibility of poly-methylation in the first step. This is likely since the product (toluene) will be more reactive than the starting material (benzene; Topic I5). One way round this problem would be to use an excess of benzene.

As a second example, let us consider the synthesis of an aromatic amine (*Fig. 3*). The alkylamine group cannot be applied to an aromatic ring directly and so must be obtained by modifying another functional group. Working backwards, the alkylamine group could be obtained by alkylation of an amino group (NH_2). An amino group cannot be directly applied to an aromatic ring either. However, an amino group could be obtained by reduction of a nitro group. A nitro group **can** be applied directly to an aromatic ring. Thus, the overall synthesis would be nitration followed by reduction, followed by alkylation.

Note that there are two methods of converting aniline ($PhNH_2$) to the final product. Alkylation is the direct method, but sometimes acylation followed by reduction gives better yields, despite the extra step. This is because it is sometimes difficult to control the alkylation to only one alkyl group (Topic O3).

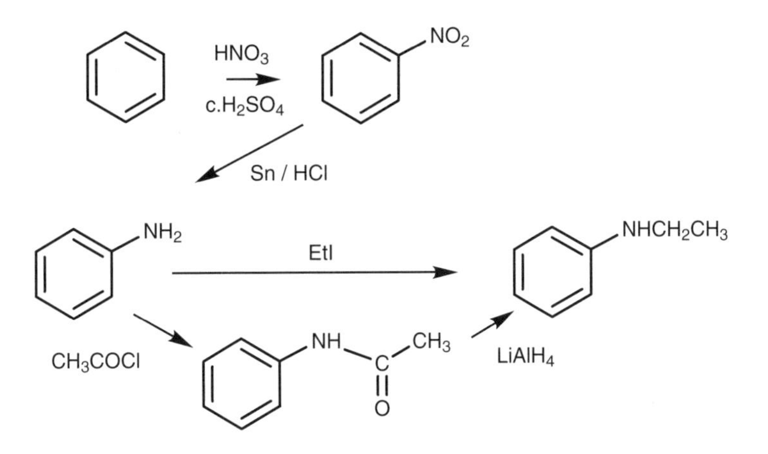

Fig. 3. *Possible synthetic routes to an aromatic amine.*

As our last example, we shall consider the synthesis of an aromatic ether (*Fig. 4*). Here an ethoxy group is attached to the aromatic ring. The ethoxy group cannot be applied directly to an aromatic ring, so we have to find a way of obtaining it from another functional group. Alkylation of a phenol group would give the desired ether, but a phenol group cannot be applied directly to the ring either. However, we can obtain the phenol from an amino group, which in turn can be obtained from a nitro group. The nitro group can be applied directly to the ring and so the synthesis involves a nitration, reduction, conversion of the amino group to a diazonium salt (see Topic O3), hydrolysis, and finally an alkylation.

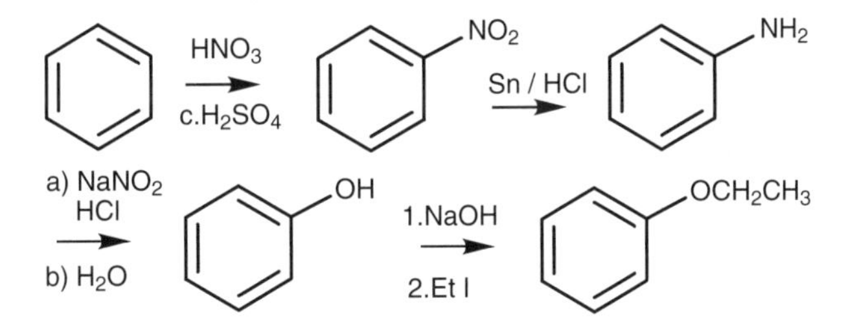

Fig. 4. *Possible synthetic route to an aromatic ether.*

15 ELECTROPHILIC SUBSTITUTIONS OF MONO-SUBSTITUTED AROMATIC RINGS

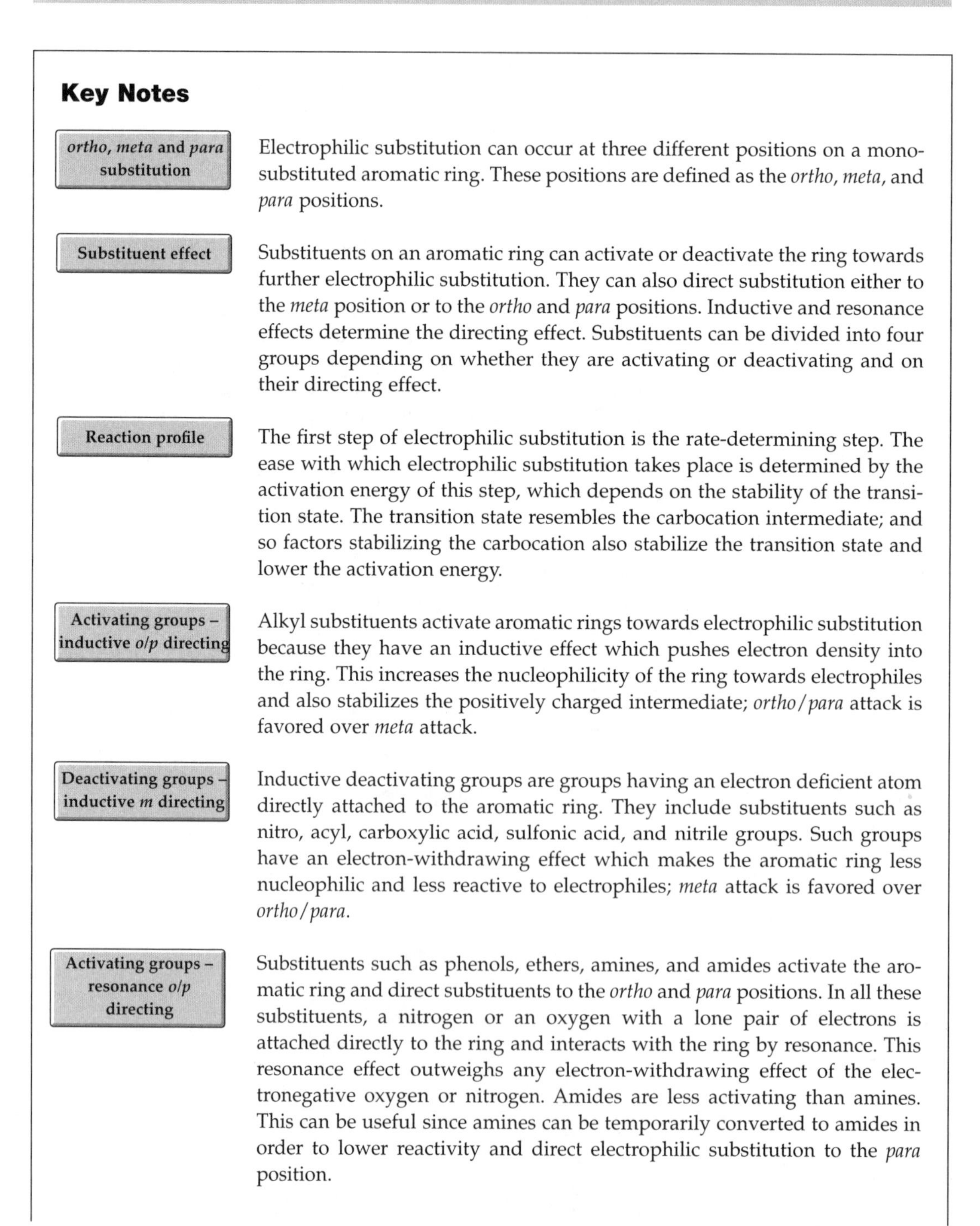

Key Notes

ortho, meta and para substitution

Electrophilic substitution can occur at three different positions on a mono-substituted aromatic ring. These positions are defined as the *ortho, meta*, and *para* positions.

Substituent effect

Substituents on an aromatic ring can activate or deactivate the ring towards further electrophilic substitution. They can also direct substitution either to the *meta* position or to the *ortho* and *para* positions. Inductive and resonance effects determine the directing effect. Substituents can be divided into four groups depending on whether they are activating or deactivating and on their directing effect.

Reaction profile

The first step of electrophilic substitution is the rate-determining step. The ease with which electrophilic substitution takes place is determined by the activation energy of this step, which depends on the stability of the transition state. The transition state resembles the carbocation intermediate; and so factors stabilizing the carbocation also stabilize the transition state and lower the activation energy.

Activating groups – inductive o/p directing

Alkyl substituents activate aromatic rings towards electrophilic substitution because they have an inductive effect which pushes electron density into the ring. This increases the nucleophilicity of the ring towards electrophiles and also stabilizes the positively charged intermediate; *ortho/para* attack is favored over *meta* attack.

Deactivating groups – inductive m directing

Inductive deactivating groups are groups having an electron deficient atom directly attached to the aromatic ring. They include substituents such as nitro, acyl, carboxylic acid, sulfonic acid, and nitrile groups. Such groups have an electron-withdrawing effect which makes the aromatic ring less nucleophilic and less reactive to electrophiles; *meta* attack is favored over *ortho/para*.

Activating groups – resonance o/p directing

Substituents such as phenols, ethers, amines, and amides activate the aromatic ring and direct substituents to the *ortho* and *para* positions. In all these substituents, a nitrogen or an oxygen with a lone pair of electrons is attached directly to the ring and interacts with the ring by resonance. This resonance effect outweighs any electron-withdrawing effect of the electronegative oxygen or nitrogen. Amides are less activating than amines. This can be useful since amines can be temporarily converted to amides in order to lower reactivity and direct electrophilic substitution to the *para* position.

Deactivating groups – resonance *o/p* directing

Halogen substituents deactivate aromatic rings towards electrophilic substitution by an inductive effect. These atoms are strongly electronegative and have an electron-withdrawing effect on the ring making it less nucleophilic and less reactive. However, once electrophilic substitution does take place, halogens direct to the *ortho* and *para* positions due to a resonance effect which helps to delocalize the positive charge onto the halogen atoms. This resonance effect is more important than the inductive effect which would normally direct substitution to the *meta* position.

Related topics

Carbocation stabilization (H5)
Electrophilic substitutions of benzene (I3)
Synthesis of di- and tri-substituted benzenes (I6)

ortho, *meta* and *para* substitution

Aromatic compounds which already contain a substituent can undergo electrophilic substitution at three different positions relative to the substituent. Consider the bromination of toluene (*Fig. 1*). Three different products are possible depending on where the bromine enters the ring. These products have the same molecular formula and are therefore constitutional isomers. The aromatic ring is said to be disubstituted and the three possible isomers are described as being *ortho*, *meta*, and *para*. The mechanisms leading to these three isomers are shown in *Fig. 2*.

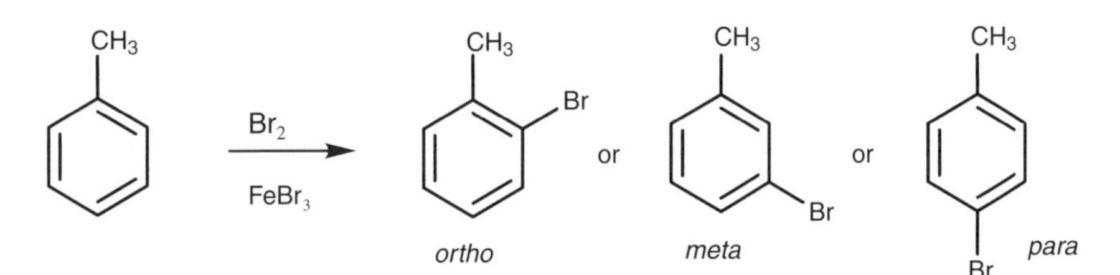

Fig. 1. ortho, meta, *and* para *isomers of bromotoluene.*

Substituent effect

Of the three possible isomers arising from the bromination of toluene, only two (the *ortho* and *para*) are formed in significant quantity. Furthermore, the bromination of toluene goes at a faster rate than the bromination of benzene. Why? The answer lies in the fact that the methyl substituent can affect the rate and the position of further substitution. A substituent can either activate or deactivate the aromatic ring towards electrophilic substitution and does so through inductive or resonance effects. A substituent can also direct the next substitution so that it goes mainly *ortho/para* or mainly *meta*.

We can classify substituents into four groups depending on the effect they have on the rate and the position of substitution, that is:

- activating groups which direct *ortho/para* by inductive effects;
- deactivating groups which direct *meta* by inductive effects;
- activating groups which direct *ortho/para* by resonance effects;
- deactivating groups which direct *ortho/para* by resonance effects.

There are no substituents which activate the ring and direct *meta*.

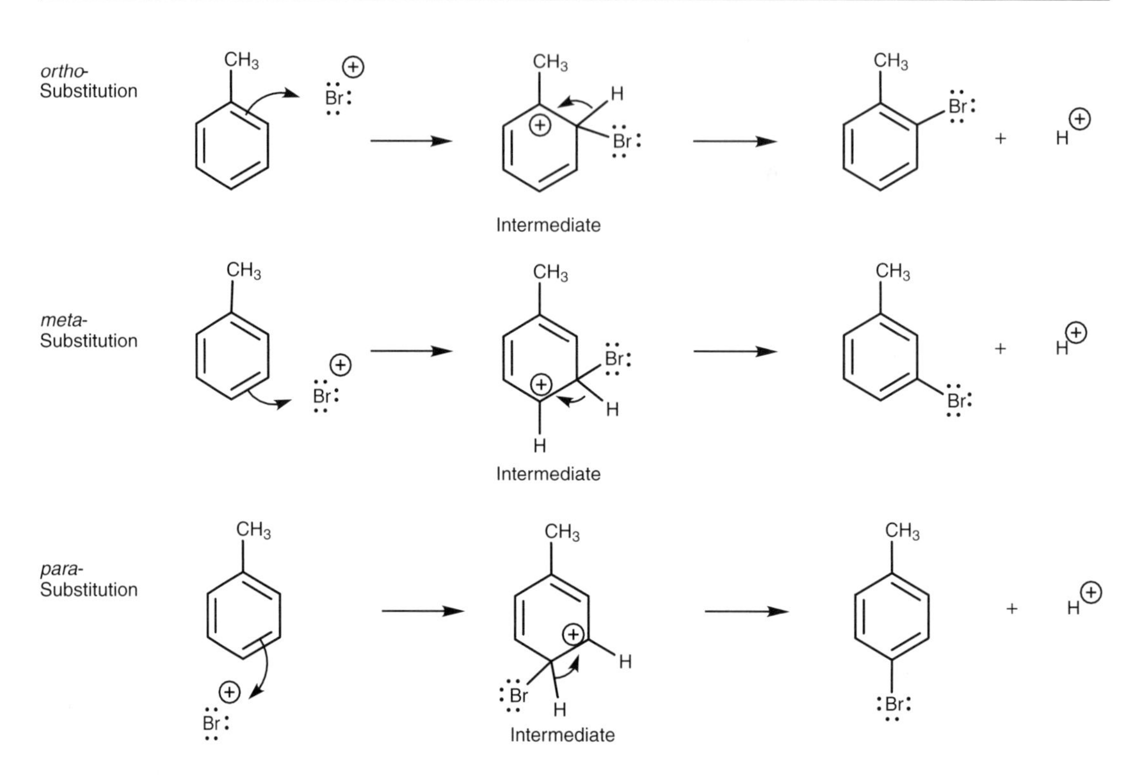

Fig. 2. Mechanisms of ortho, meta, *and* para *electrophilic substitution.*

Reaction profile

Before explaining the reasons behind the substituent effect, we have to consider the reaction profile of electrophilic substitution with respect to the relative energies of starting material, intermediate, and product. The energy diagram (*Fig. 3*) illustrates the reaction pathway for the bromination of benzene. The first stage in the mechanism is the rate-determining step and is the formation of the carbocation. This is endothermic and proceeds through a transition state which requires an activation energy ($\Delta G^{\#}$). The magnitude of $\Delta G^{\#}$ determines the rate at which the reaction will occur and this in turn is determined by the stability of the transition state. The transition state resembles the carbocation intermediate and so any factor which stabilizes the intermediate also stabilizes the transition state and favors the reaction. Therefore, in the discussions to follow we can consider the stability of relative carbocations in order to determine which reaction is more favorable.

Activating groups – inductive *o/p* directing

A methyl substituent is an example of an inductive activating group and so we shall consider again the bromination of toluene. In order to explain the directing properties of the methyl group, we need to look more closely at the mechanisms involved in generating the *ortho*, *meta*, and *para* isomers (*Fig. 2*). The preferred reaction pathway will be the one which goes through the most stable intermediate. Since a methyl group directs *ortho* and *para*, the intermediates involved in these reaction pathways are more stable than the intermediate involved in *meta* substitution. The relevant intermediates and their resonance structures are shown in *Fig. 4*.

If we compare all the resonance structures above, we can spot one *ortho* and one *para* resonance structure (boxed) where the positive charge is positioned immedi-

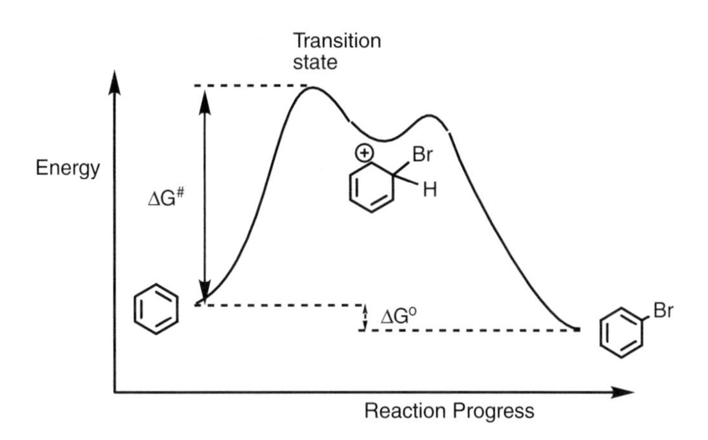

Fig. 3. Energy diagram for electrophilic substitution.

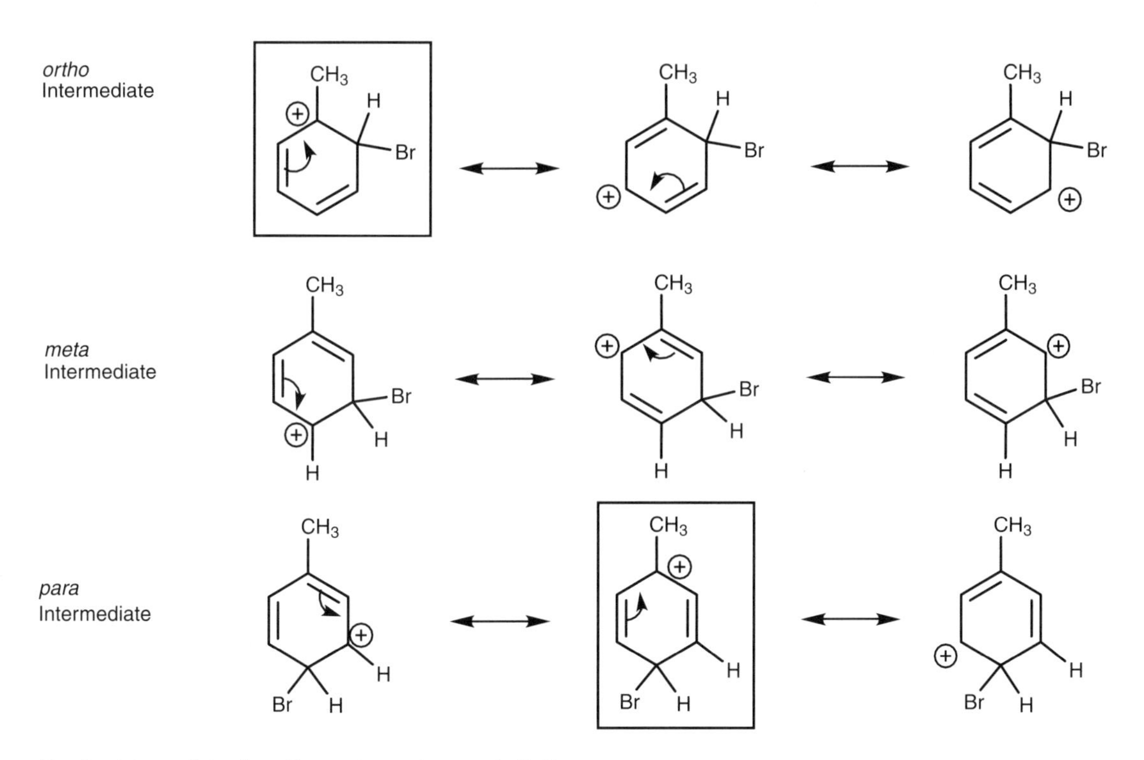

Fig. 4. Intermediates for ortho, meta, *and* para *substitution.*

ately next to the methyl substituent. An alkyl group can stabilize a neighboring positive charge by an inductive, electron-donating effect which results in some of the positive charge being spread over the alkyl group. This is an additional stabilizing effect which is only possible for the intermediates arising from *ortho* and *para* substitution. There is no such equivalent resonance structure for the *meta* intermediate and so that means that the *ortho* and *para* intermediates experience an increased stability over the *meta*, which results in a preference for these two substitution pathways.

By the same token, toluene will be more reactive than benzene. The electron-donating effect of the methyl group into the aromatic ring makes the ring inherently more nucleophilic and more reactive to electrophiles, as well as providing extra stabilization of the reaction intermediate. To sum up, alkyl groups are **activating groups** and are ***ortho, para* directing**.

The nitration of toluene also illustrates this effect (*Fig. 5*). The amount of *meta* substitution is very small as we would expect, and there is a preference for the *ortho* and *para* products. However, why is there more *ortho* substitution compared to *para* substitution? Quite simply, there are two *ortho* sites on the molecule to one *para* site and so there is double the chance of *ortho* attack to *para* attack. Based on pure statistics we would have expected the ratio of *ortho* to *para* attack to be 2:1. In fact, the ratio is closer to 1.5:1. In other words, there is **less** *ortho* substitution than expected. This is because the *ortho* sites are immediately 'next door' to the methyl substituent and the size of the substituent tends to interfere with *ortho* attack – a steric effect. The significance of this steric effect will vary according to the size of the alkyl substituent. The larger the substituent, the more *ortho* attack will be hindered.

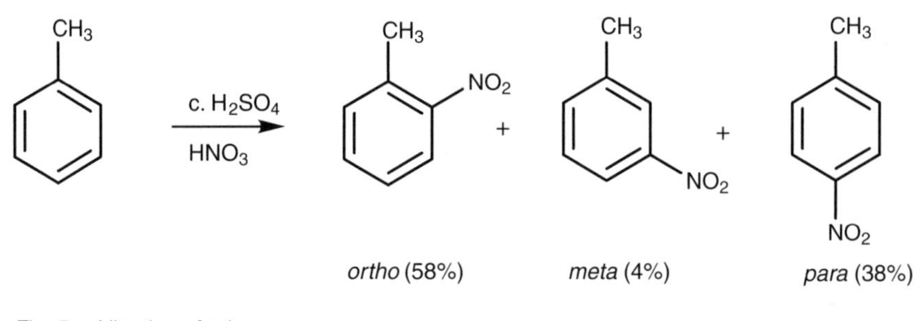

Fig. 5. Nitration of toluene.

Deactivating groups – inductive *m* directing

Alkyl groups are activating groups and direct substitution to the *ortho, para* positions. Electron withdrawing substituents (*Fig. 6*) have the opposite effect. They deactivate the ring, make the ring less nucleophilic and less likely to react with an electrophile. The electron-withdrawing effect also destabilizes the reaction intermediate and makes the reaction more difficult. This destabilization is more pronounced in the intermediates arising from *ortho/para* attack and so *meta* attack is favored.

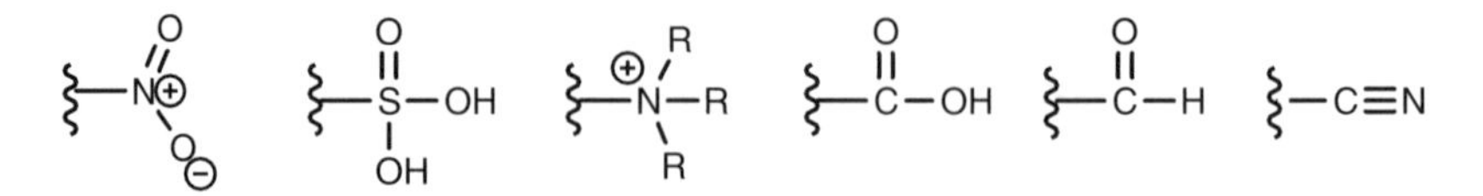

Fig. 6. Examples of electron-withdrawing groups.

All of these groups have a positively charged atom or an electron deficient atom (i.e. an electrophilic center) directly attached to the aromatic ring. Since this atom is electron deficient, it has an electron-withdrawing effect on the ring.

Deactivating groups make electrophilic substitution more difficult but the reaction will proceed under more forcing reaction conditions. However, substitution is now directed to the *meta* position. This can be explained by comparing all the possible resonance structures arising from *ortho, meta* and *para* attack. As an example, we shall consider the bromination of nitrotoluene (*Fig. 7*). Of all the possible resonance structures arising from *ortho*, *meta*, and *para* attack, there are two specific resonance structures (arising from *ortho* and *para* attack) where the positive charge is placed directly next to the electron-withdrawing nitro group (*Fig. 8*). As a result, these resonance structures are greatly destabilized. This does not occur with any of the resonance structures arising from *meta* attack and so *meta* attack is favored.

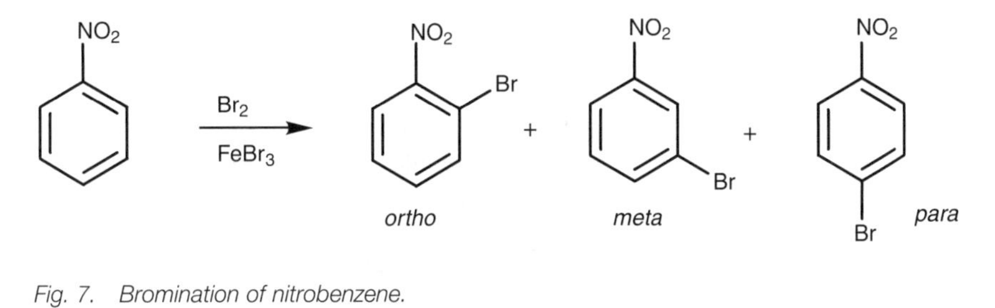

Fig. 7. Bromination of nitrobenzene.

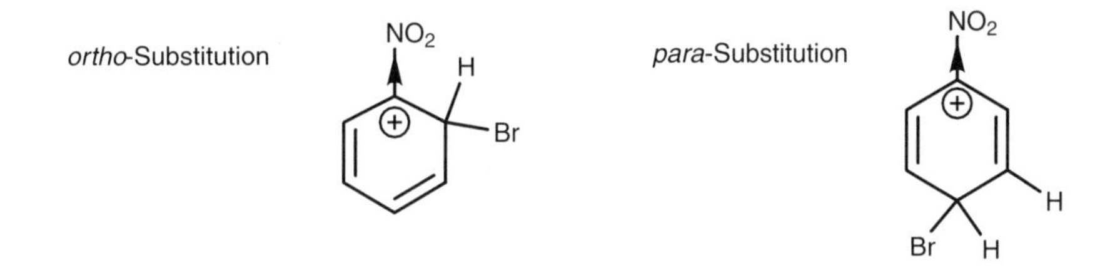

Fig. 8. Destabilizing resonance structures for the intermediate arising from ortho *and* para *substitution.*

Activating groups – resonance *o/p* directing

Phenol is an example of a substituent which activates the aromatic ring by resonance effects and which directs substitution to the *ortho* and *para* positions. In phenol, an electronegative oxygen atom is next to the aromatic ring. Since oxygen is electronegative, it should have an electron-withdrawing inductive effect and so might be expected to deactivate the ring. The fact that the phenolic group is a powerful activating group is due to the fact that oxygen is electron rich and can also act as a nucleophile, feeding electrons into the ring through a resonance process. As an example, we shall look at the nitration of phenol (*Fig. 9*).

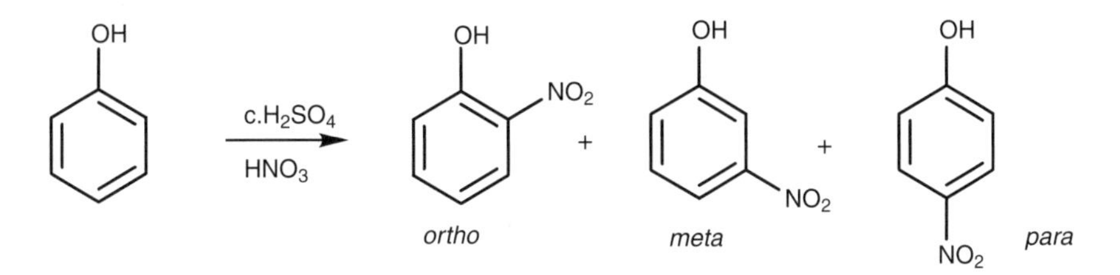

Fig. 9. Nitration of phenol.

There are three resonance structures for the intermediate formed in each form of electrophilic substitution, but there are two crucial ones to consider (*Fig. 10*), arising from *ortho* and *para* substitution. These resonance structures have the positive

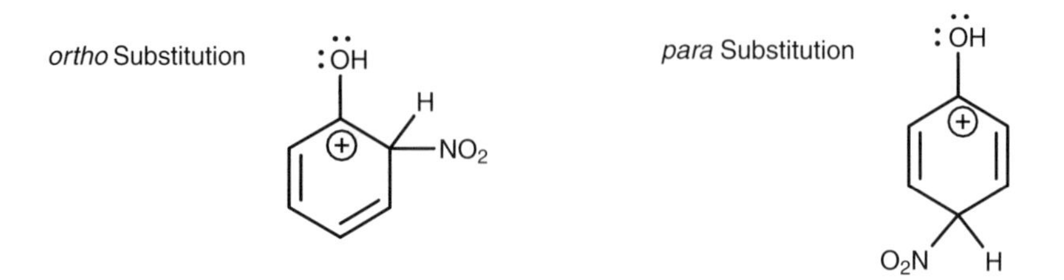

Fig. 10. Resonance structures for the intermediates arising from ortho *and* para *substitution.*

charge next to the OH substituent. If oxygen only had an inductive effect, these resonance structures would be highly unstable. However, oxygen can act as a nucleophile and can use one of its lone pairs of electrons to form a new π bond to the neighboring electrophilic center (*Fig. 11*). This results in a fourth resonance structure where the positive charge is moved out of the ring and onto the oxygen atom. Delocalizing the charge like this further stabilizes it and makes the reaction proceed more easily.

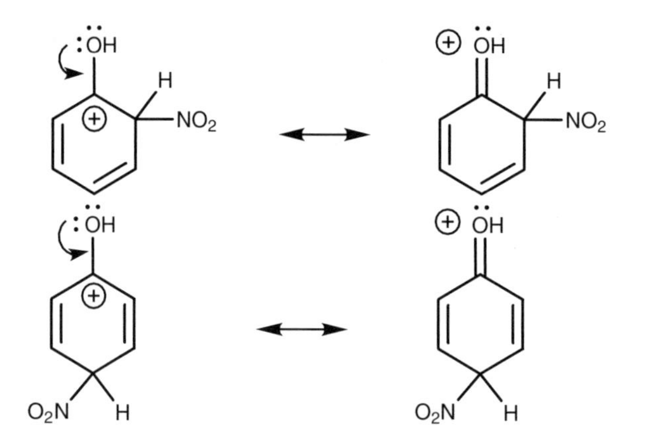

Fig. 11. Resonance interactions between the aromatic ring and oxygen.

Since none of the resonance structures arising from *meta* attack places the positive charge next to the phenol group, this fourth resonance structure is not available to the *meta* intermediate and so *meta* attack is not favored. Thus, the phenol group is an activating group which is *ortho, para* directing because of resonance effects. This resonance effect is more important than any inductive effect which the oxygen might have.

The same holds true for the following substituents: alkoxy (–OR), esters (–OCOR), amines (–NH_2, –NHR, –NR_2), and amides (–NHCOR). In all these cases, there is either a nitrogen or an oxygen next to the ring. Both these atoms are nucleophilic and have lone pairs of electrons which can be used to form an extra bond to the ring. The ease with which the group can do this depends on the nucleo-philicity of the attached atom and how well it can cope with a positive charge.

Nitrogen is more nucleophilic than oxygen since it is better able to cope with the resulting positive charge. Therefore amine substituents are stronger activating groups than ethers. On the other hand, an amide group is a weaker activating group since the nitrogen atom is less nucleophilic. This is because the nitrogen's lone pair of electrons is pulled towards the carbonyl group and is less likely to form a bond to the ring (*Fig. 12*). This property of amides can be quite

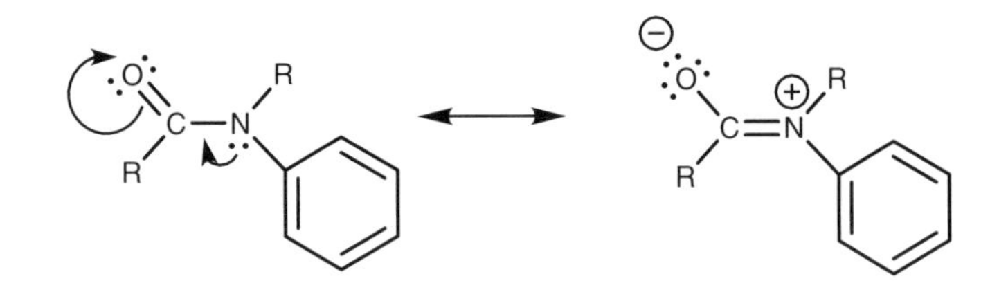

Fig. 12. Amide resonance.

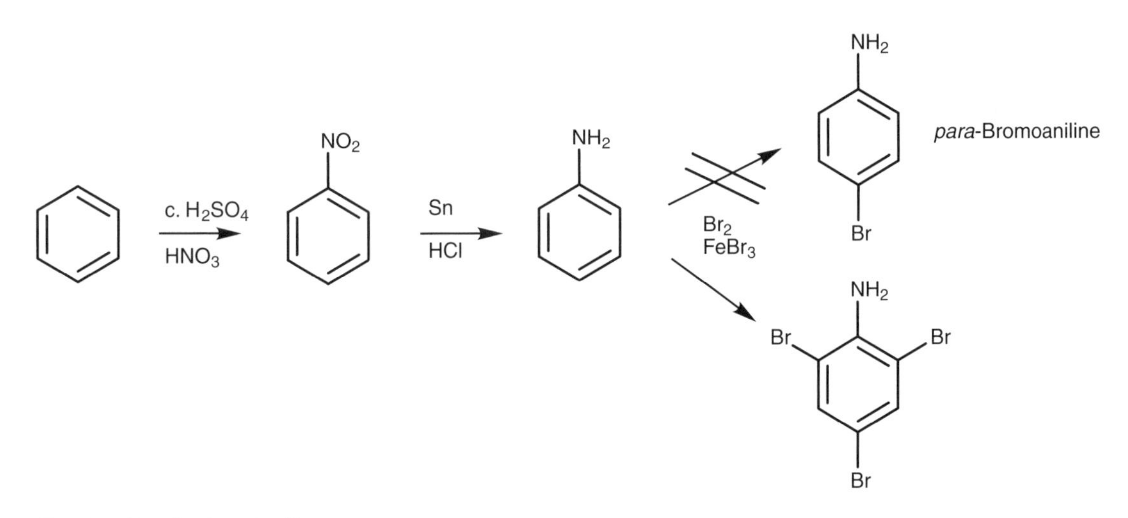

Fig. 13. Bromination of aniline.

useful. Suppose for example we wanted to make *para*-bromoaniline by brominating aniline (*Fig. 13*). In theory, this reaction scheme should give the desired product. In practice, the NH_2 group is such a strong activating group that the final bromination goes three times to give the tri-brominated product rather than the mono-brominated product.

In order to lower the activation of the amino group, we can convert it to the less activating amide group (Topic K5; *Fig. 14*). The bromination then only goes once. We also find that the bromination reaction is more selective for the *para* position than for the *ortho* position. This is because the amide group is bulkier than the NH_2 group and tends to shield the *ortho* positions from attack. Once the bromination has been completed the amide can be converted back to the amino group by hydrolysis (Topic K6).

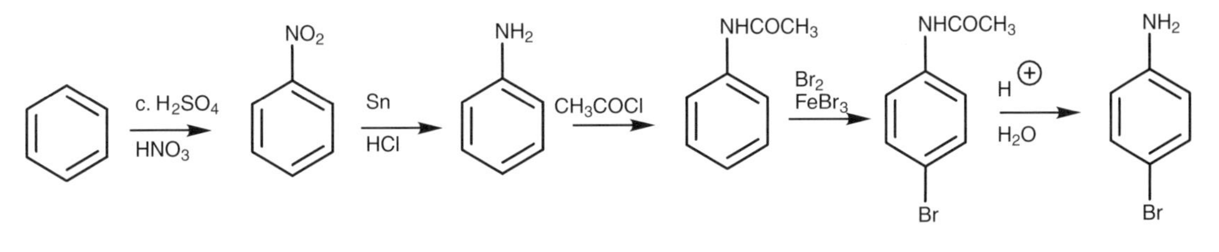

Fig. 14. Synthesis of para-*bromoaniline.*

Deactivating groups – resonance *o/p* directing

The fourth and last group of aromatic substituents are the halogen substituents which deactivate the aromatic ring and which direct substitution to the *ortho* and *para* positions. These are perhaps the trickiest to understand since they deactivate the ring by one effect, but direct substitution by a different effect. The halogen atom is strongly electronegative and therefore we would expect it to have a strong electron-withdrawing inductive effect on the aromatic ring. This would make the aromatic ring less nucleophilic and less reactive to electrophiles. It would also destabilize the required intermediate for electrophilic substitution. Halogens are also poorer nucleophiles and so any resonance effects they might have are less important than their inductive effects.

However, if halogen atoms are deactivating the ring because of inductive effects, why do they not direct substitution to the *meta* position like other electron-withdrawing groups? Let us look at a specific reaction – the nitration of bromobenzene (*Fig. 15*). There are three resonance structures for each of the three intermediates leading to these products, but the crucial ones to consider are those which position a positive charge next to the substituent. These occur with *ortho* and *para* substitution, but not *meta* substitution (*Fig. 16*). These are the crucial resonance structures as far as the directing properties of the substituent is concerned. If bromine acts inductively, it will destabilize these intermediates and direct substitution to the *meta* position. However, we know that bromine directs *ortho/para* and so it must be stabilizing the *ortho/para* intermediates rather than destabilizing them. The only way that bromine can stabilize the neighboring positive charge is by resonance in the same way as a nitrogen or oxygen atom (*Fig. 17*). Thus, the bromine acts as a nucleophile and donates one of its lone pairs to form a new bond to the electrophilic center beside it. A new π bond is formed and the positive charge is moved onto the bromine atom. This resonance effect is weak since the halogen atom is a much weaker nucleophile than oxygen or nitrogen and is less capable of stabilizing a positive charge. However, it is significant enough to direct substitution to the *ortho* and *para* positions.

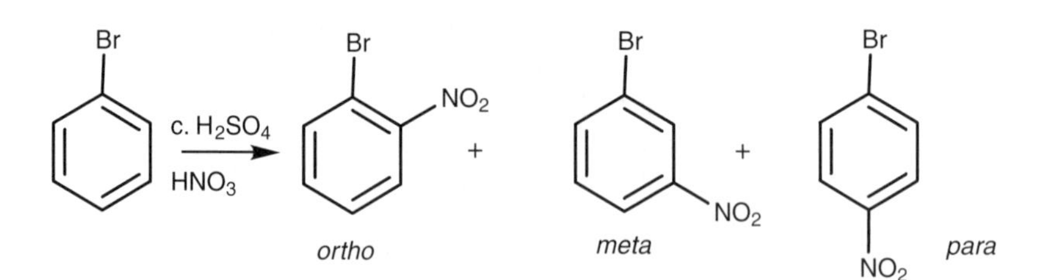

Fig. 15. Nitration of bromobenzene.

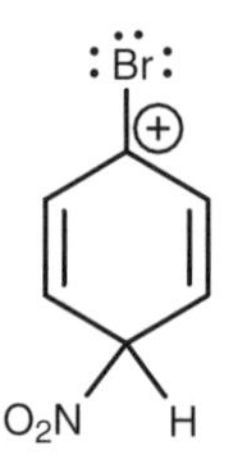

Fig. 16. Crucial resonance structures for ortho *and* para *substitution.*

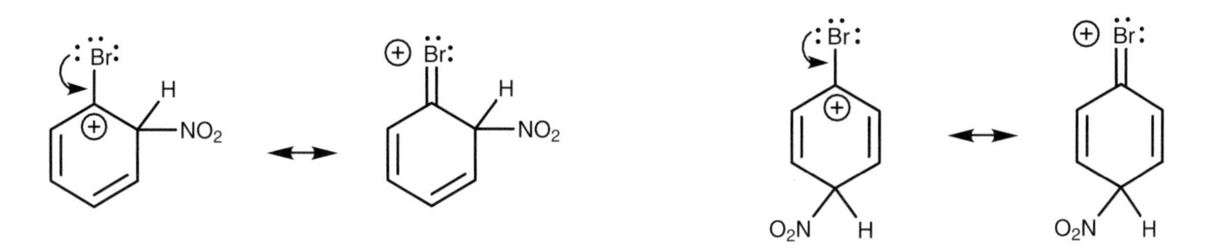

Fig. 17. Resonance interactions involving bromine.

For halogen substituents, the inductive effect is more important than the resonance effect in deactivating the ring. However, once electrophilic substitution **does** take place, resonance effects are more important than inductive effects in directing substitution.

I6 SYNTHESIS OF DI- AND TRI-SUBSTITUTED BENZENES

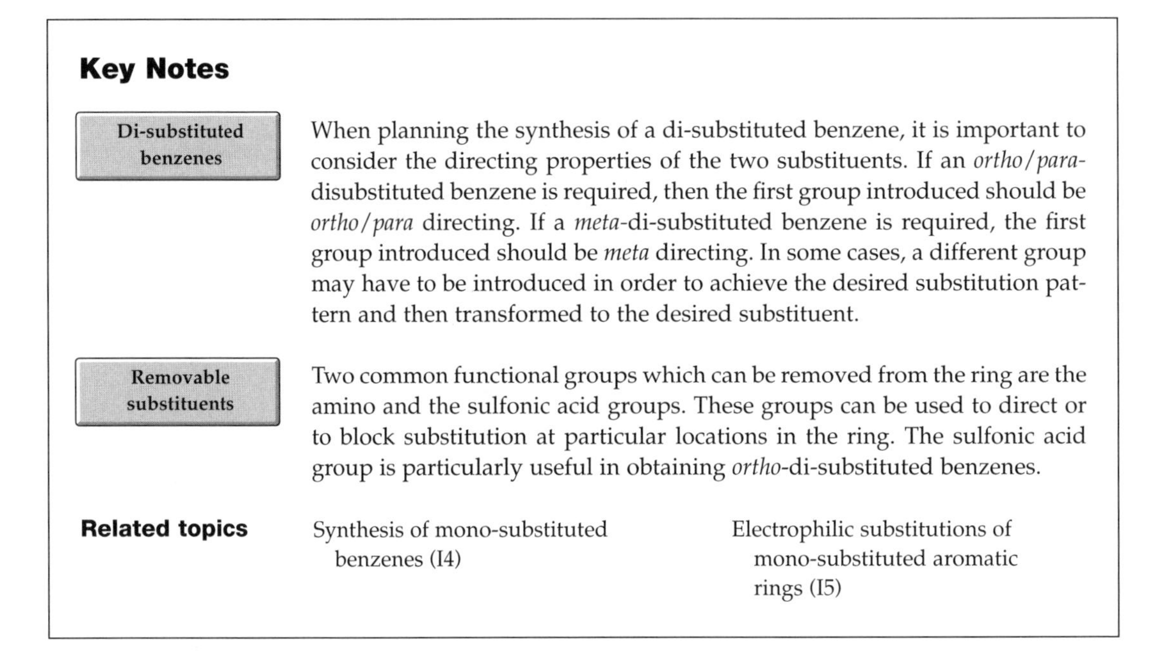

Key Notes

Di-substituted benzenes

When planning the synthesis of a di-substituted benzene, it is important to consider the directing properties of the two substituents. If an *ortho/para*-disubstituted benzene is required, then the first group introduced should be *ortho/para* directing. If a *meta*-di-substituted benzene is required, the first group introduced should be *meta* directing. In some cases, a different group may have to be introduced in order to achieve the desired substitution pattern and then transformed to the desired substituent.

Removable substituents

Two common functional groups which can be removed from the ring are the amino and the sulfonic acid groups. These groups can be used to direct or to block substitution at particular locations in the ring. The sulfonic acid group is particularly useful in obtaining *ortho*-di-substituted benzenes.

Related topics

Synthesis of mono-substituted benzenes (I4)

Electrophilic substitutions of mono-substituted aromatic rings (I5)

Di-substituted benzenes

A full understanding of how substituents direct further substitution is crucial in planning the synthesis of a di-substituted aromatic compound. For example, there are two choices which can be made in attempting the synthesis of *p*-bromonitrobenzene from benzene (*Fig. 1*). We could brominate first, then nitrate, or nitrate first then brominate. A knowledge of how substituents affect electrophilic substitution allows us to choose the most suitable route.

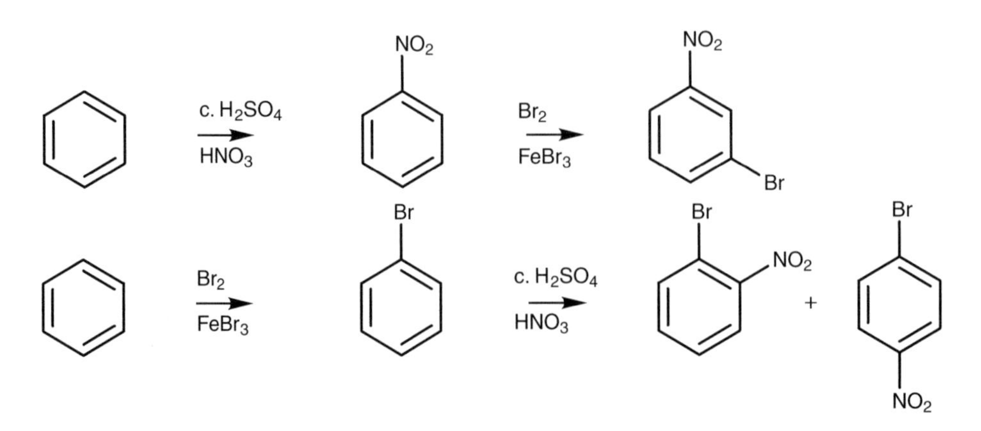

Fig. 1. Synthetic planning to di-substituted benzenes.

In the first method, nitrating first then brominating would give predominantly the *meta* isomer of the final product due to the *meta* directing properties of the nitro group. The second method is better since the directing properties of bromine are in our favor. Admittedly, we would have to separate the *para* product from the *ortho* product, but we would still get a higher yield by this route.

The synthesis of *m*-toluidine requires a little more thought (*Fig. 2*). Both the methyl and the amino substituents are activating groups and direct *ortho/para*. However, the two substituents are *meta* with respect to each other. In order to get *meta* substitution we need to introduce a substituent other than the methyl or nitro group which will direct the second substitution to the *meta* position. Moreover, once that has been achieved, the *meta* directing substituent has to be converted to one of the desired substituents. The nitro group is ideal for this since it directs *meta* and can then be converted to the required amino group.

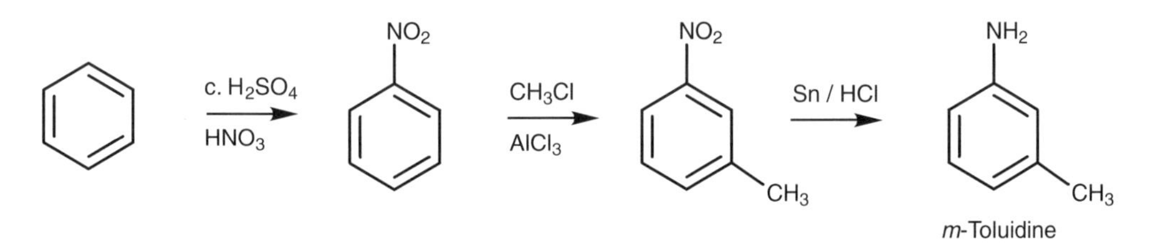

Fig. 2. Synthesis of m-*toluidine.*

This same strategy can be used for a large range of *meta*-disubstituted aromatic rings where both substituents are *ortho/para* directing since the nitro group can be transformed to an amino group which can then be transformed to a large range of different functional groups (see Topics I4 and O3). Another tricky situation is where there are two *meta*-directing substituents at *ortho* or *para* positions with respect to each other, for example, *p*-nitrobenzoic acid (*Fig. 3*). In this case, a methyl substituent is added which is *o/p* directing. Nitration is then carried out and the *para* isomer is separated from any *ortho* isomer which might be formed. The methyl group can then be oxidized to the desired carboxylic acid.

Larger alkyl groups could be used to increase the ratio of *para* to *ortho* substitution since they can all be oxidized down to the carboxylic acid (Topic I7).

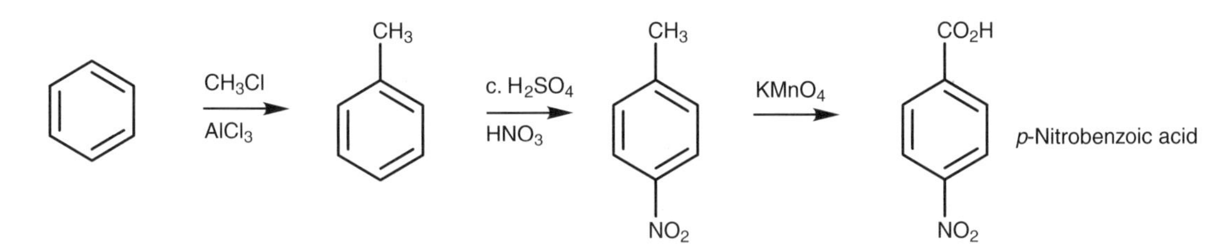

Fig. 3. Synthesis of para-*nitrobenzoic acid.*

Removable substituents

It is sometimes useful to have a substituent present which can direct or block a particular substitution, and which can then be removed once the desired substituents have been added. The reactions in *Fig. 4* are used to remove substituents from aromatic rings.

Fig. 4. Reactions which remove substituents from aromatic rings.

An example of how removable substituents can be used is in the synthesis of 1,3,5-tribromobenzene (*Fig. 5*). This structure cannot be made directly from benzene by bromination. The bromine atoms are in the *meta* positions with respect to each other, but bromine atoms direct *ortho/para*. Moreover, bromine is a deactivating group and so it would be difficult to introduce three such groups directly to benzene.

This problem can be overcome by using a strong activating group which will direct *ortho/para* and which can then be removed at the end of the synthesis. The amino group is ideal for this and the full synthesis is shown in *Fig. 5*.

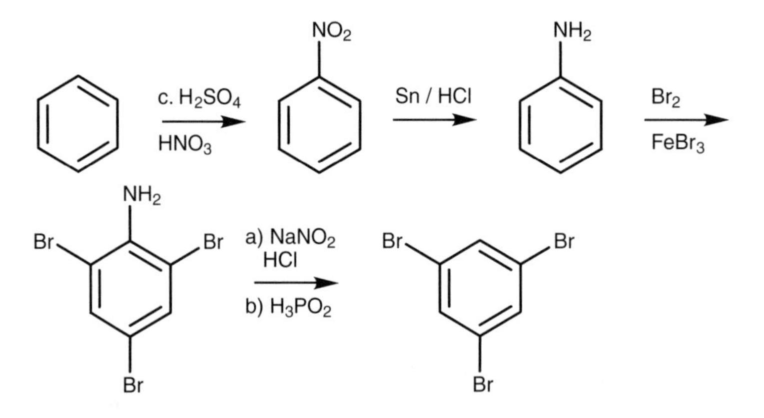

Fig. 5. Synthesis of 1,3,5-tribromobenzene.

The synthesis of *ortho*-bromotoluene illustrates how a sulfonic acid can be used in a synthesis. *o*-Bromotoluene could conceivably be synthesized by bromination of toluene or by Friedel–Crafts alkylation of bromobenzene (*Fig. 6*). However, the reaction would also give the *para*-substitution product and this is more likely if the electrophile is hindered from approaching the *ortho* position by unfavorable steric interactions. An alternative strategy would be to deliberately substitute a group at the *para* position before carrying out the bromination. This group would then act as a blocking group at the *para* position and would force the bromination to take place *ortho* to the methyl group. If the blocking group could then be removed, the desired product would be obtained. The sulfonic acid group is particularly useful in this respect since it can be easily removed once the synthesis is over (*Fig. 7*).

Note that the sulfonation of toluene could in theory take place at the *ortho* position as well as the *para* position. However, the SO_3 electrophile is bulky and so the latter position is preferred for steric reasons. Once the sulfonic acid group is present, both it and the methyl group direct bromination to the same position (*ortho* to the methyl group = *meta* to the sulfonic acid group).

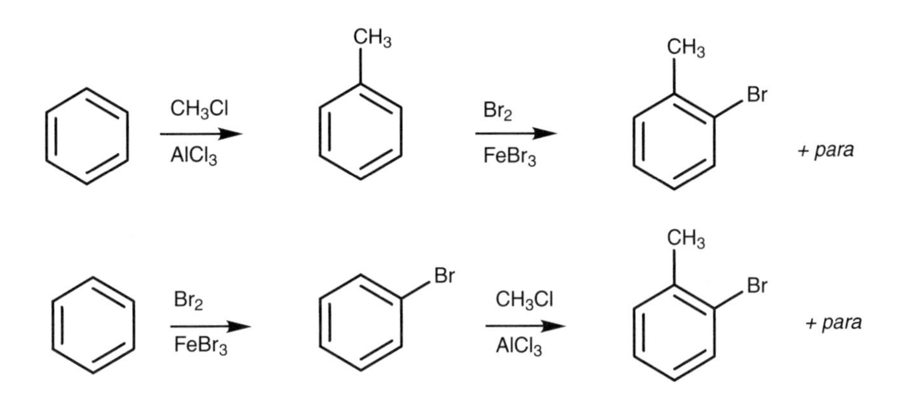

Fig. 6. Possible synthetic routes to ortho-*bromotoluene.*

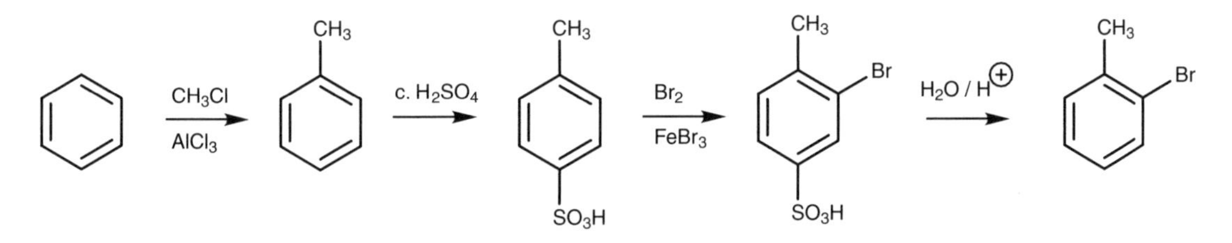

Fig. 7. Synthesis of ortho-*bromotoluene.*

I7 Oxidation and reduction

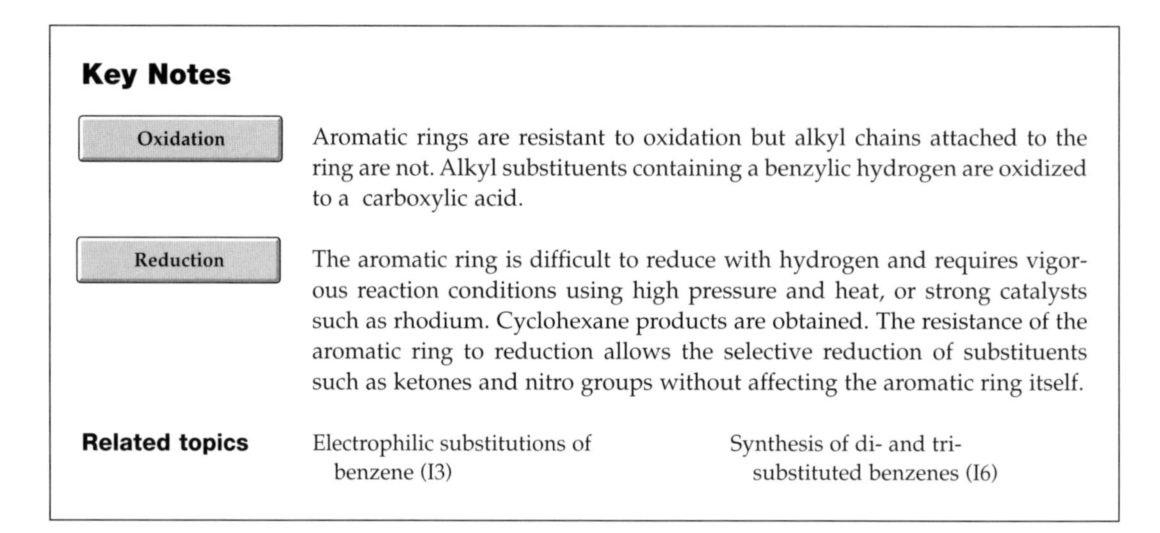

Key Notes

Oxidation	Aromatic rings are resistant to oxidation but alkyl chains attached to the ring are not. Alkyl substituents containing a benzylic hydrogen are oxidized to a carboxylic acid.
Reduction	The aromatic ring is difficult to reduce with hydrogen and requires vigorous reaction conditions using high pressure and heat, or strong catalysts such as rhodium. Cyclohexane products are obtained. The resistance of the aromatic ring to reduction allows the selective reduction of substituents such as ketones and nitro groups without affecting the aromatic ring itself.

Related topics Electrophilic substitutions of benzene (I3); Synthesis of di- and tri-substituted benzenes (I6)

Oxidation

Aromatic rings are remarkably stable to oxidation and are resistant to oxidizing agents such as potassium permanganate or sodium dichromate. However, alkyl substituents on aromatic ring are surprisingly susceptible to oxidation. This can be put to good use in the synthesis of aromatic compounds since it is possible to oxidize an alkyl chain to a carboxylic acid without oxidizing the aromatic ring (see Topic I6; *Fig. 1*). The mechanism of this reaction is not fully understood, but it is known that a **benzylic** hydrogen has to be present (i.e. the carbon directly attached to the ring must have a hydrogen). Alkyl groups lacking a benzylic hydrogen are not oxidized.

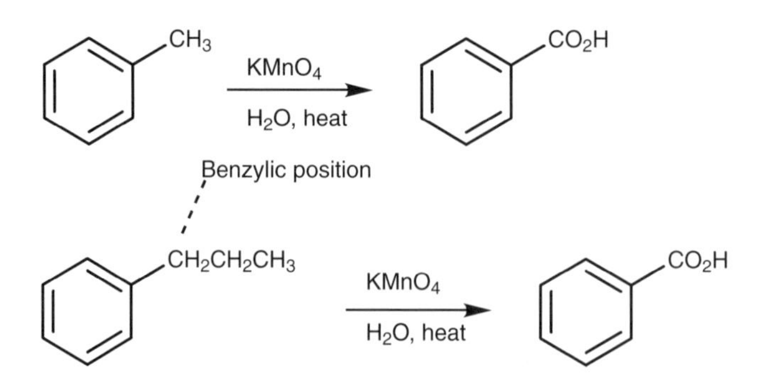

Fig. 1. Oxidation of alkyl side chains to aromatic carboxylic acids.

Reduction

Aromatic rings can be **hydrogenated** to cycloalkanes, but the reduction has to be carried out under strong conditions using a nickel catalyst, high temperature and high pressure (*Fig. 2*) – much stronger conditions than would be required for the reduction of alkenes (Topic H6). This is because of the inherent stability of aromatic rings (Topic I1). The reduction can also be carried out using hydrogen and a platinum catalyst under high pressure, or with hydrogen and a rhodium/carbon catalyst. The latter is a more powerful catalyst and the reaction can be done at room temperature and at atmospheric pressure.

The resistance of the aromatic ring to reduction is useful since it is possible to reduce functional groups which might be attached to the ring without reducing the aromatic ring itself. For example, the carbonyl group of an aromatic ketone can be reduced with hydrogen over a palladium catalyst without affecting the aromatic ring (*Fig. 3*). This allows the synthesis of primary alkylbenzenes which cannot be synthesized directly by the Friedel–Crafts alkylation (see Topic I3). It is worth noting that the aromatic ring makes the ketone group more reactive to reduction than would normally be the case. Aliphatic ketones would not be reduced under these conditions. Nitro groups can also be reduced to amino groups under these conditions without affecting the aromatic ring.

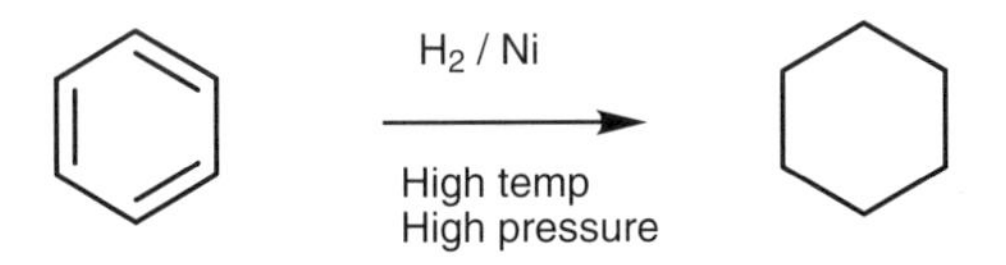

Fig. 2. Reduction of benzene to cyclohexane.

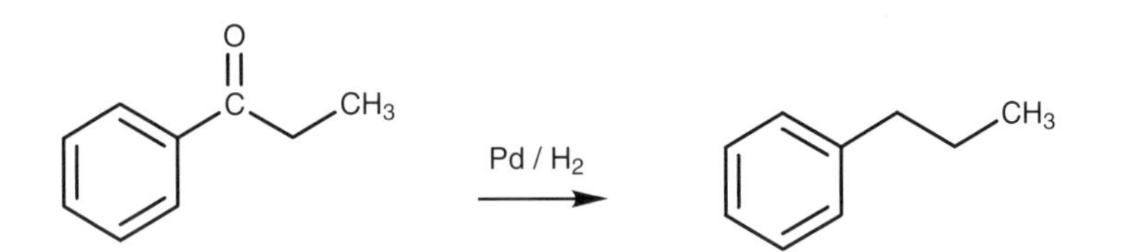

Fig. 3. Reduction of an aromatic ketone.

J1 PREPARATION

Key Notes

Functional group transformations

Functional group transformations allow the conversion of a functional group to an aldehyde or a ketone without affecting the carbon skeleton of the molecule. Aldehydes can be synthesized by the oxidation of primary alcohols, or by the reduction of esters, acid chlorides, or nitriles. Ketones can be synthesized by the oxidation of secondary alcohols. Methyl ketones can be synthesized from terminal alkynes.

C–C Bond formation

Reactions which result in the formation of aldehydes and ketones by carbon–carbon bond formation are useful in the construction of more complex carbon skeletons from simple starting materials. Ketones can be synthesized from the reaction of acid chlorides with organocuprate reagents, or from the reaction of nitriles with a Grignard or organolithium reagent. Aromatic ketones can be synthesized by the Friedel–Crafts acylation of an aromatic ring.

C–C Bond cleavage

Aldehydes and ketones can be obtained from the ozonolysis of suitably substituted alkenes.

Related topics

Reduction and oxidation of alkenes (H6)
Electrophilic additions to alkynes (H8)
Electrophilic substitutions of benzene (I3)
Reactions (K6)
Reactions of alcohols (M4)
Chemistry of nitriles (O4)

Functional group transformations

Functional group transformations allow the conversion of a functional group to an aldehyde or a ketone without affecting the carbon skeleton of the molecule. Aldehydes can be synthesized by the oxidation of primary alcohols (Topic M4), or by the reduction of esters (Topic K6), acid chlorides (Topic K6), or nitriles (Topic O4). Since nitriles can be obtained from alkyl halides (Topic L6), this is a way of adding an aldehyde unit (CHO) to an alkyl halide (*Fig. 1*).

Ketones can be synthesized by the oxidation of secondary alcohols (Topic M4). Methyl ketones can be synthesized from terminal alkynes (Topic H8).

C–C Bond formation

Reactions which result in the formation of ketones by carbon–carbon bond formation are extremely important because they can be used to construct complex

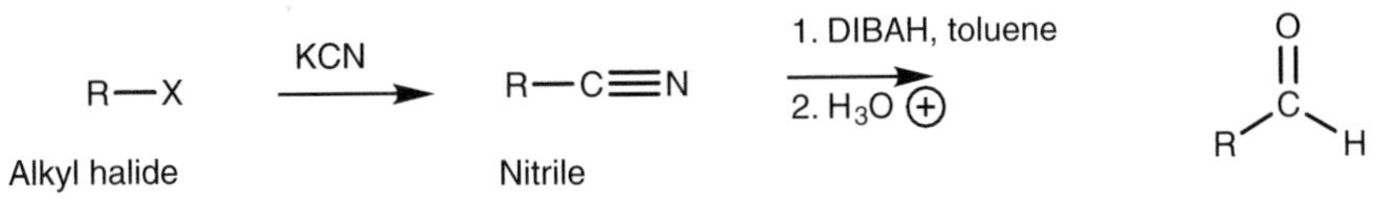

Fig. 1. Synthesis of an aldehyde from an alkyl halide with 1C chain extension.

carbon skeletons from simple starting materials. Ketones can be synthesized from the reaction of acid chlorides with organocuprate reagents (Topic K6), or from the reaction of nitriles with a Grignard or organolithium reagent (Topic O4). Aromatic ketones can be synthesized by the Friedel–Crafts acylation of an aromatic ring (Topic I3).

C–C Bond cleavage

Aldehydes and ketones can be obtained from the ozonolysis of suitably substituted alkenes (Topic H6).

J2 PROPERTIES

Key Notes

Carbonyl group

The carbonyl group is a C=O group. The carbonyl group is planar with bond angles of 120°, and consists of two sp^2 hybridized atoms (C and O) linked by a strong σ bond and a weaker π bond. The carbonyl group is polarized such that oxygen is slightly negative and carbon is slightly positive. In aldehydes and ketones, the substituents must be one or more of the following – an alkyl group, an aromatic ring, or a hydrogen.

Properties

Aldehydes and ketones have higher boiling points than alkanes of comparable molecular weight due to the polarity of the carbonyl group. However, they have lower boiling points than comparable alcohols or carboxylic acids due to the absence of hydrogen bonding. Aldehydes and ketones of small molecular weight are soluble in aqueous solution since they can participate in intermolecular hydrogen bonding with water. Higher molecular weight aldehydes and ketones are not soluble in water since the hydrophobic character of the alkyl chains or aromatic rings outweighs the polar character of the carbonyl group.

Nucleophilic and electrophilic centers

The oxygen of the carbonyl group is a nucleophilic center. The carbonyl carbon is an electrophilic center.

Keto–enol tautomerism

Ketones are in rapid equilibrium with an isomeric structure called an enol. The keto and enol forms are called tautomers and the process by which they interconvert is called keto–enol tautomerism. The mechanism can be acid or base catalyzed.

Related topics

sp^2 Hybridization (A4)
Recognition of functional groups (C1)
Intermolecular bonding (C3)
Organic structures (E4)
Enolates (G5)

Carbonyl group

Both aldehydes and ketones contain a carbonyl group (C=O). The substituents attached to the carbonyl group determine whether it is an aldehyde or a ketone, and whether it is aliphatic or aromatic (Topics C1 and C2).

The geometry of the carbonyl group is planar with bond angles of 120° (Topic A4; *Fig. 1*). The carbon and oxygen atoms of the carbonyl group are sp^2 hybridized and the double bond between the atoms is made up of a strong σ bond and a weaker π bond. The carbonyl bond is shorter than a C–O single bond (1.22 Å vs. 1.43 Å) and is also stronger since two bonds are present as opposed to one (732 kJ mol^{-1} vs. 385 kJ mol^{-1}). The carbonyl group is more reactive than a C–O single bond due to the relatively weak π bond.

The carbonyl group is polarized such that the oxygen is slightly negative and the carbon is slightly positive. Both the polarity of the carbonyl group and the

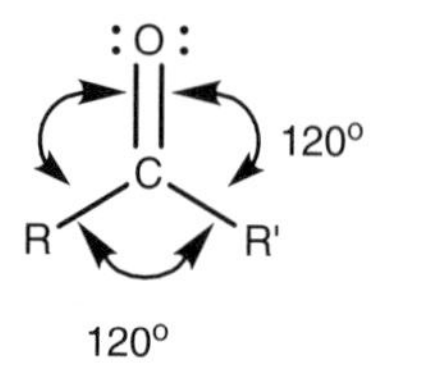

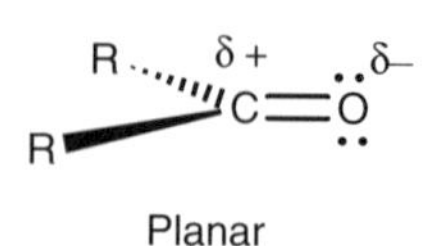

Fig. 1. Geometry of the carbonyl group.

presence of the weak π bond explains much of the chemistry and the physical properties of aldehydes and ketones. The polarity of the bond also means that the carbonyl group has a dipole moment.

Properties

Due to the polar nature of the carbonyl group, aldehydes and ketones have higher boiling points than alkanes of similar molecular weight. However, hydrogen bonding is not possible between carbonyl groups and so aldehydes and ketones have lower boiling points than alcohols or carboxylic acids.

Low molecular weight aldehydes and ketones (e.g. formaldehyde and acetone) are soluble in water. This is because the oxygen of the carbonyl group can participate in intermolecular hydrogen bonding with water molecules (Topic C3; *Fig. 2*).

As molecular weight increases, the hydrophobic character of the attached alkyl chains starts to outweigh the water solubility of the carbonyl group with the result that large molecular weight aldehydes and ketones are insoluble in water. Aromatic ketones and aldehydes are insoluble in water due to the hydrophobic aromatic ring.

Nucleophilic and electrophilic centers

Due to the polarity of the carbonyl group, aldehydes and ketones have a nucleophilic oxygen center and an electrophilic carbon center as shown for propanal (*Fig. 3*; Topic E4). Therefore, nucleophiles react with aldehydes and ketones at the carbon center, and electrophiles react at the oxygen center.

Keto–enol tautomerism

Ketones which have hydrogen atoms on their **α-carbon** (the carbon next to the carbonyl group) are in rapid equilibrium with an isomeric structure called an **enol** where the α-hydrogen ends up on the oxygen instead of the carbon. The two isomeric forms are called **tautomers** and the process of equilibration is called

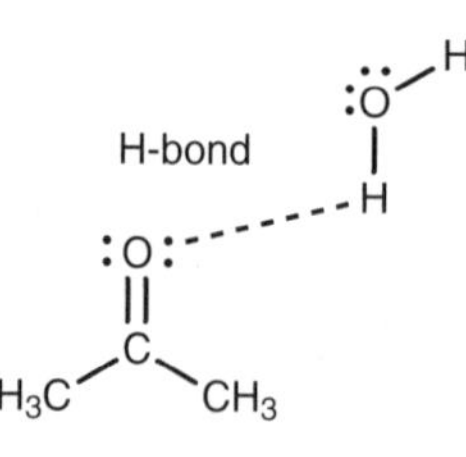

Fig. 2. Intermolecular hydrogen bonding of a ketone with water.

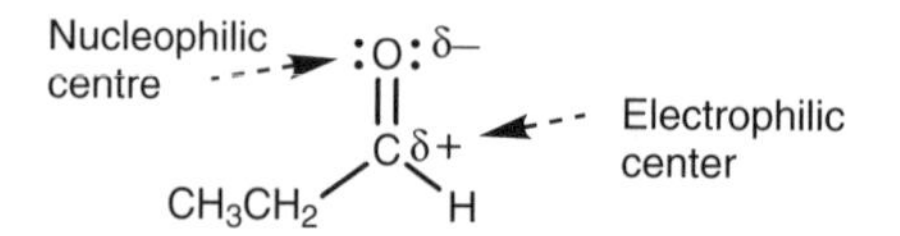

Fig. 3. Nucleophilic and electrophilic centers of the carbonyl group.

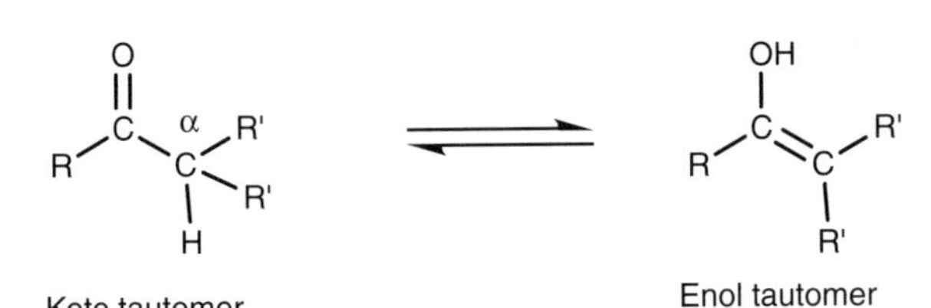

Fig. 4. Keto–enol tautomerism.

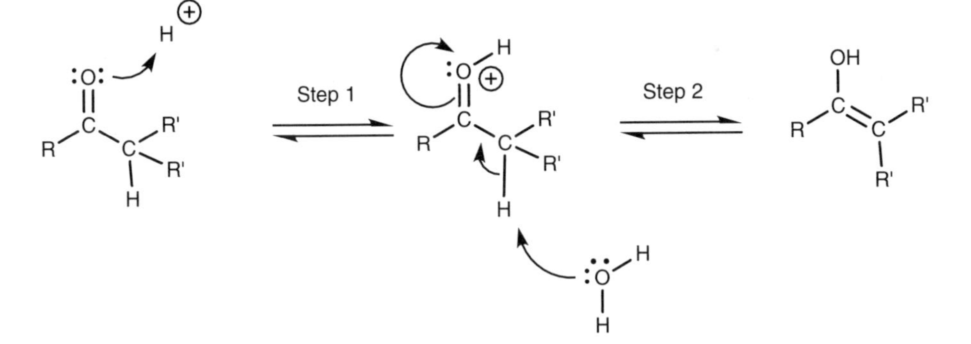

Fig. 5. Acid-catalyzed mechanism for keto–enol tautomerism.

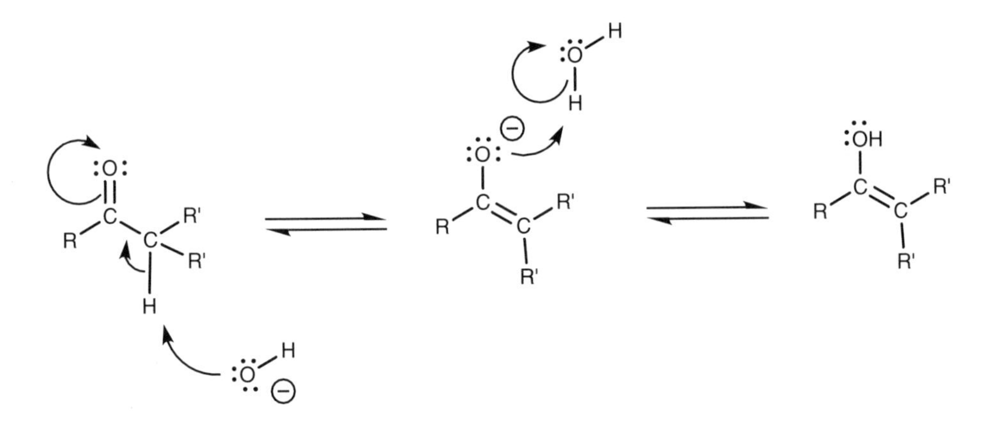

Fig. 6. Base-catalyzed mechanism for keto–enol tautomerism.

tautomerism (*Fig. 4*). In general, the equilibrium greatly favors the keto tautomer and the enol tautomer may only be present in very small quantities.

The tautomerism mechanism is catalyzed by acid or base. When catalyzed by acid (*Fig. 5*), the carbonyl group acts as a nucleophile with the oxygen using a lone pair of electrons to form a bond to a proton. This results in the carbonyl oxygen gaining a positive charge which activates the carbonyl group to attack by weak nucleophiles (Step 1). The weak nucleophile in question is a water molecule which removes the α-proton from the ketone, resulting in the formation of a new C=C double bond and cleavage of the carbonyl π bond. The enol tautomer is formed thus neutralizing the unfavorable positive charge on the oxygen (Step 2).

Under basic conditions (*Fig. 6*), an enolate ion is formed (Topic G5), which then reacts with water to form the enol.

J3 NUCLEOPHILIC ADDITION

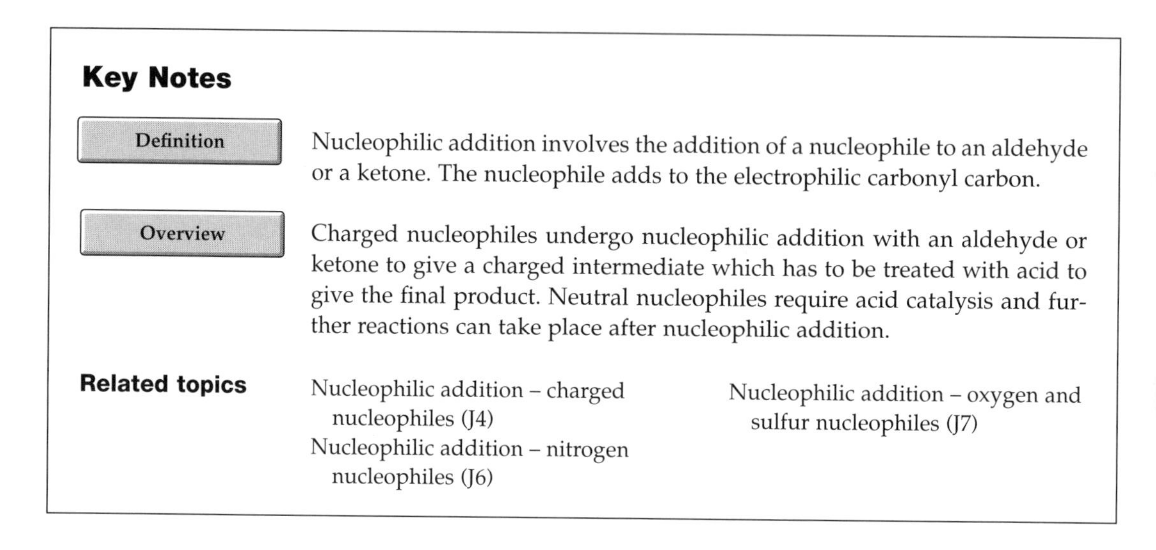

Key Notes

Definition — Nucleophilic addition involves the addition of a nucleophile to an aldehyde or a ketone. The nucleophile adds to the electrophilic carbonyl carbon.

Overview — Charged nucleophiles undergo nucleophilic addition with an aldehyde or ketone to give a charged intermediate which has to be treated with acid to give the final product. Neutral nucleophiles require acid catalysis and further reactions can take place after nucleophilic addition.

Related topics

Nucleophilic addition – charged nucleophiles (J4)
Nucleophilic addition – nitrogen nucleophiles (J6)
Nucleophilic addition – oxygen and sulfur nucleophiles (J7)

Definition

As the name of the reaction suggests, nucleophilic addition involves the addition of a nucleophile to a molecule. This is a distinctive reaction for ketones and

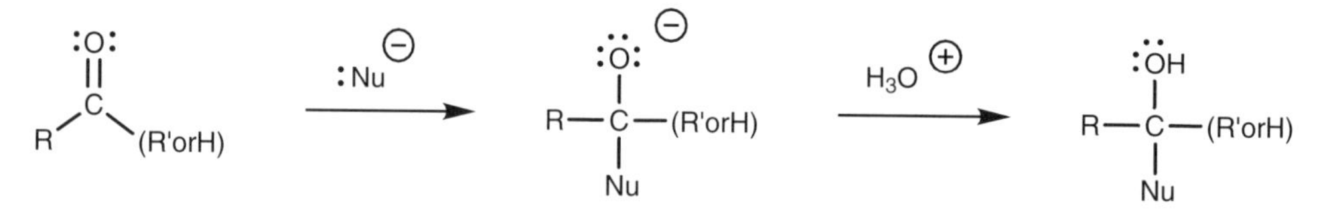

Fig. 1. Nucleophilic addition to a carbonyl group.

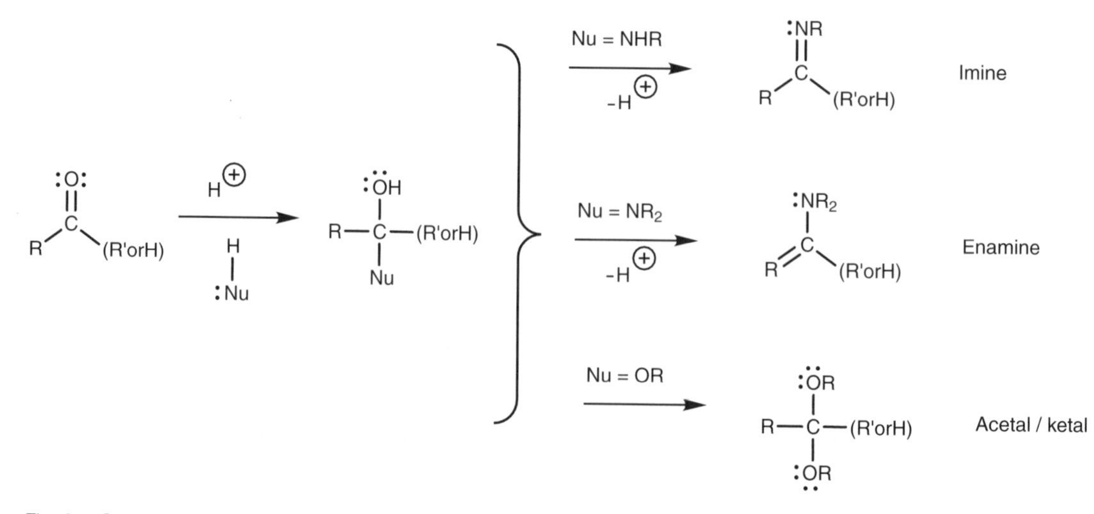

Fig. 2. Synthesis of imines, enamines, acetals, and ketals.

aldehydes and the nucleophile will add to the electrophilic carbon atom of the carbonyl group. The nucleophile can be a negatively charged ion such as cyanide or hydride, or it can be a neutral molecule such as water or alcohol.

Overview

In general, addition of charged nucleophiles results in the formation of a charged intermediate (*Fig. 1*). The reaction stops at this stage and acid has to be added to complete the reaction (Topic J4).

Neutral nucleophiles where nitrogen or oxygen is the nucleophilic center are relatively weak nucleophiles, and an acid catalyst is usually required. After nucleophilic addition has occurred, further reactions may take place leading to structures such as imines, enamines, acetals, and ketals (Topics J6 and J7; *Fig. 2*).

J4 NUCLEOPHILIC ADDITION – CHARGED NUCLEOPHILES

Key Notes

Carbanion addition

Grignard reagents (RMgX) and organolithium reagents (RLi) are used as the source of carbanions. The reaction mechanism involves nucleophilic addition of the carbanion to the aldehyde or ketone to form a negatively charged intermediate. Addition of acid completes the reaction. Both reactions are important because they involve C–C bond formation allowing the synthesis of complex molecules from simple starting materials. Primary alcohols are obtained from formaldehyde, secondary alcohols from aldehydes and tertiary alcohols from ketones.

Hydride addition

Lithium aluminum hydride ($LiAlH_4$) and sodium borohydride ($NaBH_4$) are reducing agents and the overall reaction corresponds to the nucleophilic addition of a hydride ion ($H:^-$). The reaction is a functional group transformation where primary alcohols are obtained from aldehydes and secondary alcohols are obtained from ketones.

Cyanide addition

Reaction of aldehydes and ketones with HCN and KCN produce cyanohydrins. The cyanide ion is the nucleophile and adds to the electrophilic carbonyl carbon.

Bisulfite addition

The bisulfite ion is a weakly nucleophilic anion which will only react with aldehydes and methyl ketones. The product is a water-soluble salt and so the reaction can be used to separate aldehydes and methyl ketones from larger ketones or from other water-insoluble compounds. The aldehyde and methyl ketone can be recovered by treating the salt with acid or base.

Aldol reaction

The Aldol reaction involves the nucleophilic addition of enolate ions to aldehydes and ketones to form β-hydroxycarbonyl compounds.

Related topics

Properties (J2)
Nucleophilic addition (J3)
Electronic and steric effects (J5)
Nucleophilic addition – nitrogen nucleophiles (J6)
Nucleophilic addition – oxygen and sulfur nucleophiles (J7)
Reactions of enolate ions (J8)
Organometallic reactions (L7)

Carbanion addition

Carbanions are extremely reactive species and do not occur in isolation. However, there are two reagents which can supply the equivalent of a carbanion. These are Grignard reagents and organolithium reagents. We shall look first of all at the reaction of a Grignard reagent with aldehydes and ketones (*Fig. 1*).

The Grignard reagent in this reaction is called methyl magnesium iodide

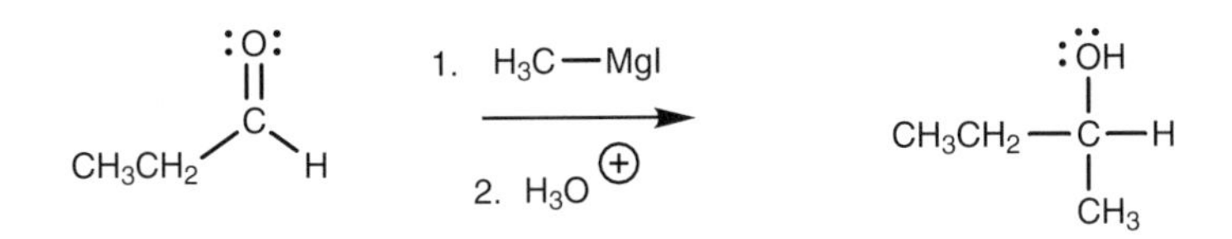

Fig. 1. Grignard reaction.

(CH_3MgI) and is the source of a methyl carbanion (Topic L7; *Fig. 2*). In reality, the methyl carbanion is never present as a separate ion, but the reaction proceeds as if it were. The methyl carbanion is the nucleophile in this reaction and the nucleophilic center is the negatively charged carbon atom. The aldehyde is the electrophile. Its electrophilic center is the carbonyl carbon atom since it is electron deficient (Topic J2).

The carbanion uses its lone pair of electrons to form a bond to the electrophilic carbonyl carbon (*Fig. 3*). At the same time, the relatively weak π bond of the carbonyl group breaks and both electrons move to the oxygen to give it a third lone pair of electrons and a negative charge (Step 1). The reaction stops at this stage, since the negatively charged oxygen is complexed with magnesium which acts as a counterion (not shown). Aqueous acid is now added to provide an electrophile in the shape of a proton. The intermediate is negatively charged and can act as a nucleophile/base. A lone pair of electrons on the negatively charged oxygen is used to form a bond to the proton and the final product is obtained (Step 2).

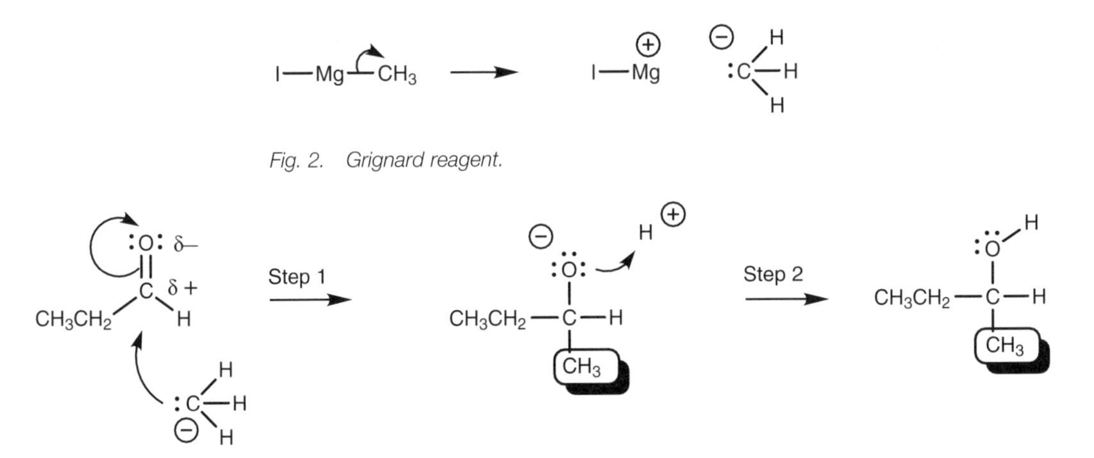

Fig. 2. Grignard reagent.

Fig. 3. Mechanism for the nucleophilic addition of a Grignard reagent.

The reaction of aldehydes and ketones with Grignard reagents is a useful method of synthesizing primary, secondary, and tertiary alcohols (*Fig. 4*). Primary alcohols can be obtained from formaldehyde, secondary alcohols can be obtained from aldehydes, and tertiary alcohols can be obtained from ketones. The reaction involves the formation of a carbon–carbon bond and so this is an important way of building up complex organic structures from simple starting materials.

The Grignard reagent itself is synthesized from an alkyl halide and a large variety of reagents are possible (Topic L7).

Organolithium reagents (Topic L7) such as CH_3Li can also be used to provide the nucleophilic carbanion and the reaction mechanism is exactly the same as that described for the Grignard reaction (*Fig. 5*).

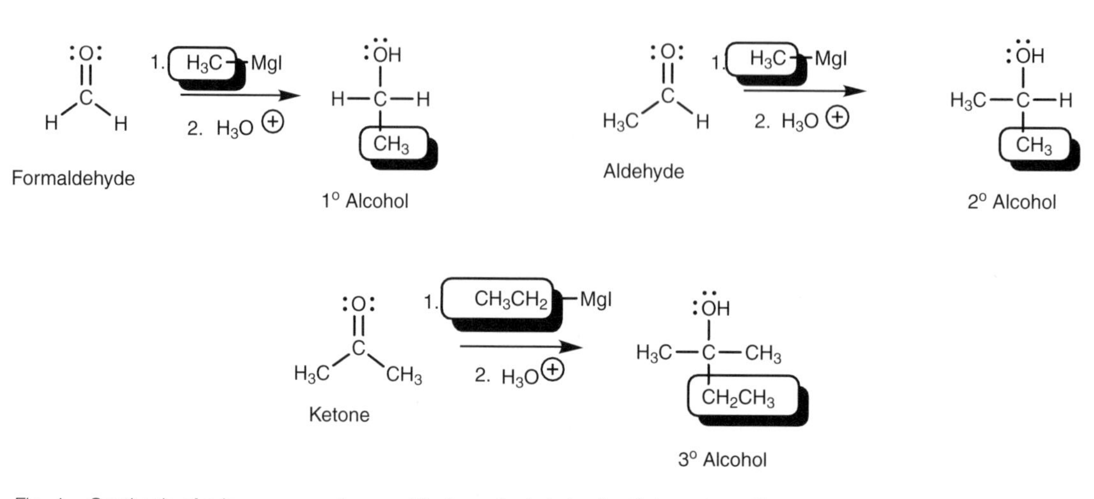

Fig. 4. Synthesis of primary, secondary, and tertiary alcohols by the Grignard reaction.

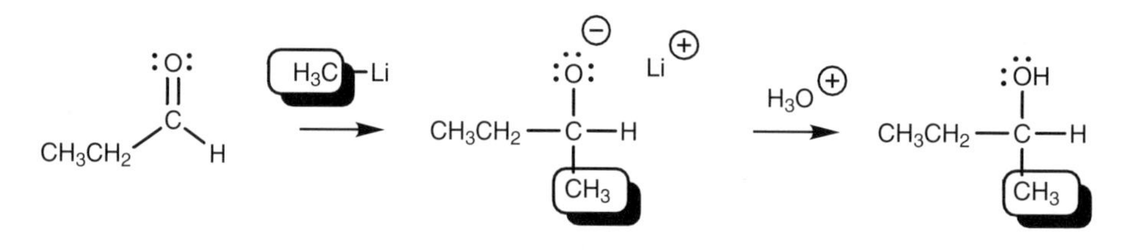

Fig. 5. Nucleophilic addition with an organolithium reagent.

Hydride addition

Reducing agents such as sodium borohydride ($NaBH_4$) and lithium aluminum hydride ($LiAlH_4$) react with aldehydes and ketones as if they are providing a hydride ion ($:H^-$; *Fig. 6*). This species is not present as such and the reaction mechanism is more complex. However, we can explain the reaction by viewing these reagents as hydride equivalents ($:H^-$). The overall reaction is an example of a functional group transformation since the carbon skeleton is unaffected. Aldehydes are converted to primary alcohols and ketones are converted to secondary alcohols.

The mechanism of the reaction is the same as that described above for the Grignard reaction (*Fig. 7*). The hydride ion equivalent adds to the carbonyl group and a negatively charged intermediate is obtained which is complexed as a lithium salt (Step 1). Subsequent treatment with acid gives the final product (Step 2). It should be emphasized again that the mechanism is actually more complex than this because the hydride ion is too reactive to exist in isolation.

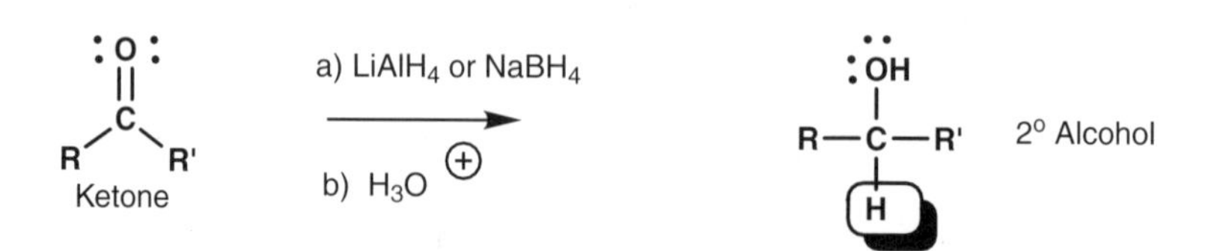

Fig. 6. Reduction of a ketone to a secondary alcohol.

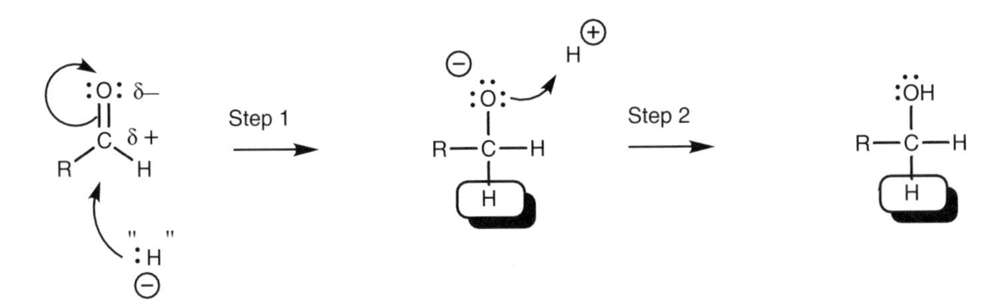

Fig. 7. Mechanism for the reaction of a ketone with $LiAlH_4$ or $NaBH_4$.

Cyanide addition

Nucleophilic addition of a cyanide ion to an aldehyde or ketone gives a **cyanohydrin** (*Fig. 8*). In the reaction, there is a catalytic amount of potassium cyanide present and this supplies the attacking nucleophile in the form of the cyanide ion (CN^-). The nucleophilic center of the nitrile group is the carbon atom since this is the atom with the negative charge. The carbon atom uses its lone pair of electrons to form a new bond to the electrophilic carbon of the carbonyl group (*Fig. 9*). As this new bond forms, the relatively weak π bond of the carbonyl group breaks and the two electrons making up that bond move onto the oxygen to give it a third lone pair of electrons and a negative charge (Step 1). The intermediate formed can now act as a nucleophile/base since it is negatively charged and it reacts with the acidic hydrogen of HCN. A lone pair of electrons from oxygen is used to form a bond to the acidic proton and the H–CN σ bond is broken at the same time such that these electrons move onto the neighboring carbon to give it a lone pair of electrons and a negative charge (Step 2). The products are the cyanohydrin and the cyanide ion. Note that a cyanide ion started the reaction and a cyanide ion is regenerated. Therefore, only a catalytic amount

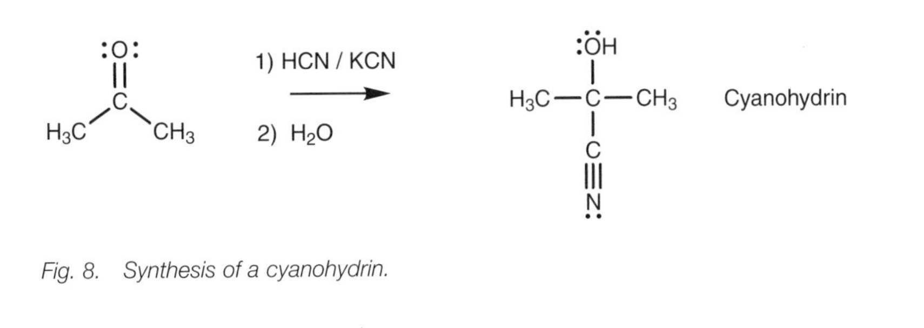

Fig. 8. Synthesis of a cyanohydrin.

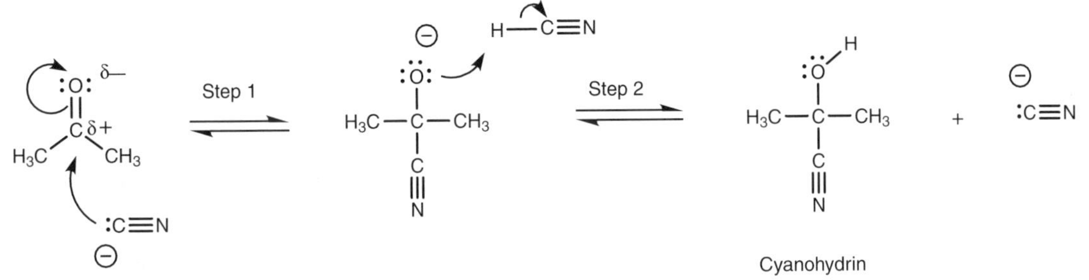

Fig. 9. Mechanism for the formation of a cyanohydrin.

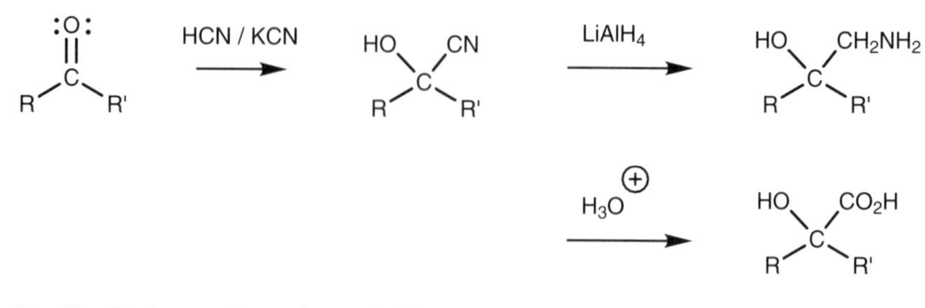

Fig. 10. Further reactions of cyanohydrins.

of cyanide ion is required to start the reaction and once the reaction has taken place, a cyanide ion is regenerated to continue the reaction with another molecule of ketone.

Cyanohydrins are useful in synthesis because the cyanide group can be converted to an amine or to a carboxylic acid (Topic O4; *Fig. 10*).

Bisulfite addition

The reaction of an aldehyde or a methyl ketone with sodium bisulfite ($NaHSO_3$) involves nucleophilic addition of a bisulfite ion ($^-{:}SO_3H$) to the carbonyl group to give a water soluble salt (*Fig. 11*). The bisulfite ion is a relatively weak nucleophile compared to other charged nucleophiles and so only the most reactive carbonyl compounds will react. Larger ketones do not react since larger alkyl groups hinder attack (Topic J5). The reaction is also reversible and so it is a useful method of separating aldehydes and methyl ketones from other ketones or from other organic molecules. This is usually done during an experimental work up where the products of the reaction are dissolved in a water immiscible organic solvent. Aqueous sodium bisulfite is then added and the mixture is shaken thoroughly in a separating funnel. Once the layers have separated, any aldehydes and methyl ketones will have undergone nucleophilic addition with the bisulfite solution and will be dissolved in the aqueous layer as the water soluble salt. The layers can

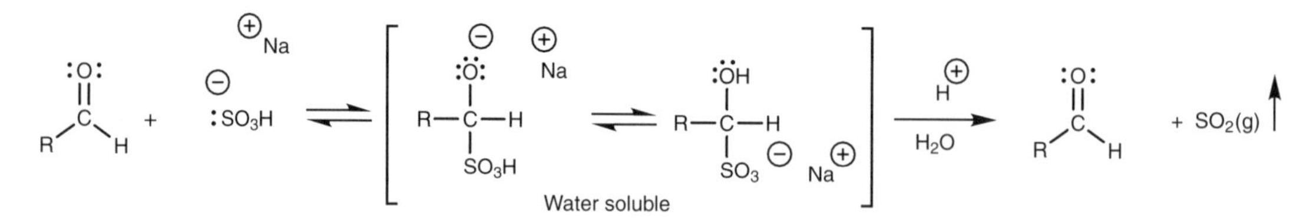

Fig. 11. Reaction of the bisulfite ion with an aldehyde.

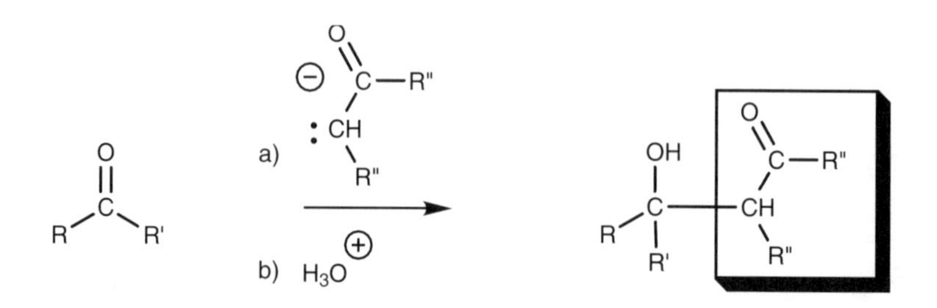

Fig. 12. The Aldol reaction.

now be separated. If the aldehyde or methyl ketone is desired, it can be recovered by adding acid or base to the aqueous layer which reverses the reaction and regenerates the carbonyl compound.

Aldol reaction

Another nucleophilic addition involving a charged nucleophile is the Aldol reaction which is covered in Topic J8. This involves the nucleophilic addition of enolate ions to aldehydes and ketones to form β-hydroxycarbonyl compounds (*Fig. 12*).

J5 Electronic and steric effects

Key Notes

Reactivity Aldehydes are more reactive to nucleophiles than ketones.

Electronic factors Alkyl groups have an inductive effect whereby they 'push' electrons towards a neighboring electrophilic center and make it less electrophilic and less reactive. Ketones have two alkyl groups and are less electrophilic than aldehydes which have only one alkyl group.

Steric factors The transition state for nucleophilic addition resembles the tetrahedral product. Therefore, any factor affecting the stability of the product will affect the stability of the transition state. Since the tetrahedral product is more crowded than the planar carbonyl compound, the presence of bulky alkyl groups will increase crowding and decrease stability. Since ketones have two alkyl groups to aldehyde's one, the transition state for ketones will be less stable than the transition state for aldehydes and the reaction will proceed more slowly. Bulky alkyl groups may also hinder the approach of the nucleophile to the reaction center – the carbonyl group.

Related topics Carbocation stabilization (H5) Nucleophilic addition – charged nucleophiles (J4)

Reactivity

Generally it is found that aldehydes are more reactive to nucleophiles than ketones. There are two factors (electronic and steric) which explain this difference in reactivity.

Electronic factors

The carbonyl carbon in aldehydes is more electrophilic than it is in ketones due to the substituents attached to the carbonyl carbon. A ketone has two alkyl groups attached whereas the aldehyde has only one. The carbonyl carbon is electron deficient and electrophilic since the neighboring oxygen has a greater share of the electrons in the double bond. However, neighboring alkyl groups have an inductive effect whereby they push electron density towards the carbonyl carbon and make it less electrophilic and less reactive to nucleophiles (*Fig. 1*).

Propanal has one alkyl group feeding electrons into the carbonyl carbon, whereas propanone has two. Therefore, the carbonyl carbon in propanal is more electrophilic than the carbonyl carbon in propanone. The more electrophilic the

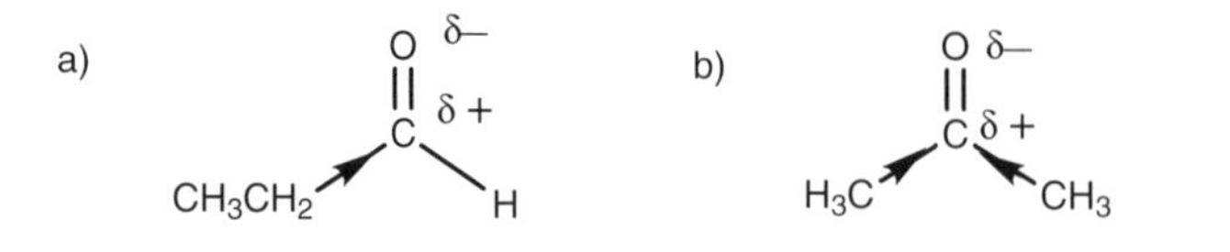

Fig. 1. Inductive effect in (a) propanal; (b) propanone.

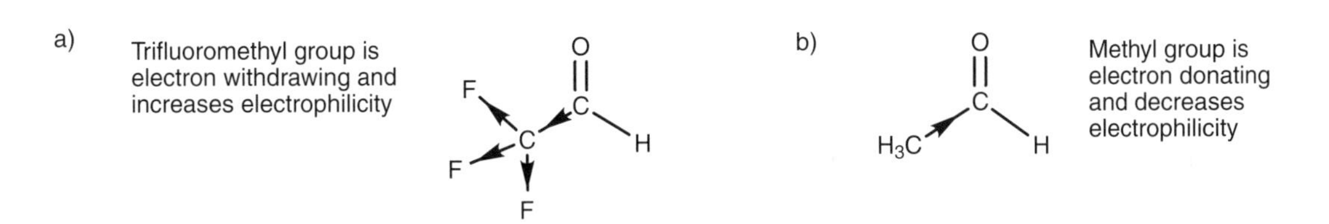

Fig. 2. Inductive effect of (a) trifluoroethanal; (b) ethanal.

carbon, the more reactive it is to nucleophiles. Therefore, propanal is more reactive than propanone.

Electron inductive effects can be used to explain differing reactivities between different aldehydes. For example the fluorinated aldehyde (*Fig. 2*) is more reactive than ethanal. The fluorine atoms are electronegative and have an electron-withdrawing effect on the neighboring carbon, making it electron deficient. This in turn has an inductive effect on the neighboring carbonyl carbon. Since electrons are being withdrawn, the electrophilicity of the carbonyl carbon is increased, making it more reactive to nucleophiles.

Steric factors

Steric factors also have a role to play in the reactivity of aldehydes and ketones. There are two ways of looking at this. One way is to look at the relative ease with which the attacking nucleophile can approach the carbonyl carbon. The other is to consider how steric factors influence the stability of the transition state leading to the final product.

Let us first consider the relative ease with which a nucleophile can approach the carbonyl carbon of an aldehyde and a ketone. In order to do that, we must consider the bonding and the shape of these functional groups (*Fig. 3*). Both molecules have a planar carbonyl group. The atoms which are in the plane are circled in white. A nucleophile will approach the carbonyl group from above or below the plane. The diagram below shows a nucleophile attacking from above. Note that the hydrogen atoms on the neighboring methyl groups are not in the plane of the carbonyl group and so these atoms can hinder the approach of a nucleophile and thus hinder the reaction. This effect will be more significant for a ketone where there are alkyl groups on either side of the carbonyl group. An aldehyde has only one alkyl group attached and so the carbonyl group is more accessible to nucleophilic attack.

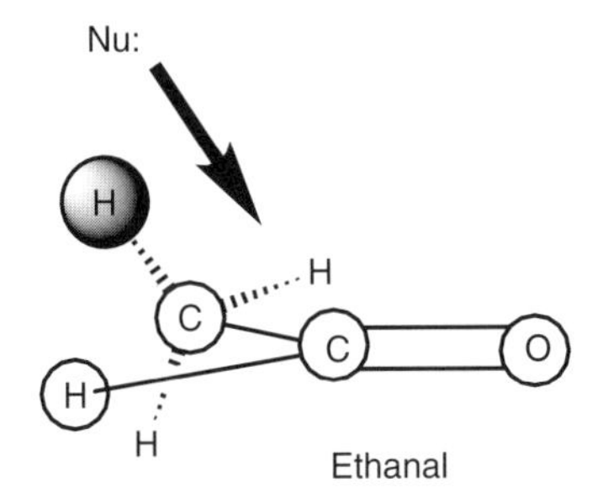

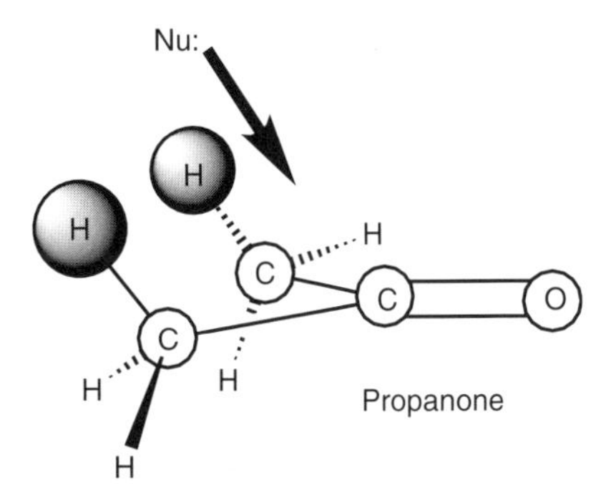

Fig. 3. Steric factors.

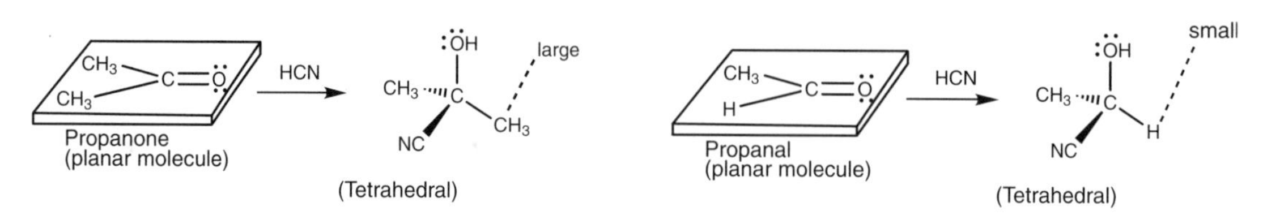

Fig. 4. Reactions of propanone and propanal with HCN.

We shall now look at how steric factors affect the stability of the transition state leading to the final product. For this we shall look at the reactions of propanone and propanal with HCN to give cyanohydrin products (*Fig. 4*).

Both propanone and propanal are planar molecules. The cyanohydrin products are tetrahedral. Thus, the reaction leads to a marked difference in shape between the starting carbonyl compound and the cyanohydrin product. There is also a marked difference in the space available to the substituents attached to the reaction site – the carbonyl carbon. The tetrahedral molecule is more crowded since there are four substituents crowded round a central carbon, whereas in the planar starting material, there are only three substituents attached to the carbonyl carbon. The crowding in the tetrahedral product arising from the ketone will be greater than that arising from the aldehyde since one of the substituents from the aldehyde is a small hydrogen atom.

The ease with which nucleophilic addition takes place depends on the ease with which the transition state is formed. In nucleophilic addition, the transition state is thought to resemble the tetrahedral product more than it does the planar starting material. Therefore, any factor which affects the stability of the product will also affect the stability of the transition state. Since crowding is a destabilizing effect, the reaction of propanone should be more difficult than the reaction of propanal. Therefore, ketones in general will be less reactive than aldehydes.

The bigger the alkyl groups, the bigger the steric effect. For example, 3-pentanone is less reactive than propanone and fails to react with the weak bisulfite nucleophile whereas propanone does (*Fig. 5*).

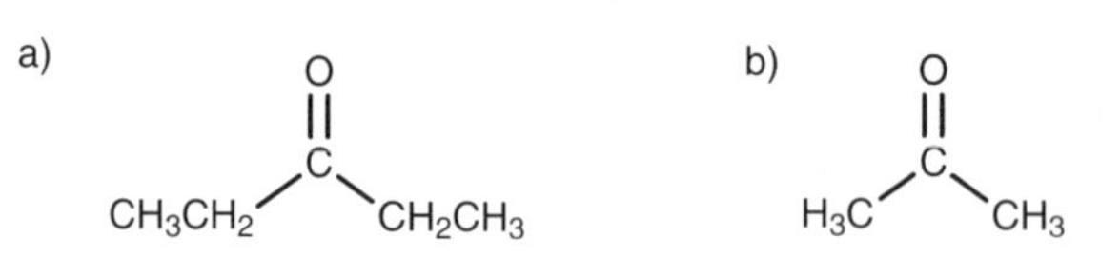

Fig. 5. (a) 3-Pentanone; (b) propanone.

J6 NUCLEOPHILIC ADDITION – NITROGEN NUCLEOPHILES

Key Notes

Imine formation

Primary amines react with aldehydes and ketones to give an imine or Schiff base. The reaction involves nucleophilic addition of the amine followed by elimination of water. Acid catalysis aids the reaction, but too much acid hinders the reaction by protonating the amine.

Enamine formation

Secondary amines undergo the same type of mechanism as primary amines, but cannot give imines as the final product. Instead, a proton is lost from a neighboring carbon and functional groups called enamines are formed.

Oximes, semicarbazones and 2,4-dinitrophenylhydrazones

Aldehydes and ketones can be converted to crystalline derivatives called oximes, semicarbazones, and 2,4-dinitrophenylhydrazone. Such derivatives were useful in the identification of liquid aldehydes and ketones.

Related topics

Nucleophilic addition (J3)
Nucleophilic addition – charged nucleophiles (J4)
Nucleophilic addition – oxygen and sulfur nucleophiles (J7)

Imine formation

The reaction of primary amines with aldehydes and ketones do not give the products expected from nucleophilic addition alone. This is because further reaction occurs once nucleophilic addition takes place. As an example, we shall consider the reaction of acetaldehyde (ethanal) with a primary amine – methylamine (*Fig. 1*). The product contains the methylamine skeleton, but unlike the previous reactions there is no alcohol group and there is a double bond between the carbon and the nitrogen. This product is called an **imine** or a **Schiff base**.

The first stage of the mechanism (*Fig. 2*) is a normal nucleophilic addition. The amine acts as the nucleophile and the nitrogen atom is the nucleophilic center. The nitrogen uses its lone pair of electrons to form a bond to the electrophilic carbonyl carbon. As this bond is being formed, the carbonyl π bond breaks with both electrons moving onto the oxygen to give it a third lone pair of electrons and a negative charge. The nitrogen also gains a positive charge, but both these charges can

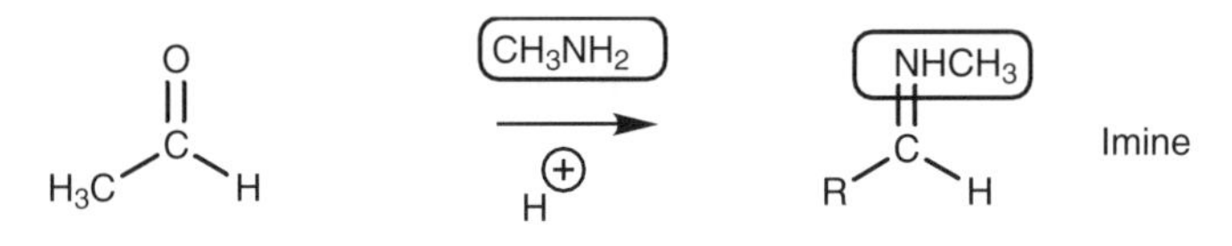

Fig. 1. Reaction of ethanal with methylamine.

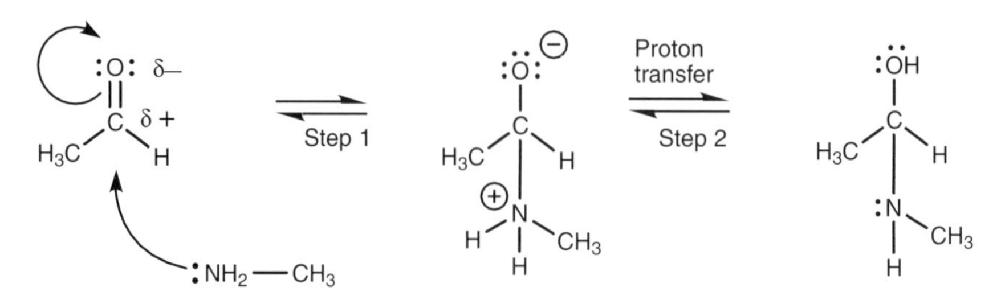

Fig. 2. Mechanism of nucleophilic addition.

be neutralized by the transfer of a proton from the nitrogen to the oxygen (Step 2). The oxygen uses up one of its lone pairs to form the new O–H bond and the electrons in the N–H bond end up on the nitrogen as a lone pair. An acid catalyst is present, but is not required for this part of the mechanism – nitrogen is a good nucleophile and although the amine is neutral, it is sufficiently nucleophilic to attack the carbonyl group without the need for acid catalysis. The intermediate obtained is the structure one would expect from nucleophilic addition alone, but the reaction does not stop there. The oxygen atom is now protonated by the acid catalyst and gains a positive charge (*Fig. 3,* Step 3). Since oxygen is electronegative, a positive charge is not favored and so there is a strong drive to neutralize the charge. This can be done if the bond to carbon breaks and the oxygen leaves as part of a water molecule. Therefore, protonation has turned the oxygen into a good leaving group. The nitrogen helps the departure of the water by using its lone pair of electrons to form a π bond to the neighboring carbon atom and a positive charged intermediate is formed (Step 4). The water now acts as a nucleophile and removes a proton from the nitrogen such that the nitrogen's lone pair is restored and the positive charge is neutralized (Step 5).

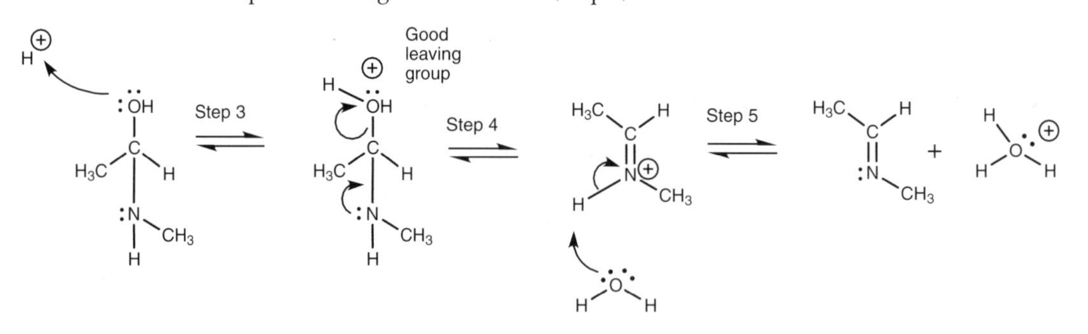

Fig. 3. Mechanism for the elimination of water.

Overall, a molecule of water has been lost in this second part of the mechanism. Acid catalysis is important in creating a good leaving group. If protonation did not occur, the leaving group would have to be the hydroxide ion which is a more reactive molecule and a poorer leaving group.

Although acid catalysis is important to the reaction mechanism, too much acid can actually hinder the reaction. This is because a high acid concentration leads to protonation of the amine, and prevents it from acting as a nucleophile.

Enamine formation

The reaction of carbonyl compounds with secondary amines cannot give imines since there is no NH proton to be lost in the final step of the mechanism. However, there is another way in which the positive charge on the nitrogen can be neutralized. This involves loss of a proton from a neighboring carbon atom (*Fig. 4*). Water acts as a base to remove the proton and the electrons which make up the C–H σ bond are used to form a new π bond to the neighboring carbon. This in turn forces the existing π bond between carbon and nitrogen to break such that both the π electrons end up on the nitrogen atom as a lone pair, thus neutralizing the charge. The final structure is known as an enamine and can prove useful in organic synthesis.

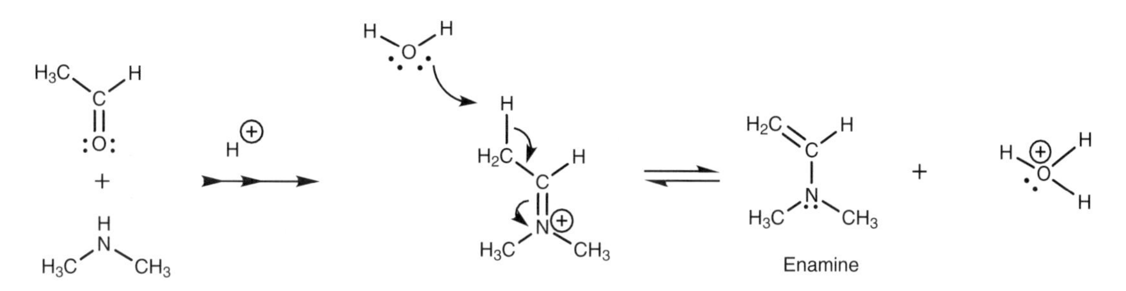

Fig. 4. Mechanism for the formation of an enamine.

Oximes, semicarbazones and 2,4-dinitrophenylhydrazones

The reaction of aldehydes and ketones with hydroxylamine (NH_2OH), semicarbazide ($NH_2NHCONH_2$) and 2,4-dinitrophenylhydrazine takes place by the same mechanism described for primary amines to give oximes, semicarbazones, and 2,4-dinitrophenylhydrazones, respectively (*Fig. 5*). These compounds were frequently synthesized in order to identify a liquid aldehyde or ketone. The products are solid and crystalline, and by measuring their melting points, the original aldehyde or ketone could be identified by looking up melting point tables of these derivatives. Nowadays, it is easier to identify liquid aldehydes and ketones spectroscopically.

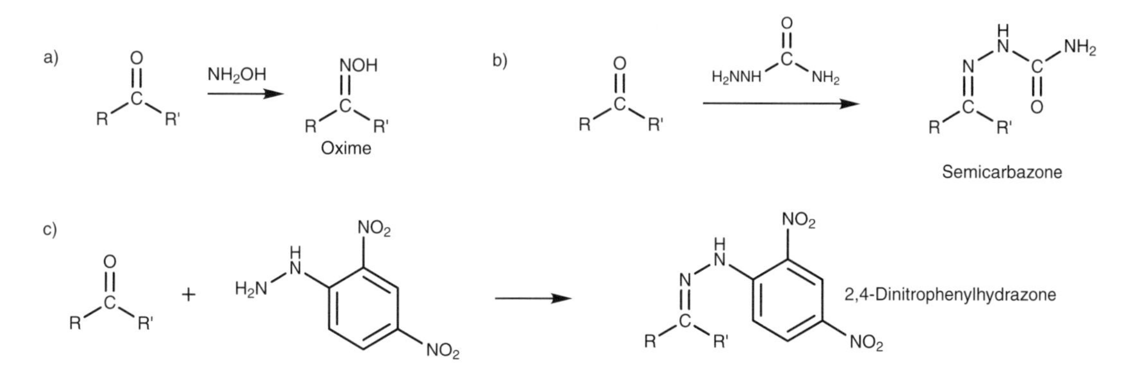

Fig. 5. Synthesis of oximes, semicarbazones, and 2,4-dinitrophenylhydrazones.

J7 Nucleophilic addition – oxygen and sulfur nucleophiles

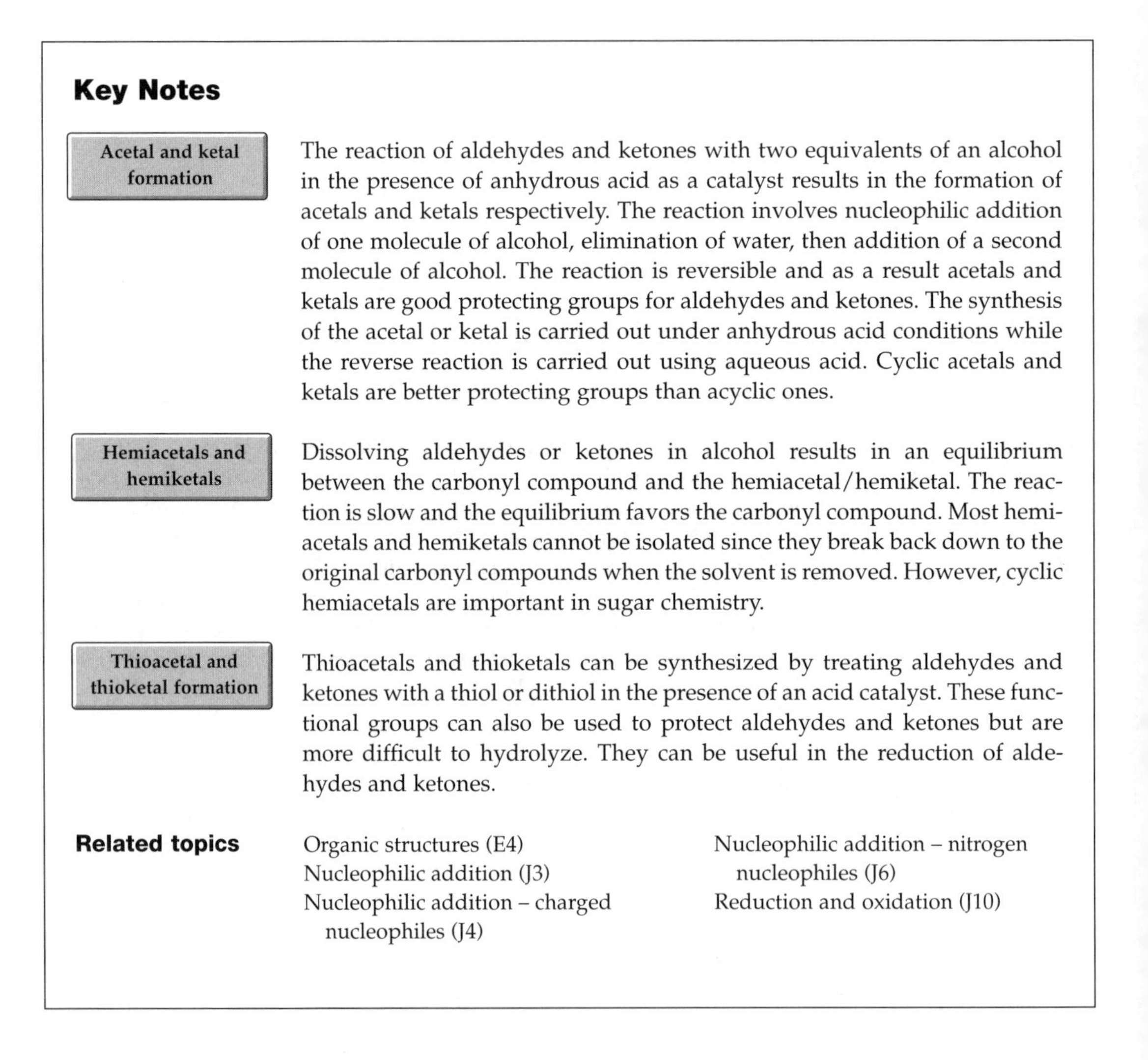

Key Notes

Acetal and ketal formation

The reaction of aldehydes and ketones with two equivalents of an alcohol in the presence of anhydrous acid as a catalyst results in the formation of acetals and ketals respectively. The reaction involves nucleophilic addition of one molecule of alcohol, elimination of water, then addition of a second molecule of alcohol. The reaction is reversible and as a result acetals and ketals are good protecting groups for aldehydes and ketones. The synthesis of the acetal or ketal is carried out under anhydrous acid conditions while the reverse reaction is carried out using aqueous acid. Cyclic acetals and ketals are better protecting groups than acyclic ones.

Hemiacetals and hemiketals

Dissolving aldehydes or ketones in alcohol results in an equilibrium between the carbonyl compound and the hemiacetal/hemiketal. The reaction is slow and the equilibrium favors the carbonyl compound. Most hemiacetals and hemiketals cannot be isolated since they break back down to the original carbonyl compounds when the solvent is removed. However, cyclic hemiacetals are important in sugar chemistry.

Thioacetal and thioketal formation

Thioacetals and thioketals can be synthesized by treating aldehydes and ketones with a thiol or dithiol in the presence of an acid catalyst. These functional groups can also be used to protect aldehydes and ketones but are more difficult to hydrolyze. They can be useful in the reduction of aldehydes and ketones.

Related topics

Organic structures (E4)
Nucleophilic addition (J3)
Nucleophilic addition – charged nucleophiles (J4)
Nucleophilic addition – nitrogen nucleophiles (J6)
Reduction and oxidation (J10)

Acetal and ketal formation

When an aldehyde or ketone is treated with an excess of alcohol in the presence of an acid catalyst, **two** molecules of alcohol are added to the carbonyl compound to give an acetal or a ketal respectively (*Fig. 1*). The final product is tetrahedral.

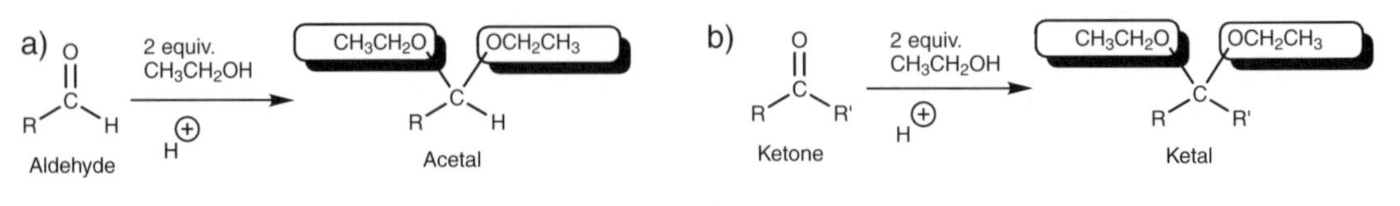

Fig. 1. Formation of an acetal and a ketal.

The reaction mechanism involves the nucleophilic addition of one molecule of alcohol to form a hemiacetal or hemiketal. Elimination of water takes place to form an oxonium ion and a second molecule of alcohol is then added (*Fig. 2*).

The mechanism is quite complex and we shall look at it in detail by considering the reaction of methanol with acetaldehyde (ethanal; *Fig. 3*). The aldehyde is the electrophile and the electrophilic center is the carbonyl carbon. Methanol is the nucleophile and the nucleophilic center is oxygen. However, methanol is a relatively weak nucleophile (Topic E4). As a result, the carbonyl group has to be activated by adding an acid catalyst if a reaction is to take place. The first step of the mechanism involves the oxygen of the carbonyl group using a lone pair of electrons to form a bond to a proton. This results in a charged intermediate where the positive charge is shared between the carbon and oxygen of the carbonyl group. Protonation increases the electrophilicity of the carbonyl group, making the carbonyl carbon even more electrophilic. As a result, it reacts better with the weakly nucleophilic alcohol. The alcoholic oxygen now uses one of its lone pairs of electrons to form a bond to the carbonyl carbon and the carbonyl π bond breaks at the same time with the π electrons moving onto the carbonyl oxygen and neutralizing the positive charge (*Fig. 4*). However, the alcoholic oxygen now has an unfavorable positive charge (which explains why methanol is a weak nucleophile in the first place). This charge is easily lost if the attached proton is lost. Both elec-

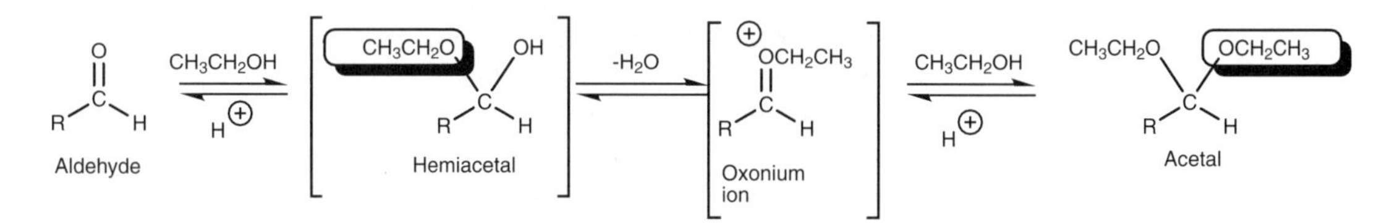

Fig. 2. Acetal formation and intermediates involved.

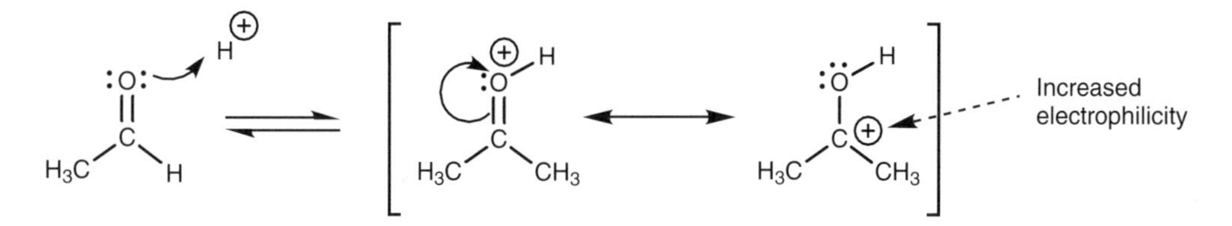

Fig. 3. Mechanism of acetal formation – step 1.

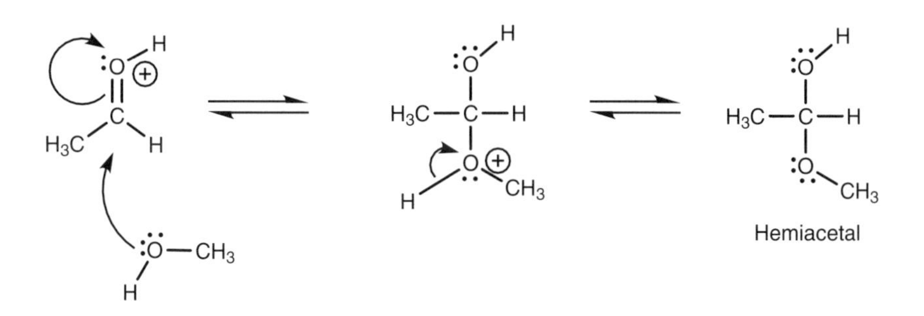

Fig. 4. Mechanism of acetal formation – steps 2 and 3.

trons in the O–H σ bond are captured by the oxygen to restore its second lone pair of electrons and neutralize the positive charge.

The intermediate formed from this first nucleophilic addition is called a **hemiacetal**. If a ketone had been the starting material, the structure obtained would have been a **hemiketal**. Once the hemiacetal is formed, it is protonated and water is eliminated by the same mechanism described in the formation of imines (Topic J6) – the only difference being that oxygen donates a lone pair of electrons to force the removal of water rather than nitrogen (*Fig. 5*). The resulting oxonium ion is extremely electrophilic and a second nucleophilic addition of alcohol takes place to give the acetal.

All the stages in this mechanism are reversible and so it is possible to convert the acetal or ketal back to the original carbonyl compound using water and an aqueous acid as catalyst. Since water is added to the molecule in the reverse mechanism, this is a process called **hydrolysis**.

Acid acts as a catalyst both for the formation and the hydrolysis of acetals and ketals, so how can one synthesize ketals and acetals in good yield? The answer lies in the reaction conditions. When forming acetals or ketals, the reaction is carried out in the absence of water using a small amount of concentrated sulfuric acid or an organic acid such as *para*-toluenesulfonic acid. The yields are further boosted if the water formed during the reaction is removed from the reaction mixture.

In order to convert the acetal or ketal back to the original carbonyl compound, an aqueous acid is used such that there is a large excess of water present and the equilibrium is shifted towards the carbonyl compounds.

Both the synthesis and the hydrolysis of acetals and ketals can be carried out in high yield and so these functional groups are extremely good as protecting groups for aldehydes and ketones. Acetals and ketals are stable to nucleophiles and basic conditions and so the carbonyl group is 'disguised' and will not react with these reagents. Cyclic acetals and ketals are best used for the protection of aldehydes and ketones. These can be synthesized by using diols rather than alcohols (*Fig. 6*).

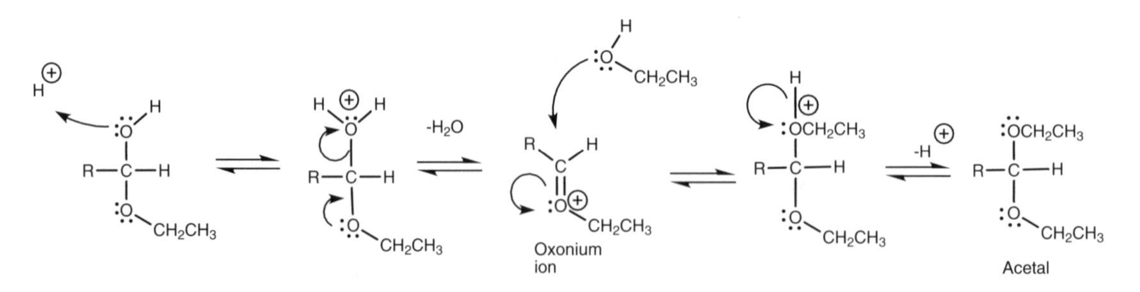

Fig. 5. Mechanism of acetal formation from a hemiacetal.

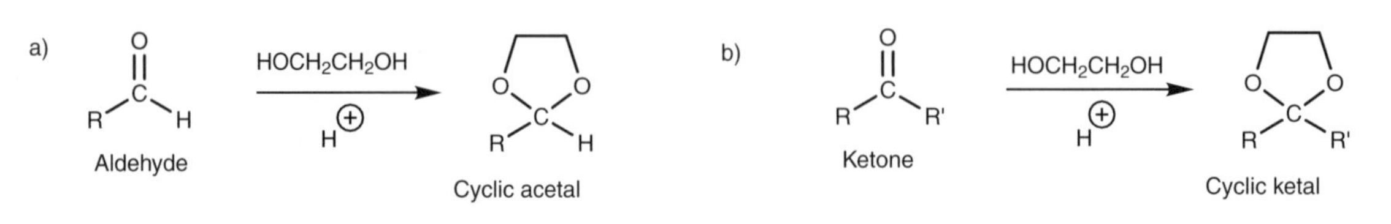

Fig. 6. Synthesis of cyclic acetals and cyclic ketals.

Hemiacetals and hemiketals

When aldehydes and ketones are dissolved in alcohol without an acid catalyst being present, only the first part of the above mechanism takes place with one alcohol molecule adding to the carbonyl group. An equilibrium is set up between the carbonyl group and the hemiacetal or hemiketal, with the equilibrium favoring the carbonyl compound (*Fig. 7*).

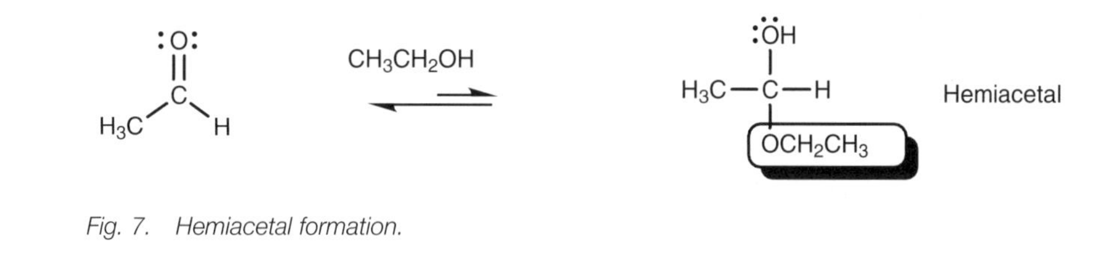

Fig. 7. Hemiacetal formation.

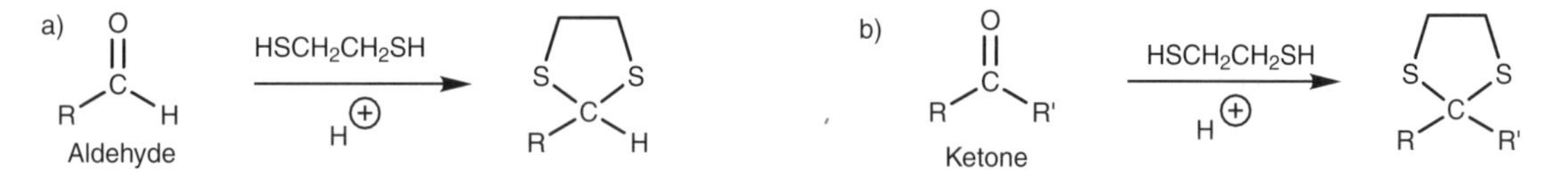

Fig. 8. Formation of (a) cyclic thioacetals and (b) cyclic thioketals.

The reaction is not synthetically useful, since it is not usually possible to isolate the products. If the solvent is removed, the equilibrium is driven back to starting materials. However, cyclic hemiacetals are important in the chemistry of sugars.

Thioacetal and thioketal formation

Thioacetals and thioketals are the sulfur equivalents of acetals and ketals and are also prepared under acid conditions (*Fig. 8*). These can also be used to protect aldehydes and ketones, but the hydrolysis of these groups is more difficult. More importantly, the thioacetals and thioketals can be removed by reduction and this provides a method of reducing aldehydes and ketones (Topic J10).

J8 Reactions of enolate ions

Key Notes

Enolate ions

Enolate ions are formed by treating aldehydes or ketones with a base. A proton has to be present on the α-carbon.

Alkylation

Enolate ions can be alkylated with an alkyl halide. *O*-Alkylation and *C*-alkylation are both possible, but the latter is more likely and more useful. The reaction allows the introduction of alkyl groups to the α-carbon of aldehydes and ketones. If there are two α-protons present, two different alkylations can be carried out in succession. β-Ketoesters are useful starting materials since the α-protons are more acidic and the alkylation is targeted to one position. The ester group is removed by decarboxylation.

Aldol reaction

The Aldol reaction involves the dimerization of an aldehyde or a ketone. In the presence of sodium hydroxide, aldehyde or ketone is converted to an enolate ion, but not all the carbonyl molecules are converted and so the enolate ion can undergo a nucleophilic addition on 'free' aldehyde or ketone. The product is a β-hydroxyaldehyde or β-hydroxyketone. Aldehydes react better than ketones in this reaction. If water is lost from the Aldol adduct, an α,β-unsaturated carbonyl structure is obtained.

Crossed Aldol reaction

The crossed Aldol reaction links two different aldehyde structures. The reaction works best if one of the aldehydes has no α-proton present and the other aldehyde is added slowly to the reaction mixture to prevent self-condensation. If a ketone is linked to an aldehyde, the reaction is known as the Claisen–Schmidt reaction. This works best if the aldehyde has no α-proton.

Related topics

sp^2 Hybridization (A4)
Organic structures (E4)
Enolates (G5)
Nucleophilic substitution (L2)
Elimination (L4)

Enolate ions

Enolate ions are formed by treating aldehydes or ketones with a base. An α-proton has to be present. The mechanism of this acid base reaction was covered in Topic G5. Enolate ions can undergo a variety of important reactions including alkylation and the Aldol reaction.

Alkylation

Treatment of an enolate ion with an alkyl halide results in a reaction known as **alkylation** (*Fig. 1*). The overall reaction involves the replacement of an α-proton with an alkyl group. The nucleophilic and electrophilic centers of the enolate ion and methyl iodide are shown (*Fig. 2*). The enolate ion has its negative charge shared between the oxygen atom and the carbon atom due to resonance (Topic G5), and so both of these atoms are nucleophilic centers. Iodomethane has a polar

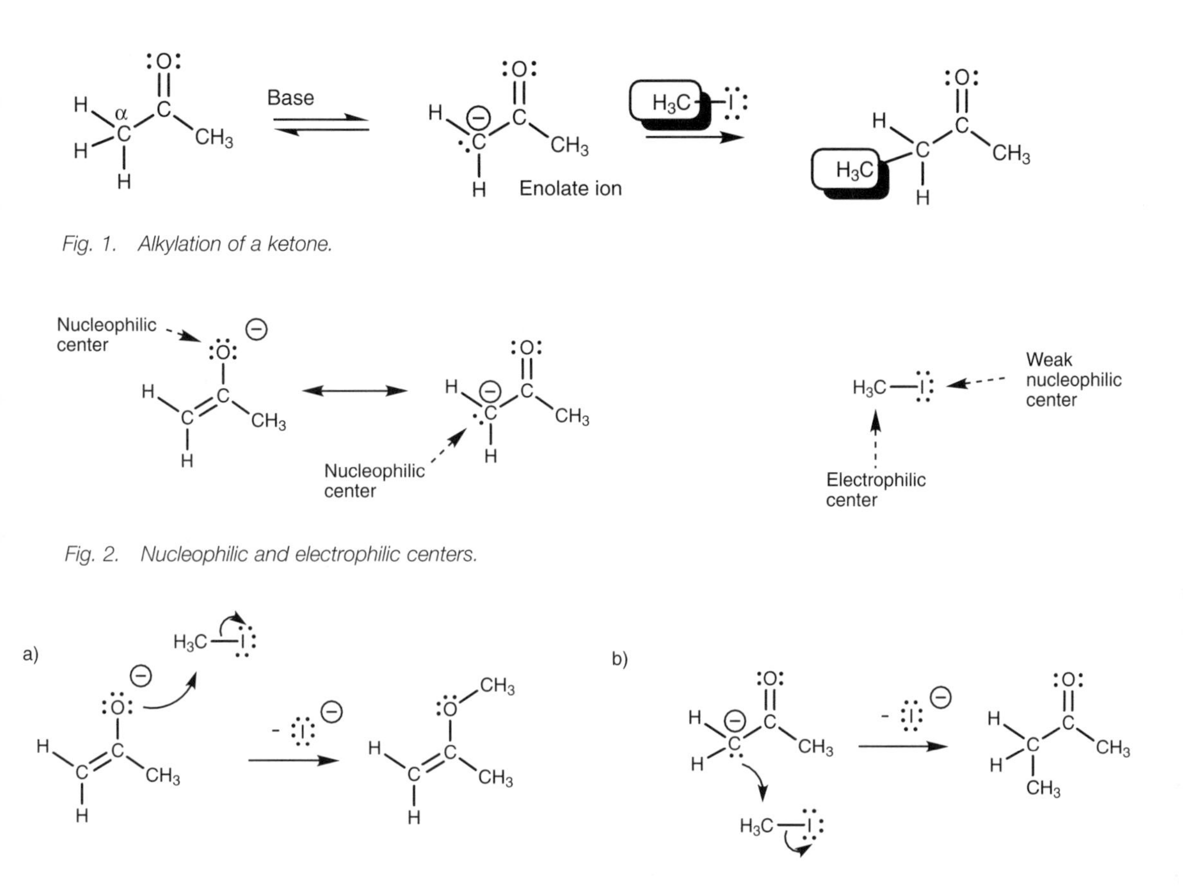

Fig. 1. Alkylation of a ketone.

Fig. 2. Nucleophilic and electrophilic centers.

Fig. 3. (a) O-Alkylation; (b) C-alkylation.

C–I bond where the iodine is a weak nucleophilic center and the carbon is a good electrophilic center (Topics E3 and E4).

One possible reaction between these molecules involves the nucleophilic oxygen using one of its lone pairs of electrons to form a new bond to the electrophilic carbon on iodomethane (*Fig. 3a*). At the same time, the C–I bond and both electrons move onto iodine to give it a fourth lone pair of electrons and a negative charge. This reaction is possible, but in practice the product obtained is more likely to arise from the reaction of the alternative carbanion structure reacting with methyl iodide (*Fig. 3b*). This is a more useful reaction since it involves the formation of a carbon–carbon bond and allows the construction of more complex carbon skeletons.

An alternative mechanism to that shown in *Fig. 3b,* but which gives the same result, starts with the enolate ion. The enolate ion is more stable than the carbanion since the charge is on the electronegative oxygen and so it is more likely that the reaction mechanism will occur in this manner (*Fig. 4*). This is a very useful reaction in organic synthesis. However, there are limitations to the type of alkyl halide which can be used in the reaction. The reaction is S_N2 with respect to the alkyl halide (see Topic L2) and so the reaction works best with primary alkyl, primary benzylic, and primary allylic halides. The enolate ion is a strong base and if it is reacted with secondary and tertiary halides, elimination of the alkyl halide takes place to give an alkene (Topic L4).

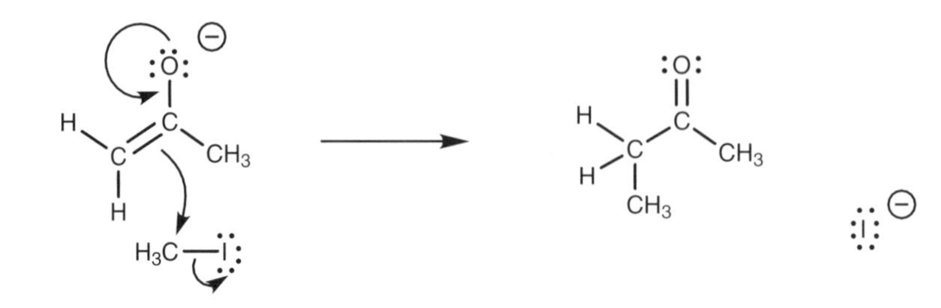

Fig. 4. Mechanism for C-alkylation of the enolate ion.

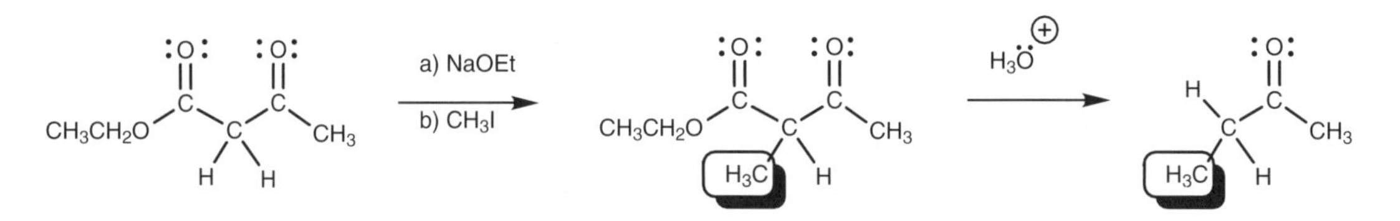

Fig. 5. Alkylation of ethyl acetoacetate.

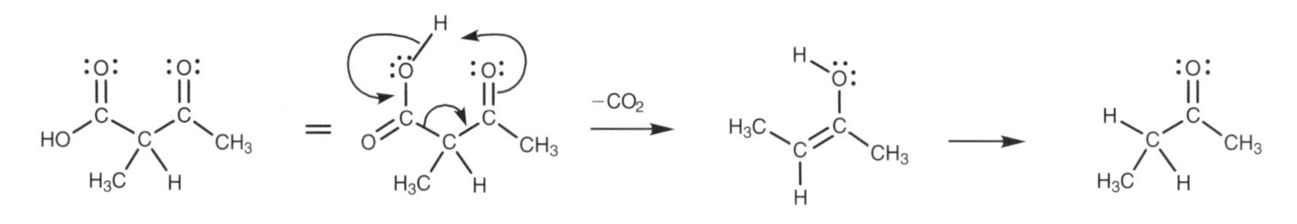

Fig. 6. Decarboxylation mechanism.

α-Alkylation works well with ketones, but not so well for aldehydes since the latter tend to undergo Aldol condensations instead (see below).

The α-protons of a ketone such as propanone are only weakly acidic and so a powerful base (e.g. lithium diisopropylamide) is required to generate the enolate ion required for the alkylation. An alternative method of preparing the same product but using a milder base is to start with ethyl acetoacetate (a β-keto ester) instead (*Fig. 5*). The α-protons in this structure are more acidic since they are flanked by two carbonyl groups (Topic G5). As a result, the enolate can be formed using a weaker base such as sodium ethoxide. Once the enolate has been alkylated, the ester group can be hydrolyzed and decarboxylated on heating with aqueous hydrochloric acid. The decarboxylation mechanism involves the β-keto group and would not occur if this group was absent (*Fig. 6*). Carbon dioxide is lost and the enol tautomer is formed. This can then form the keto tautomer by the normal keto–enol tautomerism (Topic J2).

It is possible for two different alkylations to be carried out on ethyl acetoacetate since there is more than one α-proton present (*Fig. 7*).

β-keto esters such as ethylacetoacetate are also useful in solving a problem involved in the alkylation of unsymmetrical ketones. For example, alkylating butanone with methyl iodide leads to two different products since there are α-protons on either side of the carbonyl group (*Fig. 8*). One of these products is obtained specifically by using a β-keto ester to make the target alkylation site more acidic (*Fig. 9*).

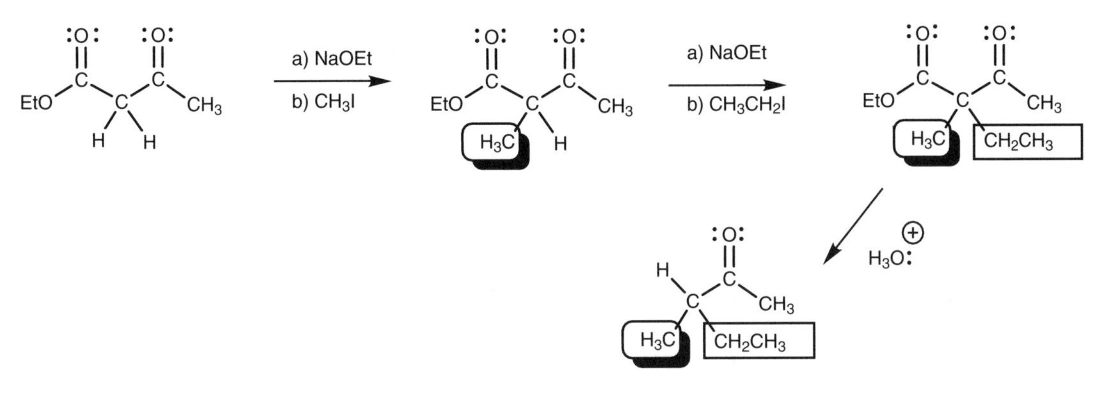

Fig. 7. Double alkylation of ethylacetoacetate.

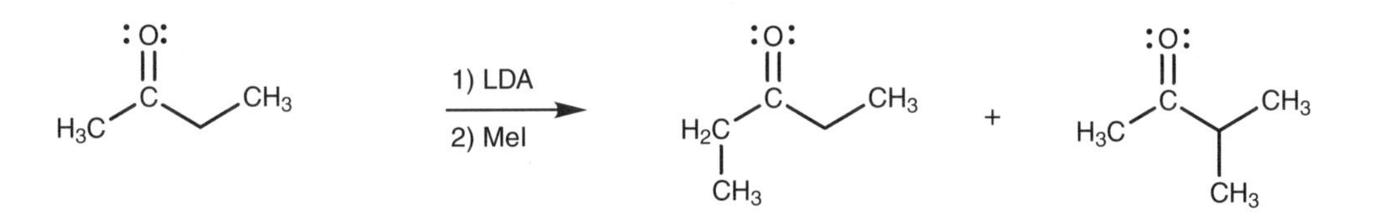

Fig. 8. Alkylation of butanone.

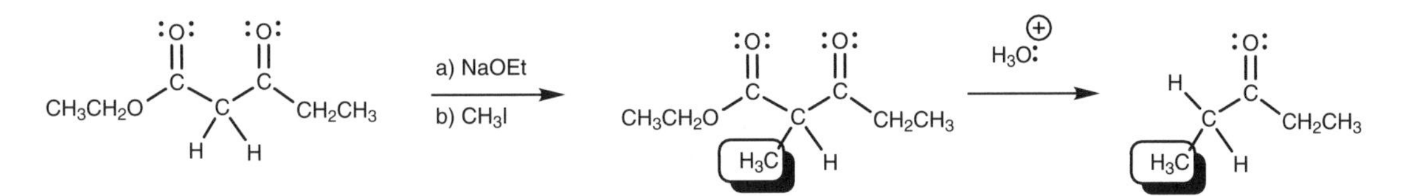

Fig. 9. Use of a β-keto ester to direct alkylation.

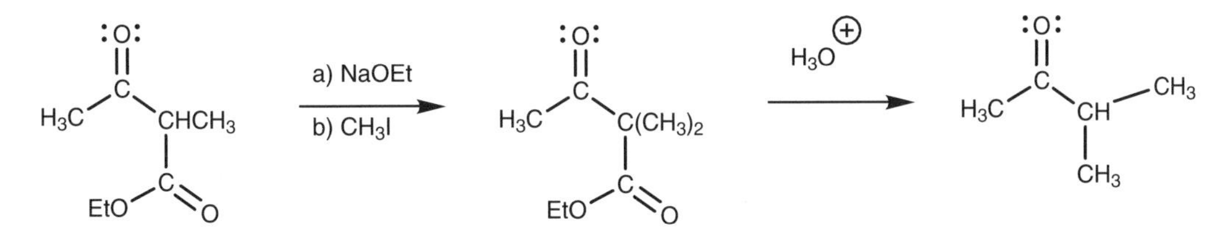

Fig. 10. Use of a β-keto ester to direct alkylation.

The alternative alkylation product could be obtained by using a different β-keto ester (*Fig. 10*).

Aldol reaction

Enolate ions can also react with aldehydes and ketones by nucleophilic addition. The enolate ion acts as the nucleophile while the aldehyde or ketone acts as an electrophile. Since the enolate ion is formed from a carbonyl compound itself, and can then react with a carbonyl compound, it is possible for an aldehyde or ketone to react with itself. We can illustrate this by looking at the reaction of acetaldehyde with sodium hydroxide (*Fig. 11*). Under these conditions, two molecules of acetaldehyde are linked together to form a β-hydroxyaldehyde.

In this reaction, two separate reactions are going on – the formation of an

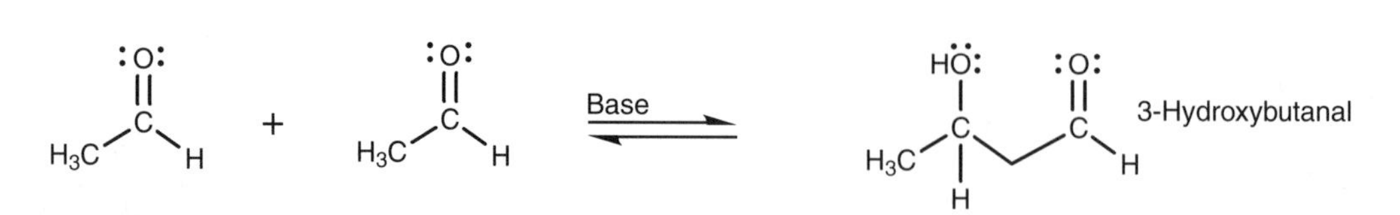

Fig. 11. Aldol reaction.

enolate ion from one molecule of acetaldehyde, and the addition of that enolate to a second molecule of acetaldehyde. The mechanism begins with the formation of the enolate ion as described in Topic G5. It is important to realize that not all of the acetaldehyde is converted to the enolate ion and so we still have molecules of acetaldehyde present in the same solution as the enolate ions. Since acetaldehyde is susceptible to nucleophilic attack, the next stage in the mechanism is the nucleophilic attack of the enolate ion on acetaldehyde (*Fig. 12*). The enolate ion has two nucleophilic centers – the carbon and the oxygen – but the preferred reaction is at the carbon atom. The first step is nucleophilic addition of the aldehyde to form a charged intermediate. The second step involves protonation of the charged oxygen. Since a dilute solution of sodium hydroxide is used in this reaction, water is available to supply the necessary proton. (Note that it would be wrong to show a free proton (H^+) since the solution is alkaline.)

If the above reaction is carried out with heating, then a different product is obtained (*Fig. 13*). This arises from elimination of a molecule of water from the Aldol reaction product. There are two reasons why this can occur. First of all, the product still has an acidic proton (i.e. there is still a carbonyl group present and an α-hydrogen next to it). This proton is prone to attack from base. Secondly, the dehydration process results in a conjugated product which results in increased stability (Topic A4). The mechanism of dehydration is shown in *Fig. 14*. First of all, the acidic proton is removed and a new enolate ion is formed. The electrons in the enolate ion can then move in such a fashion that the hydroxyl group is expelled to

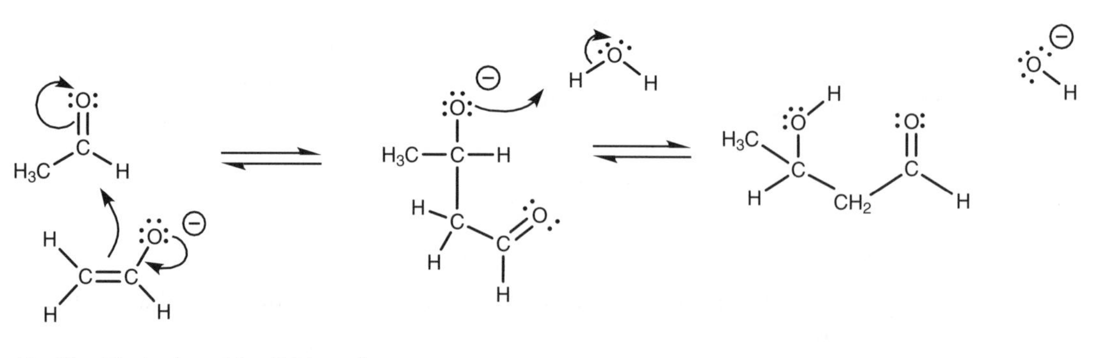

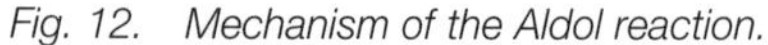
Fig. 12. Mechanism of the Aldol reaction.

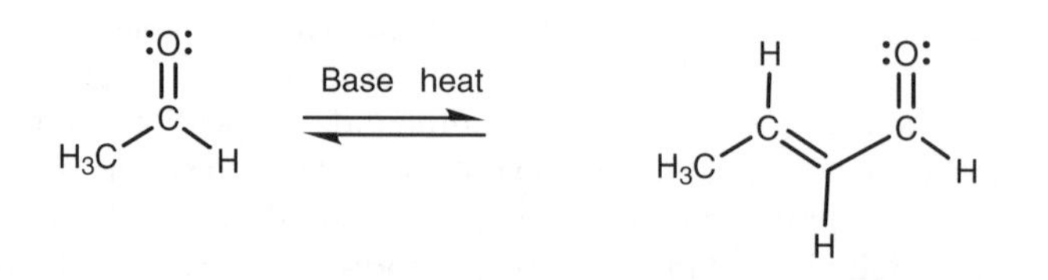

Fig. 13. Formation of 2-butenal.

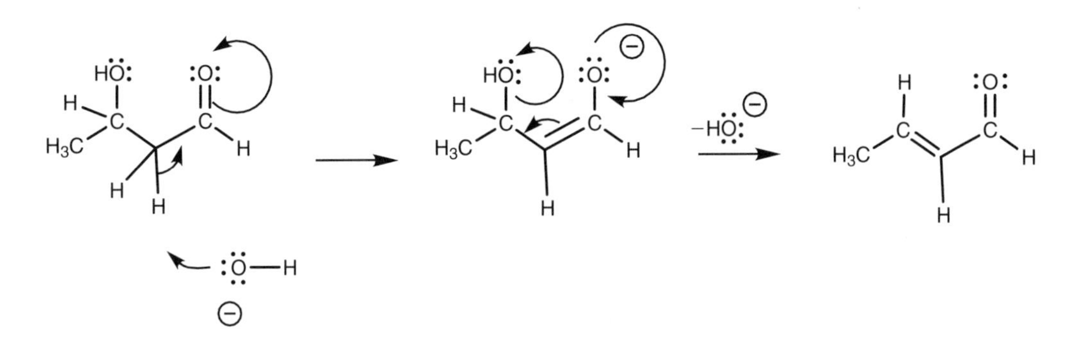

Fig. 14. Mechanism of dehydration.

give the final product – an α,β-unsaturated aldehyde. In this example, it is possible to vary the conditions such that one gets the Aldol reaction product or the α,β-unsaturated aldehyde, but in some cases only the α,β-unsaturated carbonyl product is obtained, especially when extended conjugation is possible. The Aldol reaction is best carried out with aldehydes. Some ketones will undergo an Aldol reaction, but an equilibrium is set up between the products and starting materials and it is necessary to remove the product as it is being formed in order to pull the reaction through to completion.

Crossed Aldol reaction

So far we have talked about the Aldol reaction being used to link two molecules of the same aldehyde or ketone, but it is also possible to link two different carbonyl compounds. This is known as a **crossed Aldol reaction**. For example, benzaldehyde and ethanal can be linked in the presence of sodium hydroxide (*Fig. 15*). In this example, ethanal reacts with sodium hydroxide to form the enolate ion which then reacts with benzaldehyde. Elimination of water occurs easily to give an extended conjugated system involving the aromatic ring, the double bond, and the carbonyl group.

This reaction works well because the benzaldehyde has no α-protons and cannot form an enolate ion. Therefore, there is no chance of benzaldehyde undergoing self-condensation. It can only act as the electrophile for another enolate ion. However, what is to stop the ethanal undergoing an aldol addition with itself as previously described (*Fig. 11*)?

This reaction can be limited by only having benzaldehyde and sodium hydroxide initially present in the reaction flask. Since benzaldehyde has no α-protons, no reaction can take place. A small quantity of ethanal can now be added. Reaction with excess sodium hydroxide turns most of the ethanal into its enolate ion. There will only be a very small amount of 'free' ethanal left compared to benzaldehyde and so the enolate ion is more likely to react with benzaldehyde.

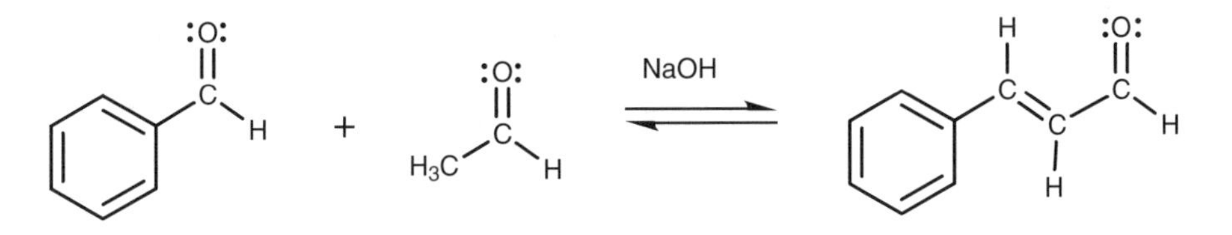

Fig. 15. Crossed Aldol reaction.

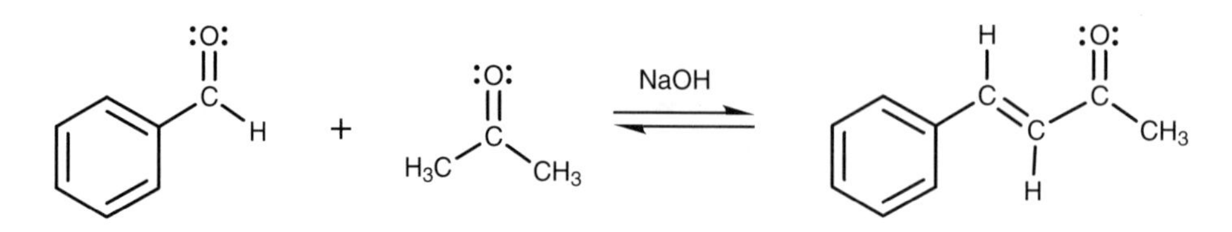

Fig. 16. Claisen–Schmidt reaction.

Once the reaction is judged to have taken place, the next small addition of ethanal can take place and the process is repeated.

Ketones and aldehydes can also be linked by the same method – a reaction known as the **Claisen–Schmidt** reaction. The most successful reactions are those where the aldehyde does not have an α-proton (*Fig. 16*).

J9 α-HALOGENATION

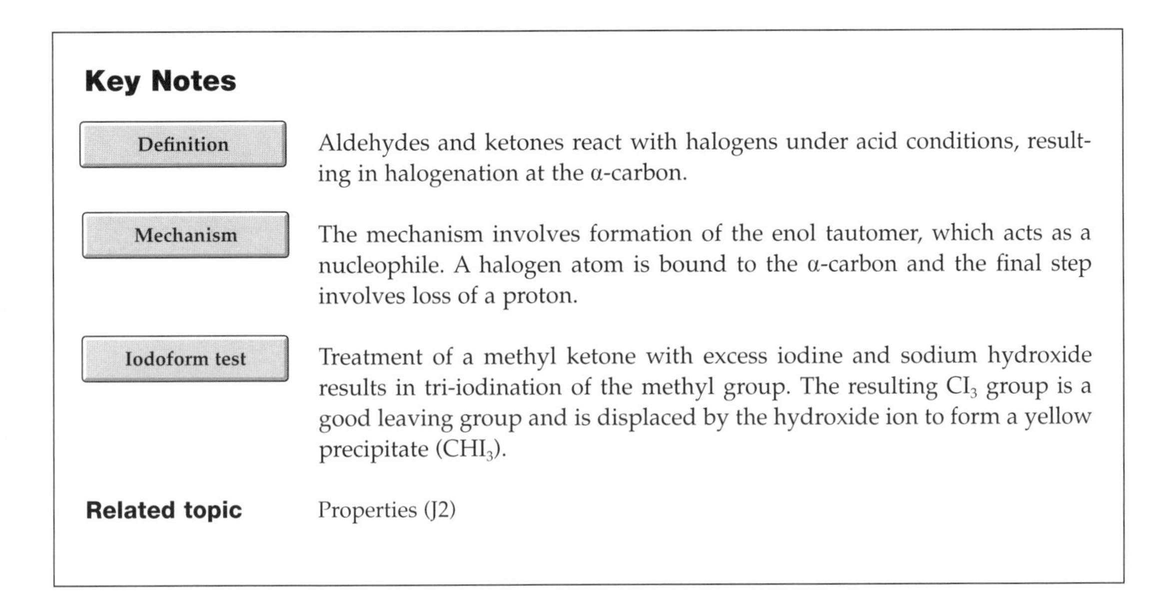

Key Notes

Definition — Aldehydes and ketones react with halogens under acid conditions, resulting in halogenation at the α-carbon.

Mechanism — The mechanism involves formation of the enol tautomer, which acts as a nucleophile. A halogen atom is bound to the α-carbon and the final step involves loss of a proton.

Iodoform test — Treatment of a methyl ketone with excess iodine and sodium hydroxide results in tri-iodination of the methyl group. The resulting CI_3 group is a good leaving group and is displaced by the hydroxide ion to form a yellow precipitate (CHI_3).

Related topic — Properties (J2)

Definition

Aldehydes and ketones react with chlorine, bromine or iodine in acidic solution, resulting in halogenation at the α-carbon (*Fig. 1*).

Mechanism

Since acid conditions are employed, this process does not involve an enolate ion. Instead, the reaction takes place through the enol tautomer of the carbonyl compound (Topic J2). The enol tautomer acts as a nucleophile with a halogen by the mechanism shown (*Fig. 2*). In the final step, the solvent acts as a base to remove the proton.

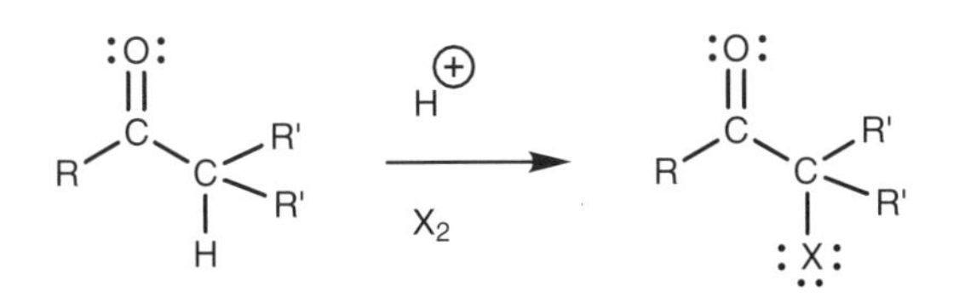

Fig. 1. α-Halogenation.

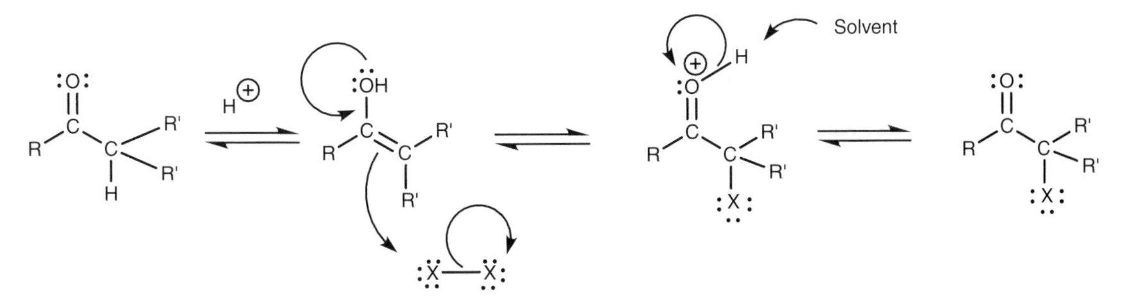

Fig. 2. Mechanism of α-halogenation.

Iodoform reaction α-Halogenation can also be carried out in the presence of base. The reaction proceeds through an enolate ion which is then halogenated (*Fig. 3*). However, it is difficult to stop the reaction at mono-halogenation since the resulting product is generally more acidic than the starting ketone due to the electron-withdrawing effect of the halogen. As a result, another enolate ion is quickly formed leading to further halogenation.

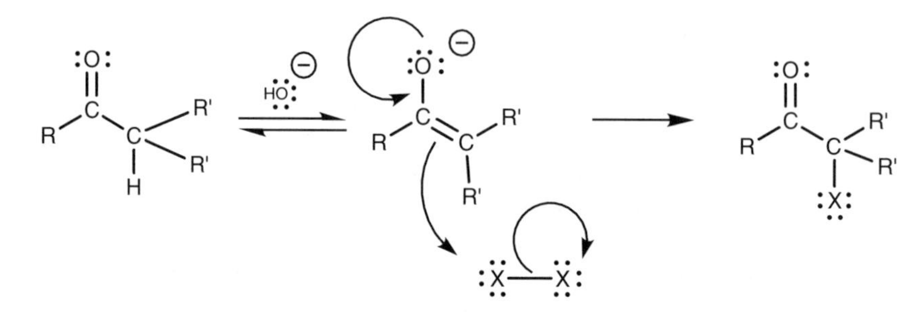

Fig. 3. α-Halogenation in the presence of base.

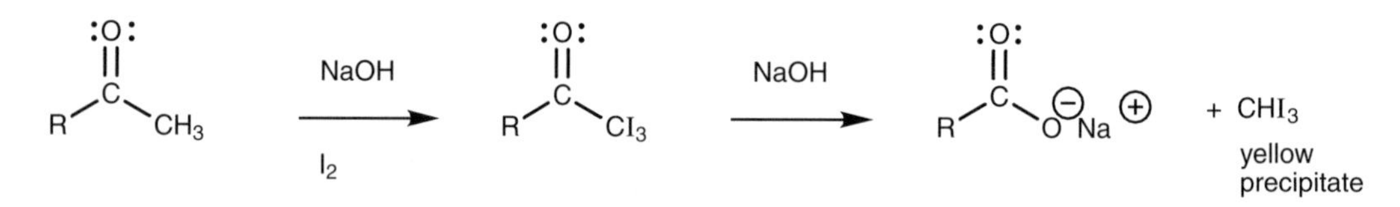

Fig. 4. The iodoform reaction.

This tendency towards multiple halogenation is the basis for a classical test called the iodoform test which is used to identify methyl ketones. The ketone to be tested is treated with excess iodine and base and if a yellow precipitate is formed, a positive result is indicated. Under these conditions, methyl ketones undergo α-halogenation three times (*Fig. 4*). The product obtained is then susceptible to nucleophilic substitution (Topic K2) whereby the hydroxide ion substitutes the tri-iodomethyl ($^{-}CI_3$) carbanion – a good leaving group due to the three electron-withdrawing iodine atoms. Tri-iodomethane is then formed as the yellow precipitate.

J10 Reduction and oxidation

Key Notes

Reduction to alcohols

Reduction of an aldehyde with sodium borohydride or lithium aluminum hydride gives a primary alcohol. Similar reduction of a ketone gives a secondary alcohol.

Reduction to alkanes

There are three methods of deoxygenating aldehydes and ketones. The method used depends on whether the compound is sensitive to acid or base. If sensitive to acid, reduction is carried out under basic conditions by the Wolff–Kishner reduction. If sensitive to base, the reaction is carried out under acid conditions – the Clemmenson reduction. If sensitive to both acid and base, the carbonyl group is converted to a dithioacetal or dithioketal then reduced with Raney nickel.

Oxidation

Aldehydes can be oxidized to carboxylic acids, but ketones are resistant to oxidation.

Related topics

Nucleophilic addition – charged nucleophiles (J4)
Nucleophilic addition – nitrogen nucleophiles (J6)
Nucleophilic addition – oxygen and sulfur nucleophiles (J7)

Reduction to alcohols

Aldehydes and ketones can be reduced to alcohols with a hydride ion – provided by reducing reagents such as sodium borohydride or lithium borohydride (Topic J4). Primary alcohols are obtained from aldehydes and secondary alcohols from ketones.

Reduction to alkanes

Aldehydes and ketones can be reduced to alkanes by three different methods which are complementary to each other. The **Wolff–Kishner** reduction is carried out under basic conditions and is suitable for compounds that might be sensitive to acid (*Fig. 1*). The reaction involves the nucleophilic addition of hydrazine followed by elimination of water to form a hydrazone. The mechanism is the same as that described for the synthesis of 2,4-dinitrophenylhydrazones (Topic J6).

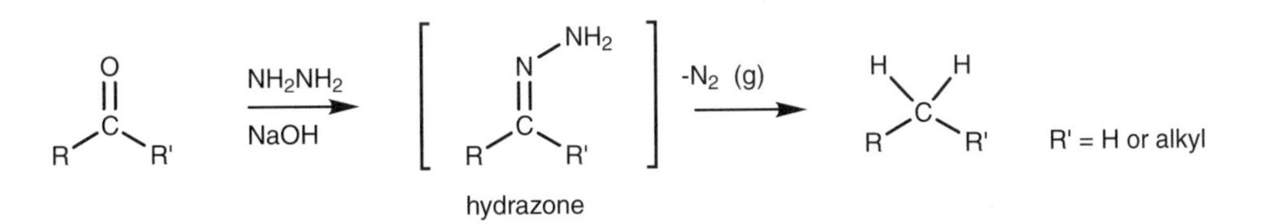

Fig. 1. Wolff–Kishner reduction.

However, the simple hydrazone formed under these reaction conditions spontaneously decomposes with the loss of nitrogen gas.

The **Clemmenson** reduction (*Fig. 2*) gives a similar product but is carried out under acid conditions and so this is a suitable method for compounds which are unstable to basic conditions.

Compounds which are sensitive to both acid and base can be reduced under neutral conditions by forming the thioacetal or thioketal (Topic J7), then reducing with Raney nickel (*Fig. 3*).

Aromatic aldehydes and ketones can also be deoxygenated with hydrogen over a palladium charcoal catalyst. The reaction takes place because the aromatic ring activates the carbonyl group towards reduction. Aliphatic aldehydes and ketones are not reduced.

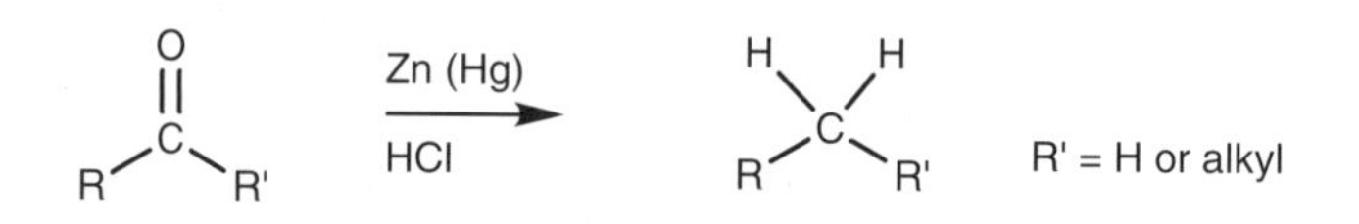

Fig. 2. Clemmenson reduction.

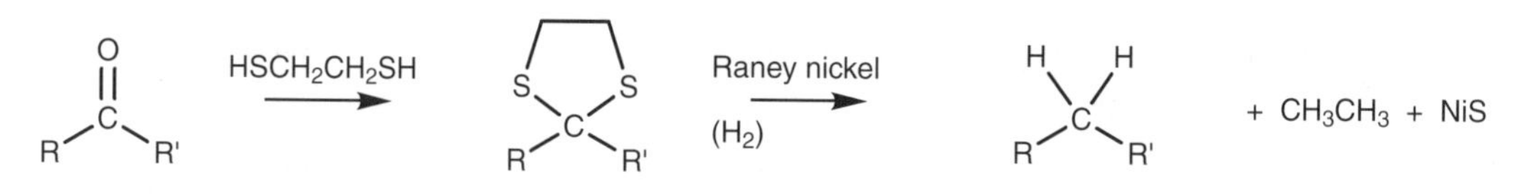

Fig. 3. Reduction via a cyclic thioketal.

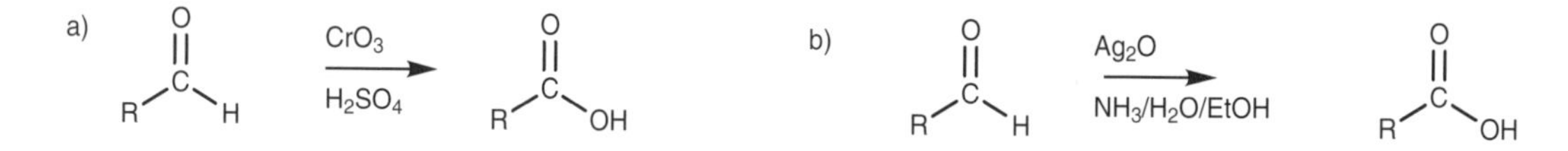

Fig. 4. (a) Oxidation of an aldehyde to form a carboxylic acid (b). Oxidation of an aldehyde using silver oxide.

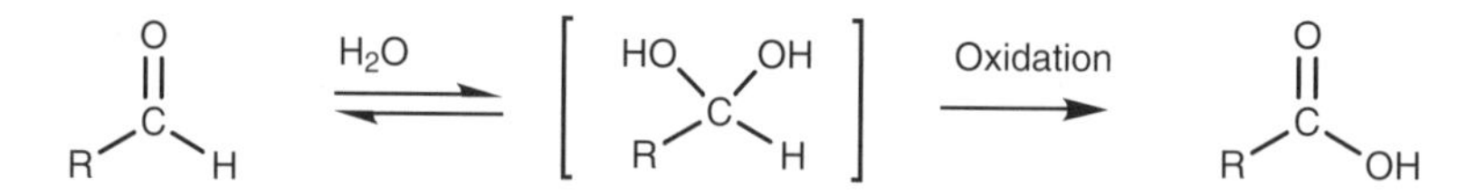

Fig. 5. 1,1-Diol intermediate.

Oxidation

Ketones are resistant to oxidation whereas aldehydes are easily oxidized. Treatment of an aldehyde with an oxidizing agent results in the formation of a carboxylic acid (*Fig. 4a*). Some compounds may be sensitive to the acid conditions used in this reaction and an alternative way of carrying out the oxidation is to use a basic solution of silver oxide (*Fig. 4b*).

Both reactions involve the nucleophilic addition of water to form a 1,1-diol or hydrate which is then oxidized in the same way as an alcohol (*Fig. 5*).

J11 α,β-UNSATURATED ALDEHYDES AND KETONES

Key Notes

Definition

α,β-Unsaturated aldehydes and ketones are aldehydes and ketones which are conjugated with a double bond.

Nucleophilic and electrophilic centers

The carbonyl oxygen of an α,β-unsaturated aldehyde or ketone is a nucleophilic center. The carbonyl carbon and the β-carbon are electrophilic centers. Nucleophilic addition can take place at either the carbonyl carbon (1,2-addition), or the β-carbon (1,4- or conjugate addition).

1,2-Addition

1,2-Addition to α,β-unsaturated aldehydes and ketones takes place with Grignard reagents and organolithium reagents.

1,4-Addition

1,4-Addition to α,β-unsaturated aldehydes and ketones takes place with organocuprate reagents, amines and the cyanide ion.

Reduction

α,β-unsaturated ketones are reduced to allylic alcohols with lithium aluminum hydride.

Related topics

sp^2 Hybridization (A4)
Conjugated dienes (H11)
Nucleophilic addition (J3)
Nucleophilic addition – charged nucleophiles (J4)

Definition

α,β-Unsaturated aldehydes and ketones are aldehydes and ketones which are conjugated with a double bond. The α-position is defined as the carbon atom next to the carbonyl group, while the β-position is the carbon atom two bonds removed (*Fig. 1*).

Nucleophilic and electrophilic centers

The carbonyl group of α,β-unsaturated aldehydes and ketones consists of a nucleophilic oxygen and an electrophilic carbon. However, α,β-unsaturated aldehydes and ketones also have another electrophilic carbon – the β-carbon. This is due to the influence of the electronegative oxygen which can result in the resonance shown (*Fig. 2*). Since two electrophilic centers are present, there are two

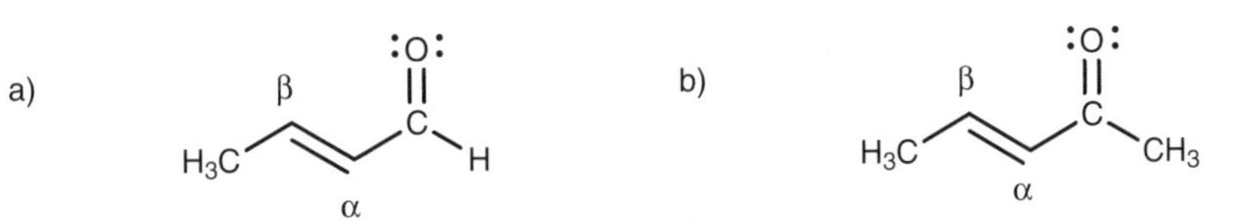

Fig. 1. (a) α,β-unsaturated aldehyde; (b) α,β-unsaturated ketone.

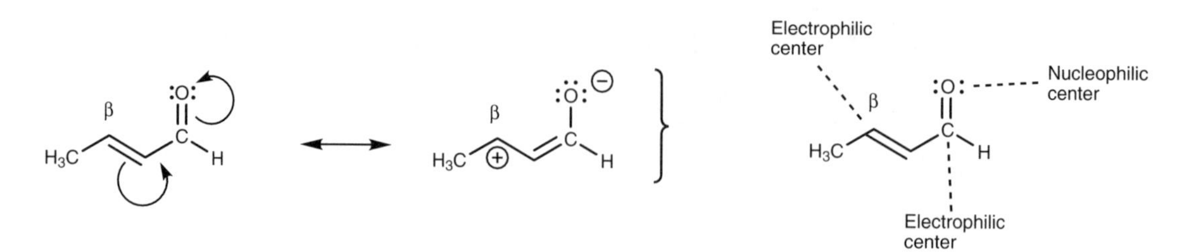

Fig. 2. Nucleophilic and electrophilic centers.

places where a nucleophile can react. In both situations, an addition reaction takes place. If the nucleophile reacts with the carbonyl carbon, this is a normal nucleophilic addition to an aldehyde or ketone and is called a **1,2-nucleophilic addition**. If the nucleophile adds to the β-carbon, this is known as a **1,4-nucleophilic addition** or a **conjugate addition**.

1,2-Addition

The mechanism of 1,2-nucleophilic addition is the same mechanism already described in Topics J3 and J4. It is found that Grignard reagents and organolithium reagents will react with α,β-unsaturated aldehydes and ketones in this way and do not attack the β-position (*Fig. 3*).

1,4-Addition

The mechanism for 1,4-addition involves two stages (*Fig. 4*). In the first stage, the nucleophile uses a lone pair of electrons to form a bond to the β-carbon. At the same time, the C=C π bond breaks and both electrons are used to form a new π bond to the carbonyl carbon. This in turn forces the carbonyl π bond to break with both of the electrons involved moving onto the oxygen as a third lone pair of electrons. The resulting intermediate is an enolate ion. Aqueous acid is now added to the reaction mixture. The carbonyl π bond is reformed, which forces open the C=C π bond. These electrons are now used to form a σ bond to a proton at the α carbon.

Conjugate addition reactions can be carried out with amines, or a cyanide ion.

Alkyl groups can also be added to the β-position by using organocuprate

Fig. 3. 1,2-Nucleophilic addition.

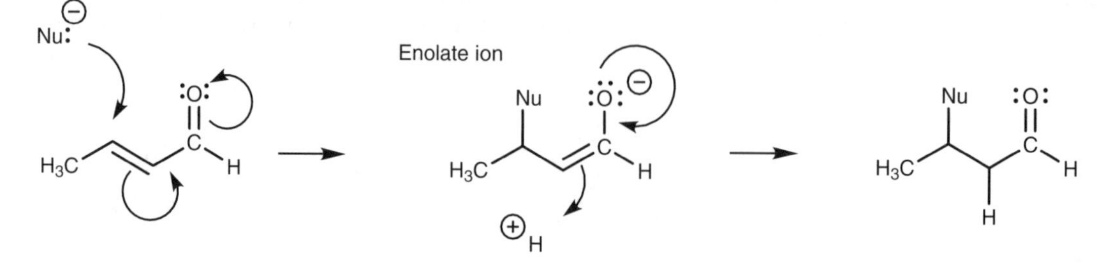

Fig. 4. Mechanism of 1,4-nucleophilic addition.

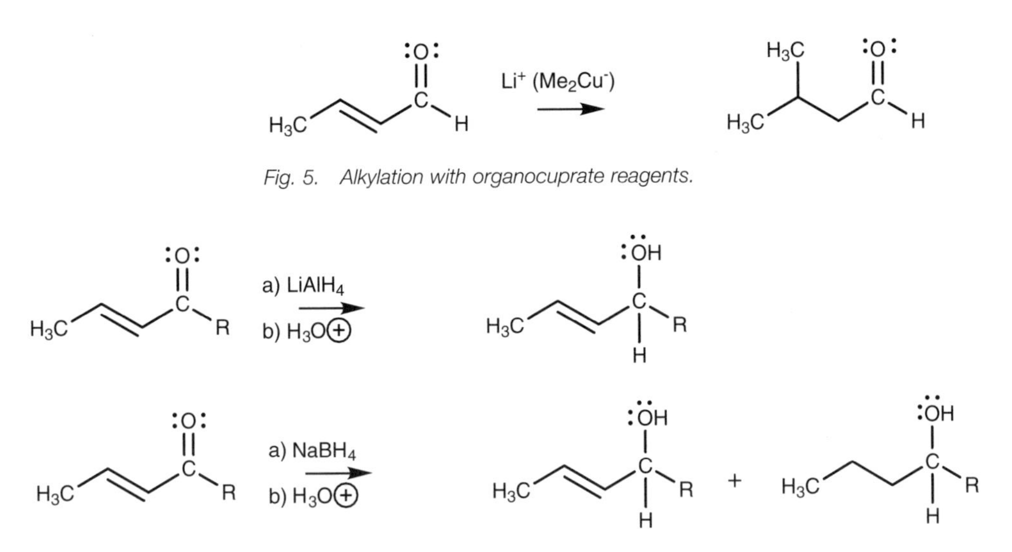

Fig. 5. Alkylation with organocuprate reagents.

Fig. 6. Reduction of α,β-unsaturated ketones.

reagents (Topic L7; *Fig. 5*). A large variety of organocuprate reagents can be prepared allowing the addition of primary, secondary and tertiary alkyl groups, aryl groups, and alkenyl groups.

Reduction

The reduction of α,β-unsaturated ketones to allylic alcohols is best carried out using lithium aluminum hydride under carefully controlled conditions (*Fig. 6*). With sodium borohydride, some reduction of the alkene also takes place.

K1 Structure and properties

Key Notes

Structure

There are four common carboxylic acid derivaties derived from a parent carboxylic acid – acid chlorides, acid anhydrides, esters, and amides. Carboxylic acids and their derivatives contain an sp^2 hybridized carbonyl group linked to a group Y where the atom directly attached to the carbonyl group is a heteroatom (Cl, O or N). The functional groups are planar with bond angles of 120°.

Bonding

The carbonyl group is made up of a strong σ bond and a weaker π bond. The carbonyl group is polarized such that the oxygen acts as a nucleophilic center and the carbon acts as an electrophilic center.

Properties

Carboxylic acids are polar and can take part in hydrogen bonding. They are soluble in water and have high boiling points. Carboxylic acids are weak acids in aqueous solution and form water soluble salts when treated with a base. Primary and secondary amides participate in hydrogen bonding and have higher boiling points than comparable aldehydes or ketones. Acid chlorides, acid anhydrides, esters, and tertiary amides are polar but are not capable of hydrogen bonding. Their boiling points are similar to aldehydes and ketones of similar molecular weight.

Reactions

Carboxylic acids and acid derivatives undergo nucleophilic substitutions.

Related topics

sp^2 Hybridization (A4)
Intermolecular bonding (C3)
Organic structures (E4)
Acid strength (G2)
Nucleophilic addition (J3)

Structure

Carboxylic acid derivatives are structures derived from a parent carboxylic acid structure. There are four common types of acid derivative – acid chlorides, acid anhydrides, esters, and amides (*Fig. 1*). These functional groups contain a carbonyl group (C=O) where both atoms are sp^2 hybridized (*Fig. 2*). The carbonyl group along with the two neighboring atoms is planar with bond angles of 120°. The carbonyl group along with the attached carbon chain is called an **acyl** group. Carboxylic acids and carboxylic acid derivatives differ in what is attached to the acyl group (i.e. Y = Cl, OCOR, OR, NR_2, or OH). Note that in all these cases, the atom in Y which is directly attached to the carbonyl group is a heteroatom (Cl, O, or N). This distinguishes carboxylic acids and their derivatives from aldehydes and ketones where the corresponding atom is hydrogen or carbon. This is important with respect to the sort of reactions which carboxylic acids and their derivatives undergo. The carboxylic acid group (COOH) is often referred to as a **carboxyl** group.

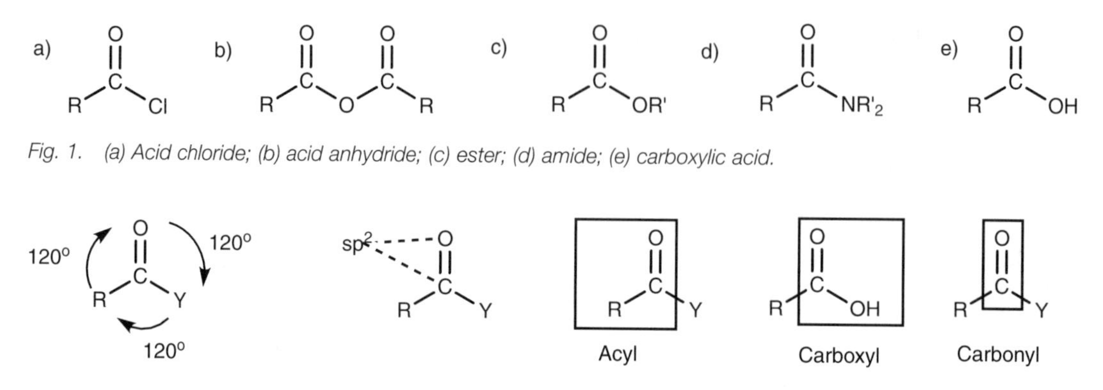

Fig. 1. *(a) Acid chloride; (b) acid anhydride; (c) ester; (d) amide; (e) carboxylic acid.*

Fig. 2. *Structure of the functional group.*

Bonding

The bonds in the carbonyl C=O group are made up of a strong σ bond and a weaker π bond (*Fig. 3*). Since oxygen is more electronegative than carbon, the carbonyl group is polarized such that the oxygen is slightly negative and the carbon is slightly positive. This means that oxygen can act as a nucleophilic center and carbon can act as an electrophilic center.

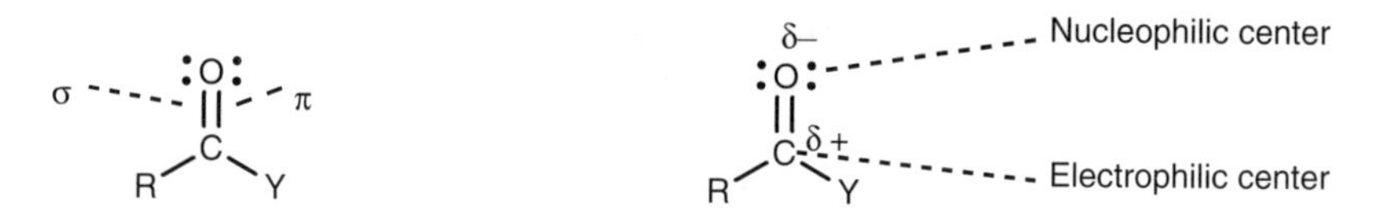

Fig. 3. *Bonding and properties.*

Properties

Carboxylic acids and their derivatives are polar molecules due to the polar carbonyl group and the presence of a heteroatom in the group Y. Carboxylic acids can associate with each other as dimers (*Fig. 4*) through the formation of two intermolecular hydrogen bonds which means that carboxylic acids have higher boiling points than alcohols of comparable molecular weight. It also means that low molecular weight carboxylic acids are soluble in water. However, as the molecular weight of the carboxylic acid increases, the hydrophobic character of the alkyl portion eventually outweighs the polar character of the functional group such that higher molecular weight carboxylic acids are insoluble in water.

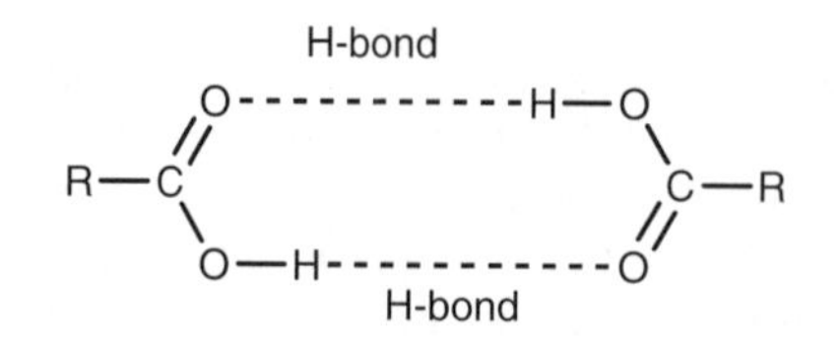

Fig. 4. *Intermolecular H-bonding.*

Primary amides and secondary amides also have a hydrogen capable of hydrogen bonding (i.e. RCON<u>H</u>R′, RCON<u>H</u>$_2$), resulting in higher boiling points for these compounds compared to aldehydes and ketones of similar molecular weight. Acid chlorides, acid anhydrides, esters, and tertiary amides are polar, but lack a hydrogen atom capable of participating in hydrogen bonding. As a result, they have lower boiling points than carboxylic acids or alcohols of similar molecular weight, and similar boiling points to comparable aldehydes and ketones.

Carboxylic acids are weak acids in aqueous solution (Topic G2), forming an equilibrium between the free acid and the carboxylate ion. In the presence of a base such as sodium hydroxide or sodium hydrogen carbonate, they ionize to form water-soluble salts and this provides a method of separating carboxylic acids from other organic compounds.

Reactions

Carboxylic acids and carboxylic acid derivatives commonly react with nucleophiles in a reaction known as **nucleophilic substitution** (*Fig. 5*). The reaction involves replacement of one nucleophile with another. Nucleophilic substitution is possible because the displaced nucleophile contains an electronegative heteroatom (Cl, O, or N) which is capable of stabilizing a negative charge.

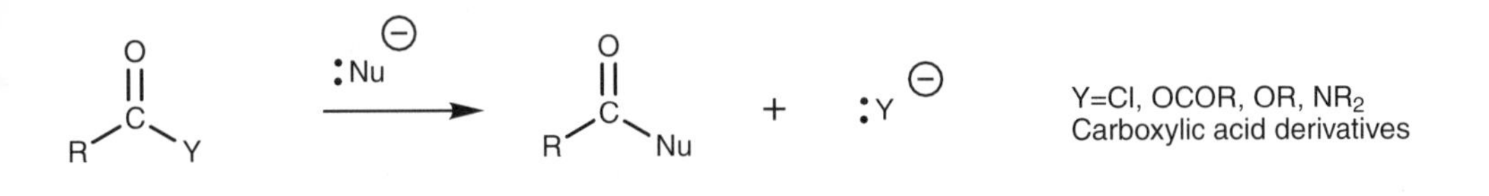

Fig. 5. Nucleophilic substitution.

K2 NUCLEOPHILIC SUBSTITUTION

Key Notes

Definition

Nucleophilic substitutions are reactions which involve the substitution of one nucleophile for another nucleophile. Alkyl halides, carboxylic acids, and carboxylic acid derivatives undergo nucleophilic substitution, but the mechanisms for alkyl halides are quite different from those of carboxylic acids and carboxylic acid derivatives.

Mechanism – charged nucleophiles

There are two steps in the nucleophilic substitution of a carboxylic acid derivative with a charged nucleophile. The first step is the same as the first step of nucleophilic addition to aldehydes and ketones. The second step involves reformation of the carbonyl group and expulsion of the leaving group.

Mechanism – neutral nucleophiles

The mechanism of nucleophilic substitution with neutral nucleophiles is the same as the mechanism for charged nucleophiles, except that an extra step is required in order to remove a proton.

Addition vs substitution

Aldehydes and ketones undergo nucleophilic addition rather than nucleophilic substitution since the latter mechanism would require the cleavage of a strong C–H or C–C bond with the generation of a highly reactive hydride ion or carbanion.

Related topics

Organic structures (E4)
Base strength (G3)
Nucleophilic addition (J3)
Preparation of carboxylic acid derivatives (K5)
Reactions (K6)
Nucleophilic substitution (L2)

Definition

Nucleophilic substitutions are reactions which involve the substitution of one nucleophile for another. Alkyl halides, carboxylic acids, and carboxylic acid derivatives undergo nucleophilic substitution. However, the mechanisms involved for alkyl halides (Topic L2) are quite different from those involved for carboxylic acids and their derivatives. The reaction of a methoxide ion with ethanoyl chloride is an example of nucleophilic substitution (*Fig. 1*), where one nucleophile (the methoxide ion) substitutes another nucleophile (Cl^-).

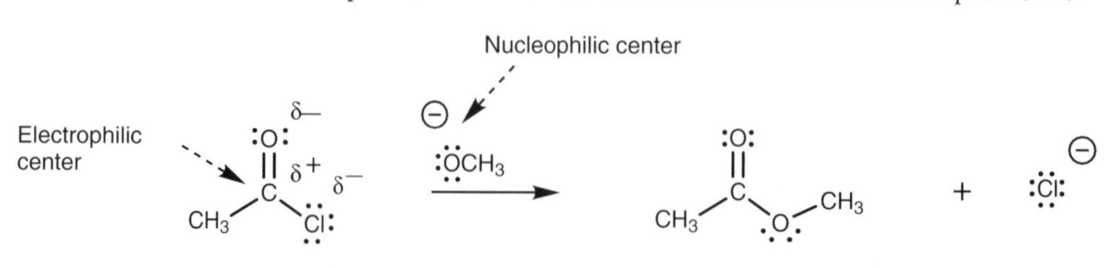

Fig. 1. Nucleophilic substitution.

Mechanism – charged nucleophiles

We shall use the reaction in *Fig. 1* to illustrate the mechanism of nucleophilic substitution (*Fig. 2*). The methoxide ion uses one of its lone pairs of electrons to form a bond to the electrophilic carbonyl carbon of the acid chloride. At the same time, the relatively weak π bond of the carbonyl group breaks and both of the π electrons move onto the carbonyl oxygen to give it a third lone pair of electrons and a negative charge. This is exactly the same first step involved in nucleophilic addition to aldehydes and ketones. However, with an aldehyde or a ketone, the tetrahedral structure is the final product. With carboxylic acid derivatives, the lone pair of electrons on oxygen return to reform the carbonyl π bond (Step 2). As this happens, the C–Cl bond breaks with both electrons moving onto the chlorine to form a chloride ion which departs the molecule. This explains how the products are formed, but why should the C–Cl σ bond break in preference to the C–OMe σ bond or the C–CH_3 σ bond? The best explanation for this involves looking at the leaving groups which would be formed from these processes (*Fig. 3*). The leaving groups would be a chloride ion, a methoxide ion and a carbanion, respectively. The chloride ion is the best leaving group because it is the most stable. This is because chlorine is more electronegative than oxygen or carbon and can stabilize the negative charge. This same mechanism is involved in the nucleophilic substitutions of all the other carboxylic acid derivatives and a general mechanism can be drawn (*Fig. 4*).

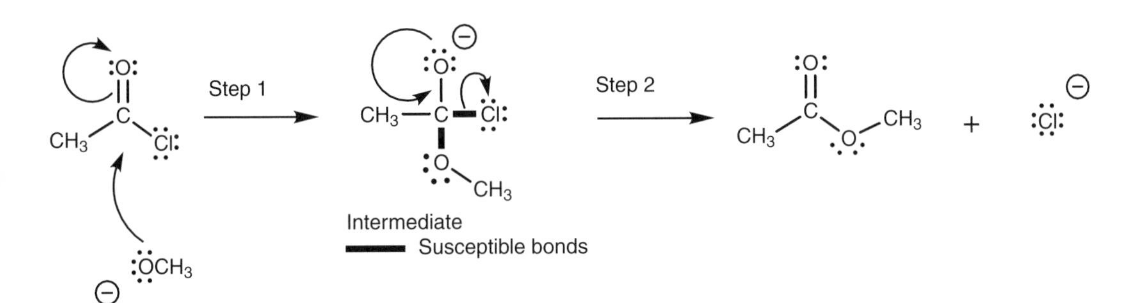

Fig. 2. Mechanism of the nucleophilic substitution.

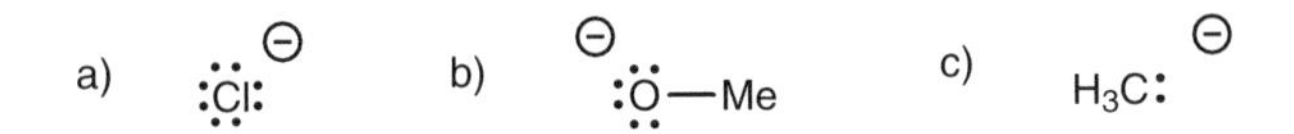

Fig. 3. Leaving groups; (a) chloride; (b) methoxide; (c) carbanion.

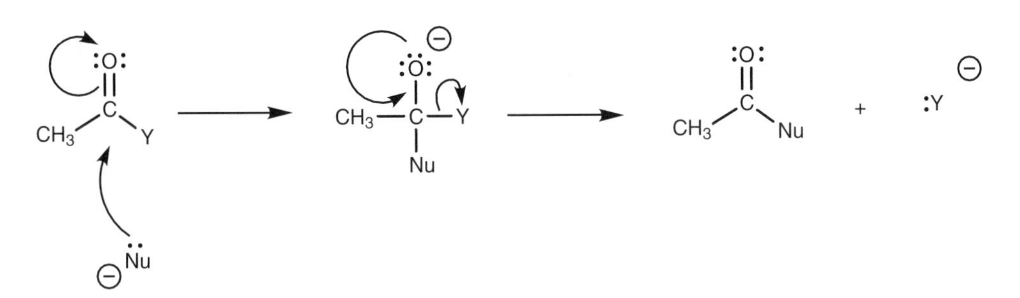

Fig. 4. General mechanism for nucleophilic substitution.

Mechanism – neutral nucleophiles

Acid chlorides are sufficiently reactive to react with uncharged nucleophiles. For example, ethanoyl chloride will react with methanol to give an ester (*Fig. 5*). Oxygen is the nucleophilic center in methanol and uses one of its lone pairs of electrons to form a new bond to the electrophilic carbon of the acid chloride (*Fig. 6*). As this new bond forms, the carbonyl π bond breaks and both electrons move onto the carbonyl oxygen to give it a third lone pair of electrons and a negative charge (Step 1). Note that the methanol oxygen gains a positive charge since it has effectively lost an electron by sharing its lone pair with carbon in the new bond. A positive charge on oxygen is not very stable and so the second stage in the mechanism is the loss of a proton. Both electrons in the O–H bond move onto the oxygen to restore a second lone pair of electrons and thus neutralize the charge. Methanol can aid the process by acting as a base. The final stage in the mechanism is the same as before. The carbonyl π bond is reformed and as this happens, the C–Cl σ bond breaks with both electrons ending up on the departing chloride ion as a fourth lone pair of electrons. Finally, the chloride anion can remove a proton from $CH_3OH_2^+$ to form HCl and methanol (not shown).

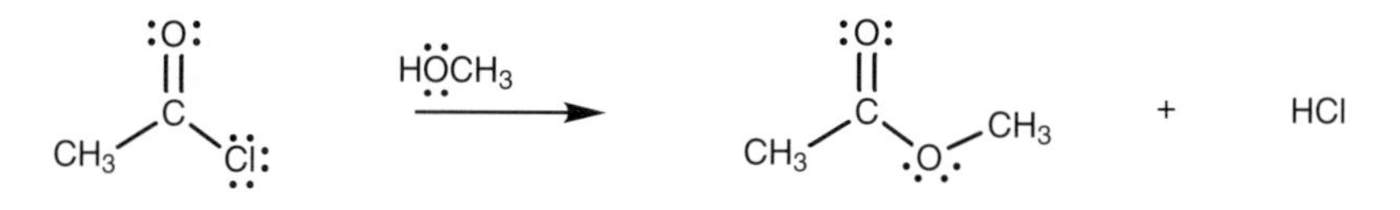

Fig. 5. Ethanoyl chloride reacting with methanol to form methyl ethanoate.

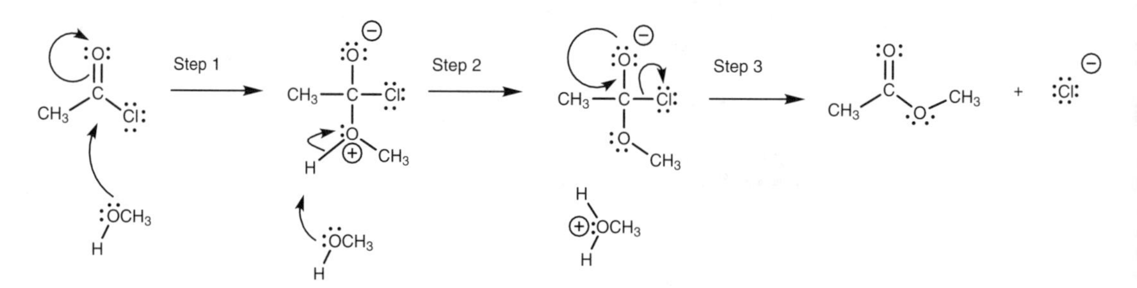

Fig. 6. Mechanism for the reaction of an alcohol with an acid chloride.

The above mechanism is essentially the same mechanism involved in the reaction of ethanoyl chloride with sodium methoxide, the only difference being that we have to remove a proton during the reaction mechanism.

The same mechanism holds true for nucleophilic substitutions of other carboxylic acid derivatives with neutral nucleophiles and we can write a general mechanism (*Fig. 7*). In practice, acids or bases are often added to improve yields. Specific examples are described in Topic K5.

Addition vs substitution

Carboxylic acids and carboxylic acid derivatives undergo nucleophilic substitution whereas aldehydes and ketones undergo nucleophilic addition. This is because nucleophilic substitution of a ketone or an aldehyde would generate a carbanion or a hydride ion respectively (*Fig. 8*). These ions are unstable and highly reactive, so they are only formed with difficulty. Furthermore, C–C and C–H σ bonds are not easily broken. Therefore, nucleophilic substitutions of aldehydes or ketones are not feasible.

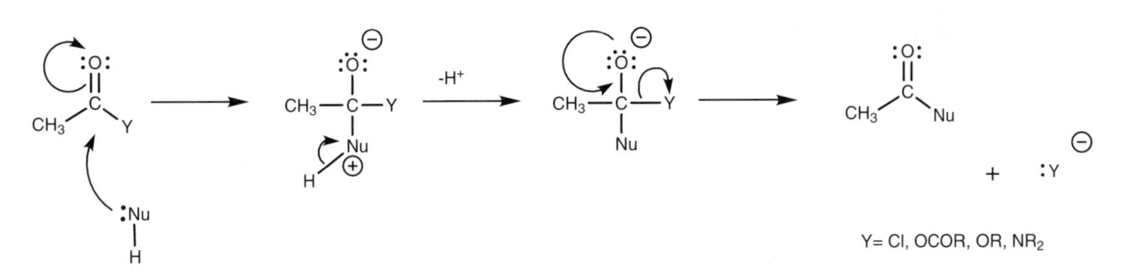

Fig. 7. General mechanism for the nucleophilic substitution of a neutral nucleophile with a carboxylic acid derivative.

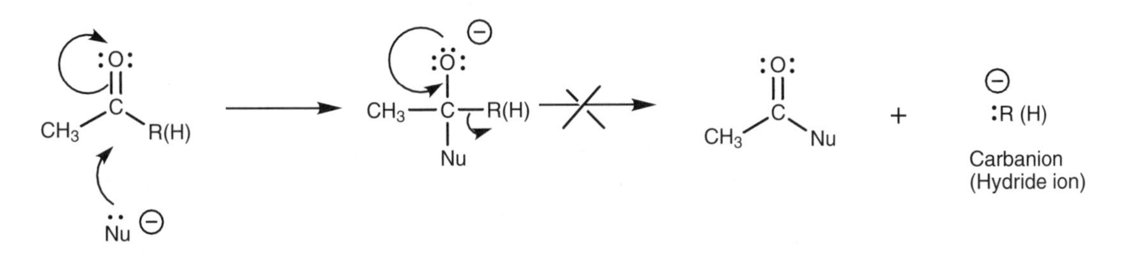

Fig. 8. Unfavorable formation of an unstable carbanion or hydride ion.

K3 REACTIVITY

Key Notes

Reactivity order

Acid chlorides are more reactive than acid anhydrides towards nucleophilic substitution. Acid anhydrides are more reactive than esters, and esters are more reactive than amides. It is possible to convert a reactive acid derivative to a less reactive acid derivative, but not the other way round.

Electronic factors

The relative reactivity of the four different acid derivatives is determined by the relative electrophilicities of the carbonyl carbon atom. Neighboring electronegative atoms increase the electrophilicity of the carbonyl group through an inductive effect. The greater the electronegativity of the neighboring atom, the greater the effect. Chlorine is more electronegative than oxygen, and oxygen is more electronegative than nitrogen. Thus, acid chlorides are more reactive than acid anhydrides and esters, while amides are the least reactive of the acid derivatives. Resonance effects play a role in diminishing the electrophilic character of the carbonyl carbon. Neighboring atoms containing a lone pair of electrons can feed these electrons into the carbonyl center to form a resonance structure where the carbonyl π bond is broken. This resonance is significant in amines where nitrogen is a good nucleophile, but is insignificant in acid chlorides where chlorine is a poor nucleophile. Resonance involving oxygen is weak but significant enough to explain the difference in reactivity between acid anhydrides and esters. Since the resonance in acid anhydrides is split between two carbonyl groups, the decrease in reactivity is less significant than in esters.

Steric factors

Bulky groups attached to the carbonyl group can hinder the approach of nucleophiles and result in lowered reactivity. Bulky nucleophiles will also react more slowly.

Carboxylic acids

Carboxylic acids are more likely to undergo acid–base reactions with nucleophiles rather than nucleophilic substitution. Nucleophilic substitution requires prior activation of the carboxylic acid.

Related topics

Nucleophilic substitution (K2)
Preparations of carboxylic acid derivatives (K5)
Reactions (K6)

Reactivity order

Acid chlorides can be converted to acid anhydrides, esters, or amides (Topic K5). These reactions are possible because acid chlorides are the most reactive of the four carboxylic acid derivatives. Nucleophilic substitutions of the other acid derivatives are more limited because they are less reactive. For example, acid anhydrides can be used to synthesize esters and amides, but cannot be used to synthesize acid chlorides. The possible nucleophilic reactions for each carboxylic acid derivative depends on its reactivity with respect to the other acid derivatives (*Fig. 1*). Reactive acid derivatives can be converted to less reactive (more stable) acid derivatives, but not the other way round. For example, an ester can be converted to an amide, but not to an acid anhydride.

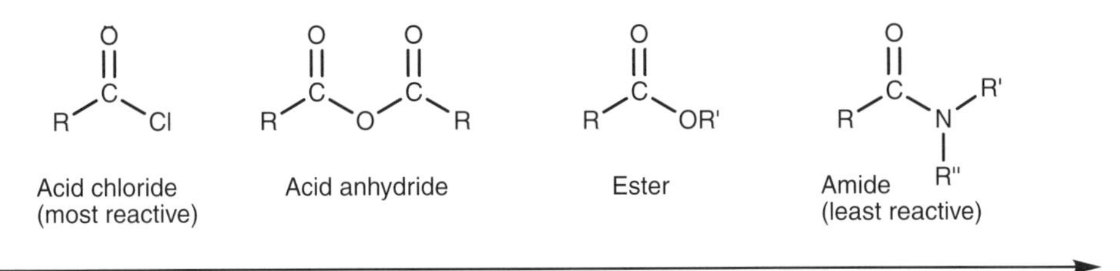

Fig. 1. Relative reactivity of carboxylic acid derivatives.

Electronic factors

But why is there this difference in reactivity? The first step in the nucleophilic substitution mechanism (involving the addition of a nucleophile to the electrophilic carbonyl carbon) is the rate-determining step. Therefore, the more electrophilic this carbon is, the more reactive it will be. The nature of Y has a significant effect in this respect (*Fig. 2*).

Y is linked to the acyl group by an electronegative heteroatom (Cl, O, or N) which makes the carbonyl carbon more electrophilic. The extent to which this happens depends on the electronegativity of Y. If Y is strongly electronegative (e.g. chlorine), it has a strong electron-withdrawing effect on the carbonyl carbon making it more electrophilic and more reactive to nucleophiles. Since chlorine is more electronegative than oxygen, and oxygen is more electronegative than nitrogen, acid chlorides are more reactive than acid anhydrides and esters, while acid anhydrides and esters are more reactive than amides.

The electron-withdrawing effect of Y on the carbonyl carbon is an inductive effect. With amides, there is an important resonance contribution which **decreases** the electrophilicity of the carbonyl carbon (*Fig. 3*). The nitrogen has a lone pair of electrons which can form a bond to the neighboring carbonyl carbon. As this new bond is formed, the weak π bond breaks and both electrons move onto oxygen to give it a third lone pair of electrons and a negative charge. Since the nitrogen's lone pair of electrons is being fed into the carbonyl group, the carbonyl carbon becomes less electrophilic and is less prone to attack by an incoming nucleophile.

In theory, this resonance could also occur in acid chlorides, acid anhydrides, and esters to give resonance structures (*Fig. 4*). However, the process is much less important since oxygen and chlorine are less nucleophilic than nitrogen. In these structures, the positive charge ends up on an oxygen or a chlorine atom. These atoms are more electronegative than nitrogen and less able to stabilize a positive charge. These resonance structures might occur to a small extent with esters and acid anhydrides, but are far less likely in acid chlorides. This trend also matches the trend in reactivity.

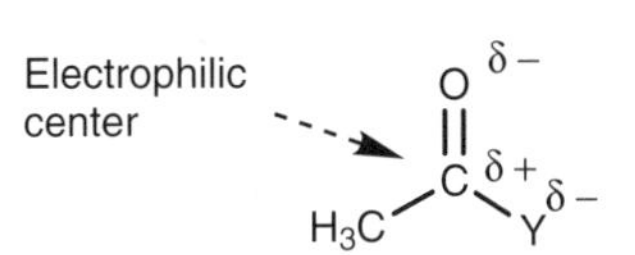

Y= Cl, OCOR, OR, NR_2

Fig. 2. The electrophilic center of a carboxylic acid derivative.

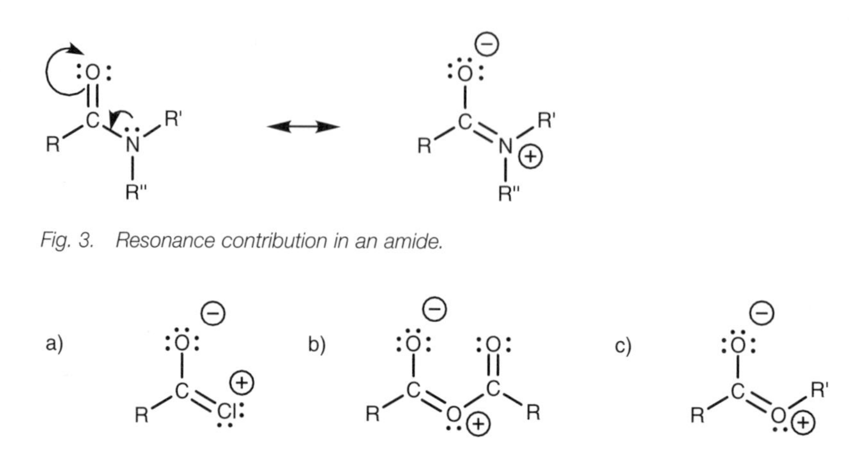

Fig. 3. Resonance contribution in an amide.

Fig. 4. Resonance structures for (a) an acid chloride; (b) an acid anhydride; (c) an ester.

Although the resonance effect is weak in esters and acid anhydrides, it can explain why acid anhydrides are more reactive than esters. Acid anhydrides have two carbonyl groups and so resonance can take place with either carbonyl group (*Fig. 5*). As a result, the lone pair of the central oxygen is 'split' between both groups which means that the resonance effect is split between both carbonyl groups. This means that the effect of resonance at any one carbonyl group is diminished and it will remain strongly electrophilic. With an ester, there is only one carbonyl group and so it experiences the full impact of the resonance effect. Therefore, its electrophilic strength will be diminished relative to an acid anhydride.

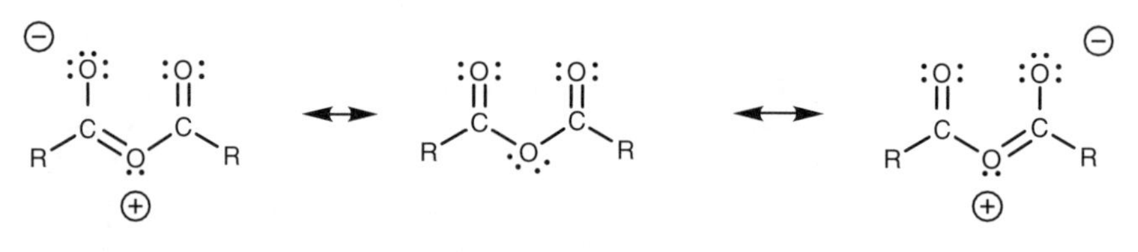

Fig. 5. Resonance structures for an acid anhydride.

Steric factors

Steric factors can play a part in the reactivity of acid derivatives. For example, a bulky group attached to the carbonyl group can hinder the approach of nucleophiles and hence lower reactivity. The steric bulk of the nucleophile can also have an influence in slowing down the reaction. For example, acid chlorides react faster with primary alcohols than they do with secondary or tertiary alcohols. This allows selective esterification if a molecule has more than one alcohol group present (*Fig. 6*).

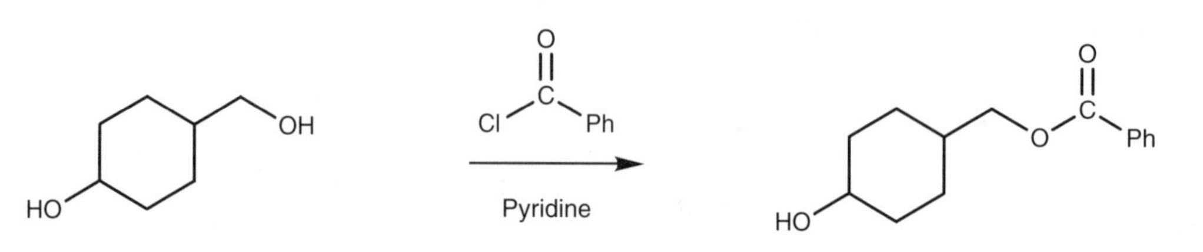

Fig. 6. Selective esterification of a primary alcohol.

Carboxylic acids

Where do carboxylic acids fit into the reactivity order described above? The nucleophilic substitution of carboxylic acids is complicated by the fact that an acidic proton is present. Since most nucleophiles can act as bases, the reaction of a carboxylic acid with a nucleophile results in an acid–base reaction rather than nucleophilic substitution.

However, carboxylic acids can undergo nucleophilic substitution if they are activated in advance (Topic K5).

K4 PREPARATIONS OF CARBOXYLIC ACIDS

Key Notes

Functional group transformations

Primary alcohols and aldehydes are converted to carboxylic acids by oxidation. Acid chlorides, acid anhydrides, esters, and amides can be hydrolyzed to their parent carboxylic acids, but only the hydrolysis of esters serves a useful synthetic role if the ester is being used as a protecting group.

C–C bond formation

Aromatic carboxylic acids are obtained by the oxidation of alkyl benzenes. Alkyl halides can be converted to carboxylic acids where the carbon chain has been extended by one carbon unit. Two methods are possible. The alkyl halide can be converted to a cyanide which is then hydrolyzed. Alternatively, the alkyl halide can be converted to a Grignard reagent then treated with carbon dioxide. A range of carboxylic acids can be prepared by alkylating diethyl malonate, then hydrolyzing and decarboxylating the product.

Bond cleavage

Alkenes can be cleaved across the double bond by potassium permanganate. Carboxylic acids are formed if a vinylic proton is present.

Related topics

Oxidation and reduction (I7)
Nucleophilic addition – charged nucleophiles (J4)
Reduction and oxidation (J10)
Reactions (K6)
Enolate reactions (K7)
Chemistry of nitriles (O4)

Functional group transformations

Carboxylic acids can be obtained by the oxidation of primary alcohols or aldehydes (Topic J10), the hydrolysis of nitriles (Topic O4), or the hydrolysis of esters (Topic K6) which can be used as protecting groups for carboxylic acids. Amides can also be hydrolyzed to carboxylic acids. However, fiercer reaction conditions are required due to the lower reactivity of amides and so amides are less useful as carboxylic acid protecting groups.

Although acid chlorides and anhydrides are easily hydrolyzed to carboxylic acids, the reaction serves no synthetic purpose since acid chlorides and acid anhydrides are synthesized from carboxylic acids in the first place. They are also too reactive to be used as protecting groups.

C–C bond formation

Aromatic carboxylic acids can be obtained by oxidation of alkyl benzenes (Topic I7). It does not matter how large the alkyl group is, since they are all oxidized to a benzoic acid structure.

There are two methods by which alkyl halides can be converted to a carboxylic acid and in both cases, the carbon chain is extended by one carbon. One method involves substituting the halogen with a cyanide ion (Topic L6) , then hydrolyzing the cyanide group (Topic O4; *Fig. 1a*). This works best with primary alkyl halides. The other method involves the formation of a Grignard reagent (Topic L7) which is then treated with carbon dioxide (*Fig. 1b*).

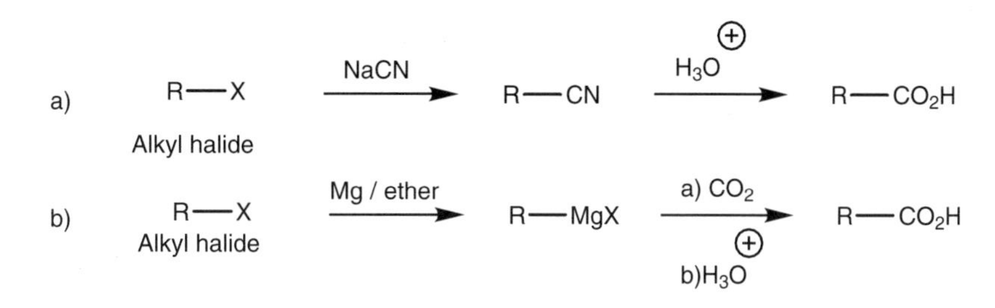

Fig. 1. Synthetic routes from an alkyl halide to a carboxylic acid.

The mechanism for the Grignard reaction is similar to the nucleophilic addition of a Grignard reagent to an aldehyde or ketone (Topic J4; *Fig. 2*).

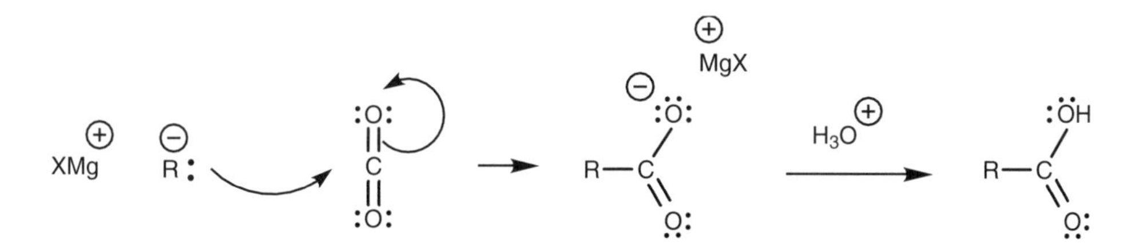

Fig. 2. Mechanism for the Grignard reaction with carbon dioxide.

A range of carboxylic acids can be prepared by alkylating diethyl malonate, then hydrolyzing and decarboxylating the product (Topic K7; *Fig. 3*).

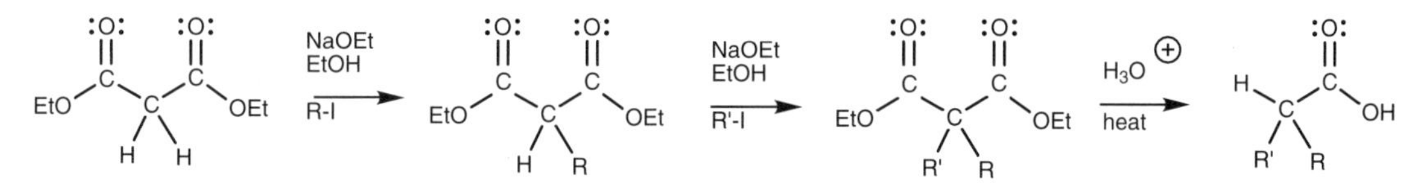

Fig. 3. Synthesis of carboxylic acids from diethyl malonate.

Bond cleavage

Alkenes can be cleaved with potassium permanganate to produce carboxylic acids (*Fig. 4*). A vinylic proton has to be present, that is a proton directly attached to the double bond.

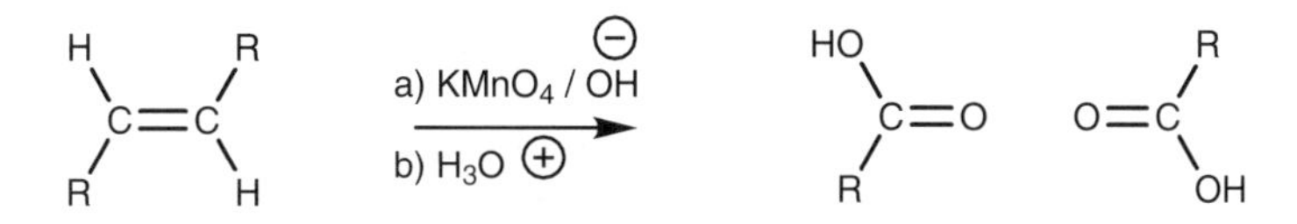

Fig. 4. Synthesis of carboxylic acids from alkenes.

K5 PREPARATIONS OF CARBOXYLIC ACID DERIVATIVES

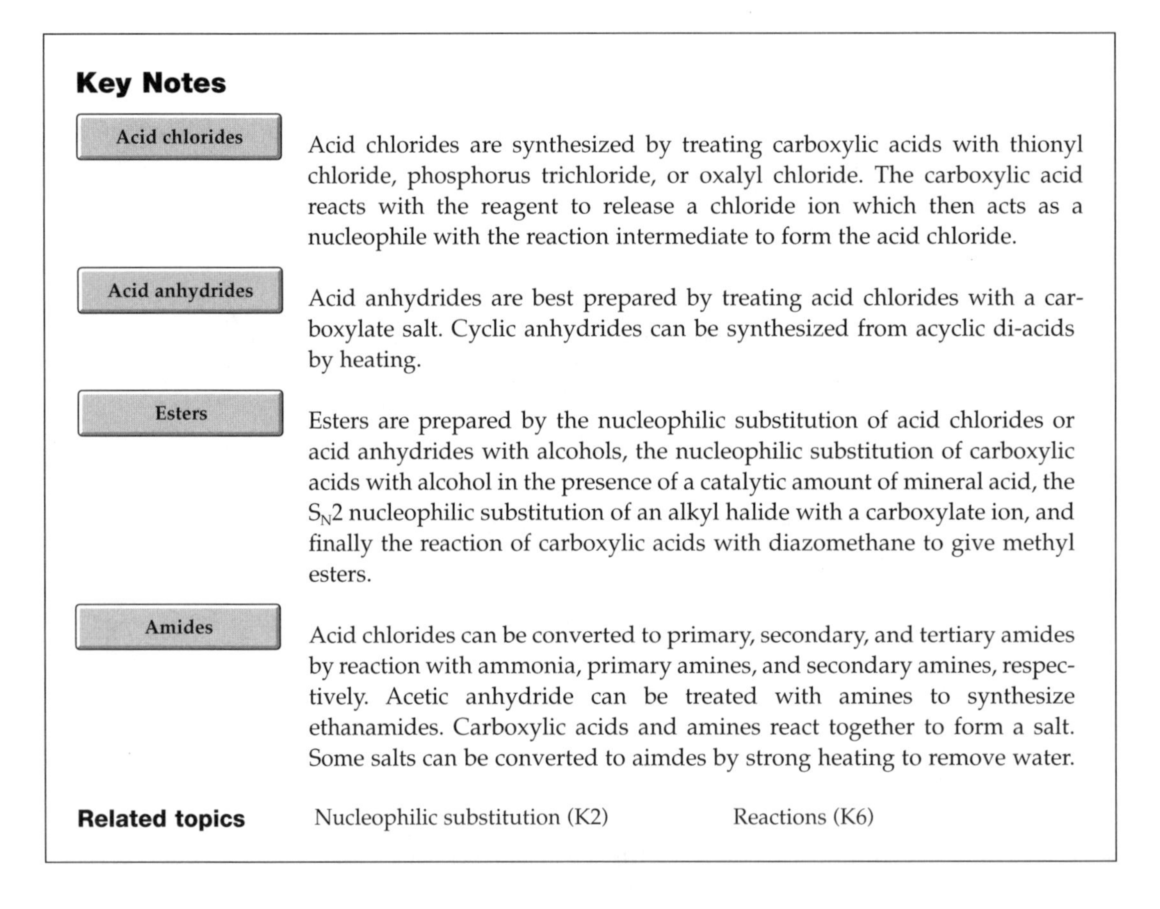

Key Notes

Acid chlorides

Acid chlorides are synthesized by treating carboxylic acids with thionyl chloride, phosphorus trichloride, or oxalyl chloride. The carboxylic acid reacts with the reagent to release a chloride ion which then acts as a nucleophile with the reaction intermediate to form the acid chloride.

Acid anhydrides

Acid anhydrides are best prepared by treating acid chlorides with a carboxylate salt. Cyclic anhydrides can be synthesized from acyclic di-acids by heating.

Esters

Esters are prepared by the nucleophilic substitution of acid chlorides or acid anhydrides with alcohols, the nucleophilic substitution of carboxylic acids with alcohol in the presence of a catalytic amount of mineral acid, the S_N2 nucleophilic substitution of an alkyl halide with a carboxylate ion, and finally the reaction of carboxylic acids with diazomethane to give methyl esters.

Amides

Acid chlorides can be converted to primary, secondary, and tertiary amides by reaction with ammonia, primary amines, and secondary amines, respectively. Acetic anhydride can be treated with amines to synthesize ethanamides. Carboxylic acids and amines react together to form a salt. Some salts can be converted to aimdes by strong heating to remove water.

Related topics Nucleophilic substitution (K2) Reactions (K6)

Acid chlorides

Acid chlorides can be prepared from carboxylic acids using thionyl chloride ($SOCl_2$), phosphorus trichloride (PCl_3), or oxalyl chloride (ClCOCOCl *Fig. 1*).

The mechanism for these reactions is quite involved, but in general involves the OH group of the carboxylic acid acting as a nucleophile to form a bond to the reagent and displacing a chloride ion. This has three important consequences. First of all, the chloride ion can attack the carbonyl group to introduce the required chlorine atom. Secondly, the acidic proton is no longer present and so an

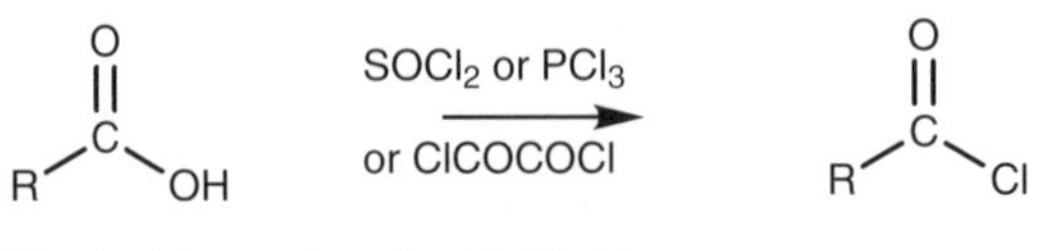

Fig. 1. Preparation of acid chlorides.

acid–base reaction is prevented. Thirdly, the original OH group is converted into a good leaving group and is easily displaced once the chloride ion makes its attack. The reaction of a carboxylic acid with thionyl chloride follows the general pathway shown in *Fig. 2*.

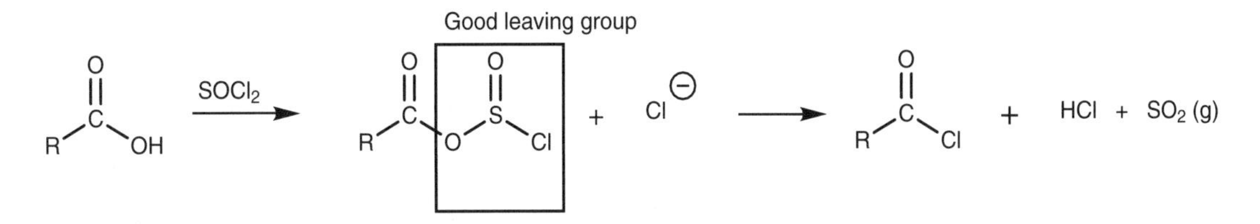

Fig. 2. Intermediate involved in the thionyl chloride reaction to form an acid chloride.

The leaving group (SO_2Cl) spontaneously fragments to produce hydrochloric acid and sulfur dioxide. The latter is lost as a gas which helps to drive the reaction to completion. The detailed mechanism is shown in *Fig. 3*.

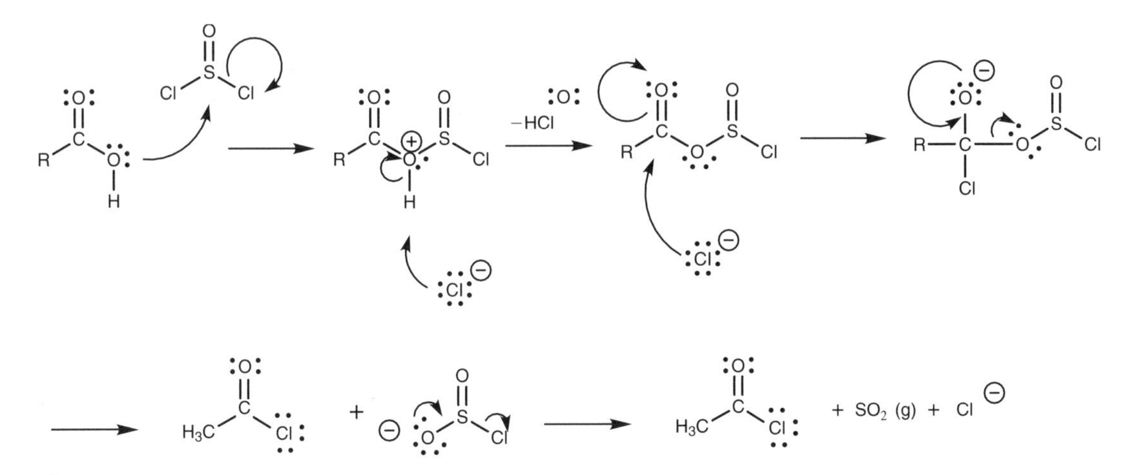

Fig. 3. Mechanism for the thionyl chloride reaction with a carboxylic acid to form an acid chloride.

Acid anhydrides

Acid anhydrides are best prepared by treating acid chlorides with a carboxylate salt (Topic K6). Carboxylic acids are not easily converted to acid anhydrides directly. However five-membered and six-membered cyclic anhydrides can be synthesized from diacids by heating the acyclic structures to eliminate water (*Fig. 4*).

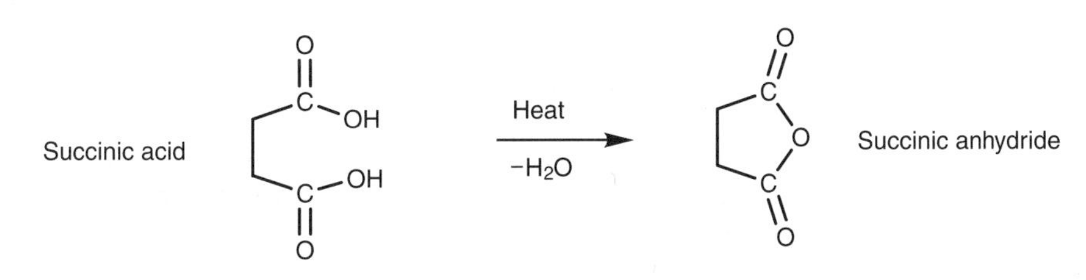

Fig. 4. Synthesis of cyclic acid anhydrides from acyclic di-acids.

Esters

There are many different ways in which esters can be synthesized. A very effective method is to react an acid chloride with an alcohol in the presence of pyridine (Topic K6). Acid anhydrides also react with alcohols to give esters, but are less reactive (Topic K6). Furthermore, the reaction is wasteful since half of the acyl content in the acid anhydride is wasted as the leaving group (i.e. the carboxylate ion). This is not a problem if the acid anhydride is cheap and readily available. For example, acetic anhydride is useful for the synthesis of a range of acetate esters (*Fig. 5*).

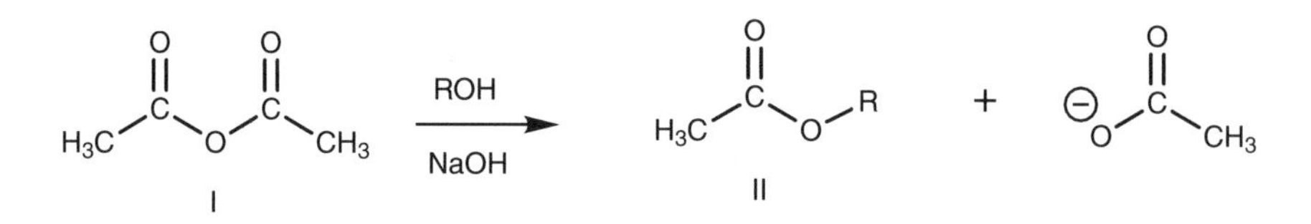

Fig. 5. Synthesis of alkyl ethanoates (II) from acetic anhydride (I).

A very common method of synthesizing simple esters is to treat a carboxylic acid with a simple alcohol in the presence of a catalytic amount of mineral acid (*Fig. 6*). The acid catalyst is required since there are two difficult steps in the reaction mechanism. First of all, the alcohol molecule is not a good nucleophile and so the carbonyl group has to be activated. Secondly, the OH group of the carboxylic acid is not a good leaving group and this has to be converted into a better leaving group. The mechanism (*Fig. 7*) is another example of nucleophilic substitution. In the first step, the carbonyl oxygen forms a bond to the acidic proton. This results in the carbonyl oxygen gaining a positive charge. This makes the carbonyl carbon more electrophilic and activates it to react with the weakly nucleophilic alcohol. In the second step, the alcohol uses a lone pair of electrons to form a bond to the carbonyl carbon. At the same time, the carbonyl π bond breaks and both electrons move onto the carbonyl oxygen to form a lone pair of electrons and thus neutralize the positive charge. Activation of the carbonyl group is important since the incoming alcohol gains an unfavorable positive charge during this step. In the third stage, a proton is transferred from the original alcohol portion to the OH group which we want to remove. By doing so, the latter moiety becomes a much better leaving group. Instead of a hydroxide ion, we can now remove a neutral water molecule. This is achieved in the fourth step where the carbonyl π bond is reformed and the water molecule is expelled.

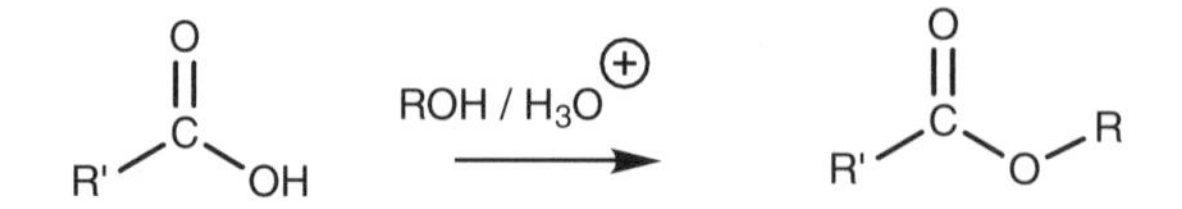

Fig. 6. Esterification of a carboxylic acid.

All the steps in the reaction mechanism are in equilibrium and so it is important to use the alcohol in large excess (i.e. as solvent) in order to drive the equilibrium to products. This is only practical with cheap and readily available alcohols such as methanol and ethanol. On the other hand, if the carboxylic acid is cheap and readily available it could be used in large excess instead.

An excellent method of preparing methyl esters is to treat carboxylic acids with diazomethane (*Fig. 8*). Good yields are obtained because nitrogen is formed as one

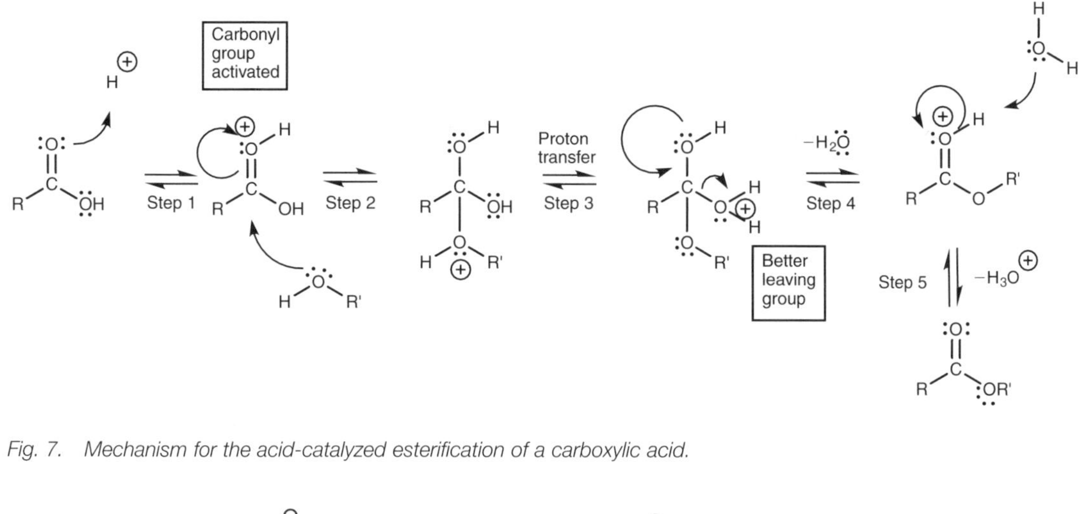

Fig. 7. Mechanism for the acid-catalyzed esterification of a carboxylic acid.

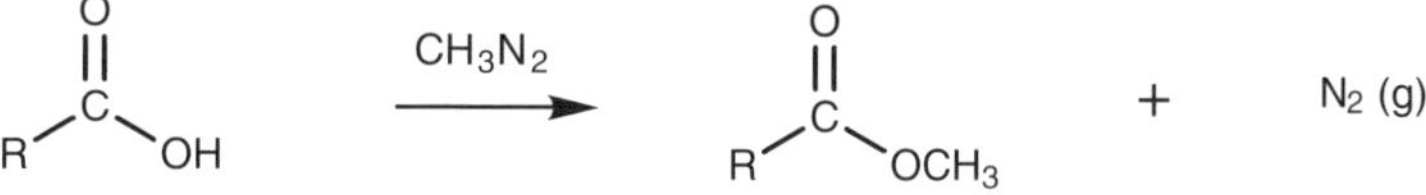

Fig. 8. Synthesis of methyl esters using diazomethane.

of the products and since it is lost from the reaction mixture, the reaction is driven to completion. However, diazomethane is an extremely hazardous chemical which can explode, and strict precautions are necessary when using it.

Lastly, the carboxylic acid can be converted to a carboxylate ion and then treated with an alkyl halide (*Fig. 9*). The reaction involves the S_N2 nucleophilic substitution of an alkyl halide (Topics L2 & L6) and so the reaction works best with primary alkyl halides.

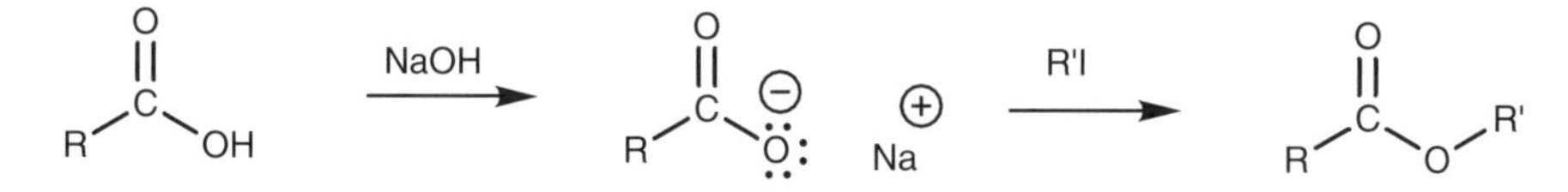

Fig. 9. Synthesis of an ester by nucleophilic substitution of an alkyl halide.

Amides

Amides can be prepared from acid chlorides by nucleophilic substitution (Topic K6). Treatment with ammonia gives a primary amide, treatment with a primary amine gives a secondary amide, and treatment with a secondary amine gives a tertiary amide. Tertiary amines cannot be used in this reaction because they do not give a stable product.

Two equivalents of amine are required for the above reactions since one equivalent of the amine is used up in forming a salt with the hydrochloric acid which is produced in the reaction. This is clearly wasteful on the amine, especially if the amine is valuable. To avoid this, one equivalent of sodium hydroxide can be added to the reaction in order to neutralize the HCl.

Amides can also be synthesized from acid anhydrides and esters (Topic K6), but in general these reactions offer no advantage over acid chlorides since acid anhydrides and esters are less reactive. Furthermore, with acid anhydrides, half of the parent carboxylic acid is lost as the leaving group. This is wasteful and so acid anhydrides are only used for the synthesis of amides if the acid anhydride is cheap and freely available (e.g. acetic anhydride).

The synthesis of amides directly from carboxylic acids is not easy since the reaction of an amine with a carboxylic acid is a typical acid–base reaction resulting in the formation of a salt (*Fig. 10*). Some salts can be converted to an amide by heating strongly to expel water, but there are better methods available as previously described.

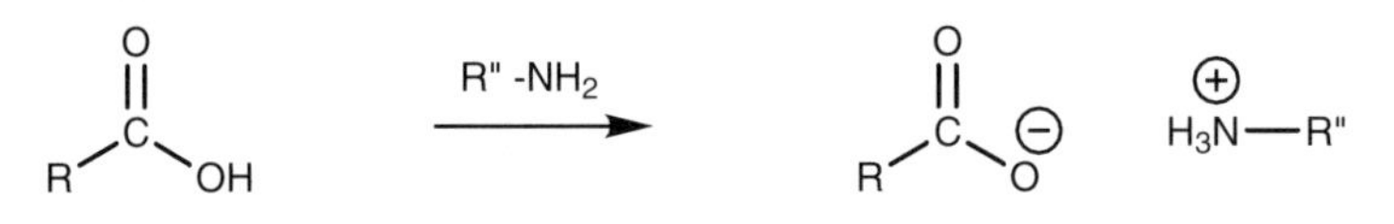

Fig. 10. Salt formation.

K6 Reactions

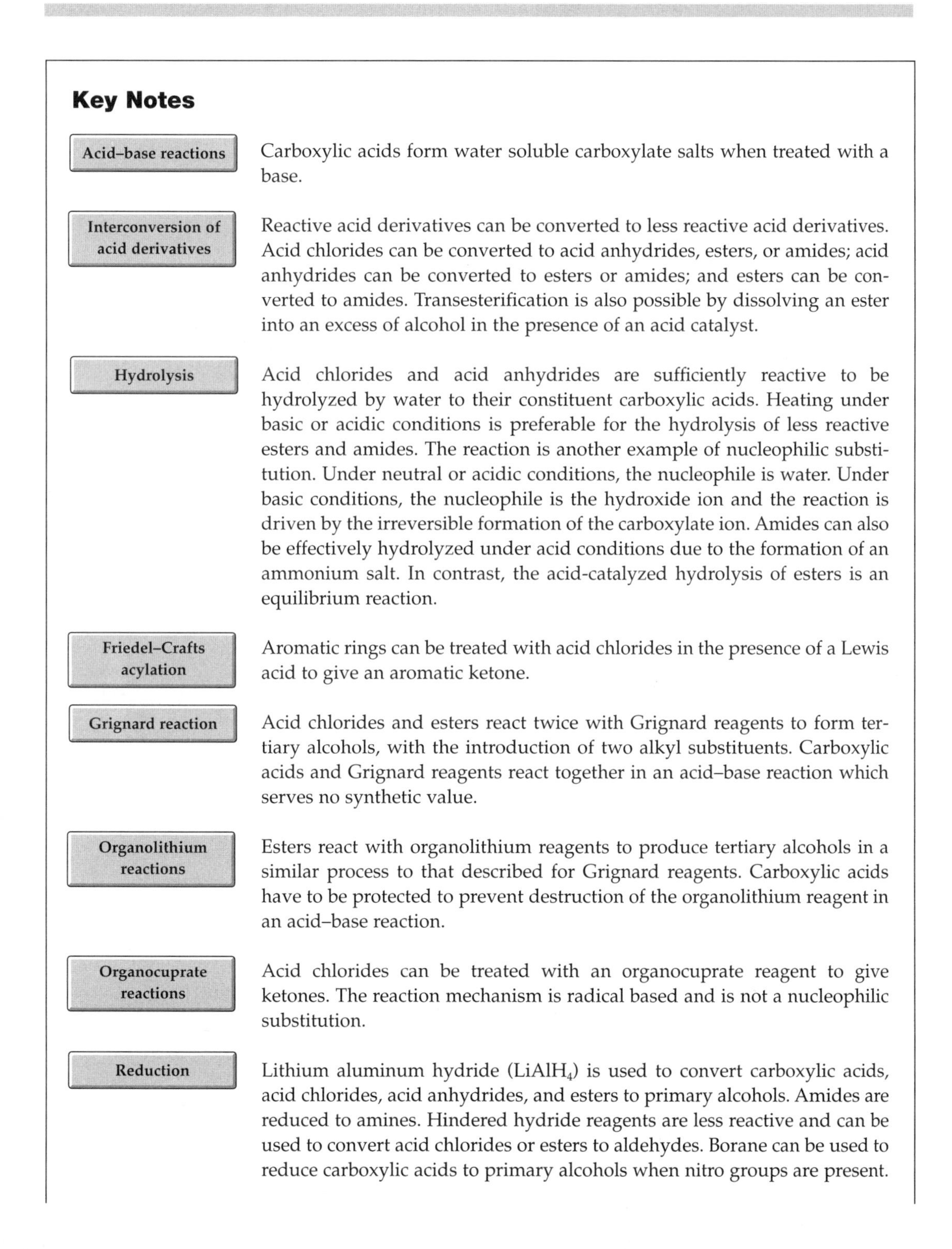

Key Notes

Acid–base reactions

Carboxylic acids form water soluble carboxylate salts when treated with a base.

Interconversion of acid derivatives

Reactive acid derivatives can be converted to less reactive acid derivatives. Acid chlorides can be converted to acid anhydrides, esters, or amides; acid anhydrides can be converted to esters or amides; and esters can be converted to amides. Transesterification is also possible by dissolving an ester into an excess of alcohol in the presence of an acid catalyst.

Hydrolysis

Acid chlorides and acid anhydrides are sufficiently reactive to be hydrolyzed by water to their constituent carboxylic acids. Heating under basic or acidic conditions is preferable for the hydrolysis of less reactive esters and amides. The reaction is another example of nucleophilic substitution. Under neutral or acidic conditions, the nucleophile is water. Under basic conditions, the nucleophile is the hydroxide ion and the reaction is driven by the irreversible formation of the carboxylate ion. Amides can also be effectively hydrolyzed under acid conditions due to the formation of an ammonium salt. In contrast, the acid-catalyzed hydrolysis of esters is an equilibrium reaction.

Friedel–Crafts acylation

Aromatic rings can be treated with acid chlorides in the presence of a Lewis acid to give an aromatic ketone.

Grignard reaction

Acid chlorides and esters react twice with Grignard reagents to form tertiary alcohols, with the introduction of two alkyl substituents. Carboxylic acids and Grignard reagents react together in an acid–base reaction which serves no synthetic value.

Organolithium reactions

Esters react with organolithium reagents to produce tertiary alcohols in a similar process to that described for Grignard reagents. Carboxylic acids have to be protected to prevent destruction of the organolithium reagent in an acid–base reaction.

Organocuprate reactions

Acid chlorides can be treated with an organocuprate reagent to give ketones. The reaction mechanism is radical based and is not a nucleophilic substitution.

Reduction

Lithium aluminum hydride ($LiAlH_4$) is used to convert carboxylic acids, acid chlorides, acid anhydrides, and esters to primary alcohols. Amides are reduced to amines. Hindered hydride reagents are less reactive and can be used to convert acid chlorides or esters to aldehydes. Borane can be used to reduce carboxylic acids to primary alcohols when nitro groups are present.

Sodium borohydride does not reduce carboxylic acids or their derivatives and can be used to reduce aldehydes and ketones without affecting carboxylic acids or acid derivatives.

Dehydration of primary amides

Primary amides are dehydrated to nitriles on treatment with a dehydrating agent such as thionyl chloride.

Related topics

Acid strength (G2)
Electrophilic substitutions of benzene (I3)
Nucleophilic addition – charged nucleophiles (J4)
Reduction and oxidation (J10)
Nucleophilic substition (K2)
Preparations of carboxylic acid derivatives (K5)

Acid–base reactions

Since carboxylic acids have an acidic proton ($CO_2\underline{H}$), they form water soluble carboxylate salts on treatment with a base (e.g. sodium hydroxide or sodium bicarbonate, Topic G2; *Fig. 1*).

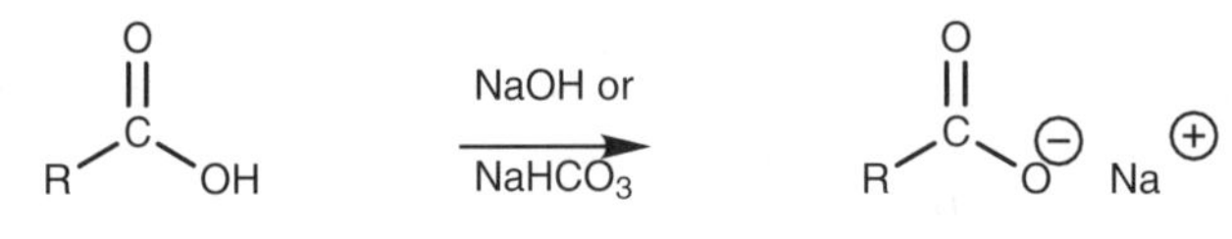

Fig. 1. Salt formation.

Interconversion of acid derivatives

Reactive acid derivatives can be converted to less reactive acid derivatives by nucleophilic substitution. This means that acid chlorides can be converted to acid anhydrides, esters, and amides (*Fig. 2*). Hydrochloric acid is released in these reactions and this may lead to side reactions. As a result, pyridine or sodium hydroxide may be added in order to mop up the hydrochloric acid (*Fig. 3*).

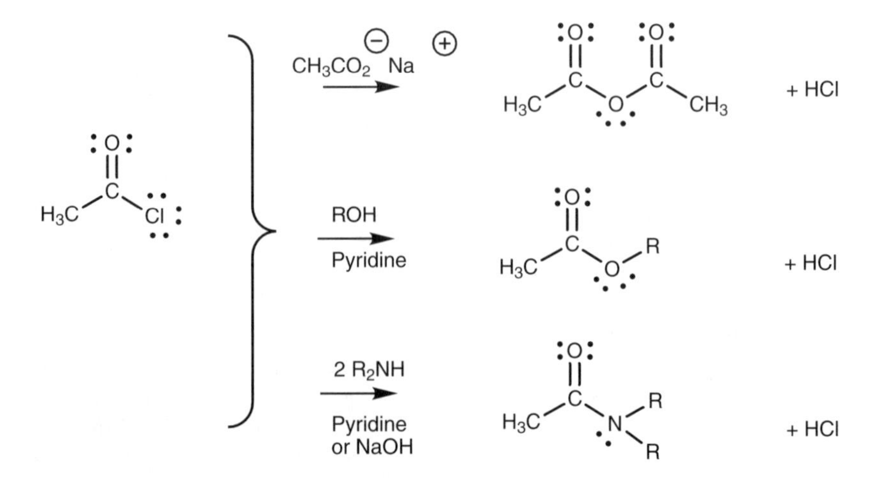

Fig. 2. Nucleophilic substitutions of an acid chloride.

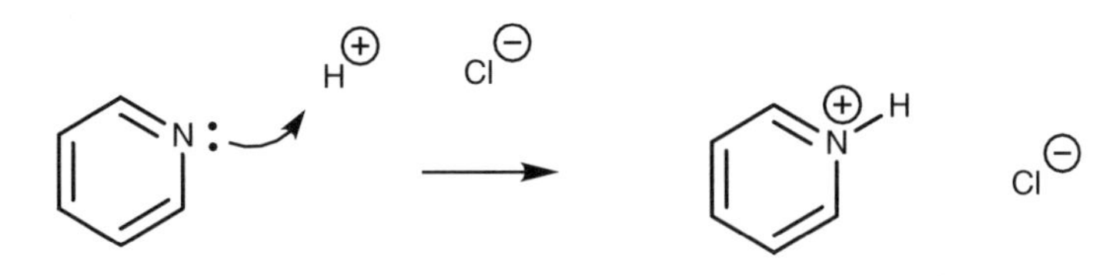

Fig. 3. Role of pyridine in 'mopping up' protons.

Acid anhydrides can be converted to esters and amides but not to acid chlorides (*Fig. 4*).

Esters can be converted to amides but not to acid chlorides or acid anhydrides (*Fig. 5*).

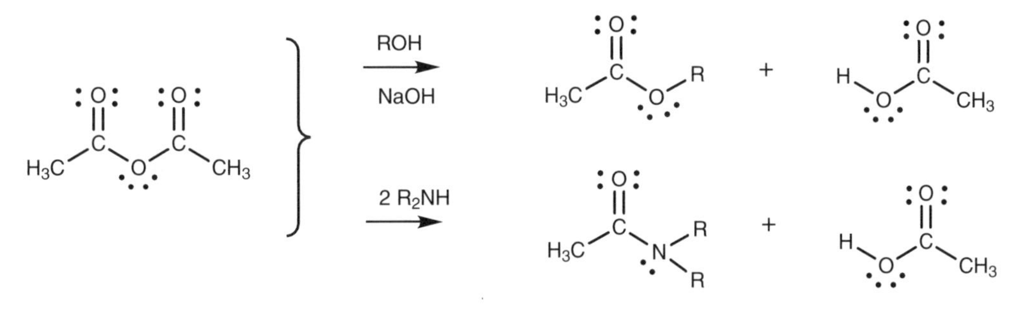

Fig. 4. Nucleophilic substitutions of acid anhydrides.

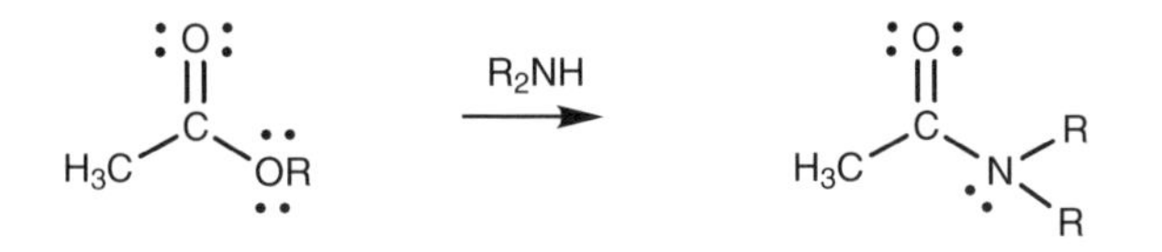

Fig. 5. Nucleophilic substitutions of an ester.

Esters can also be converted by nucleophilic substitution from one type of ester to another – a process called **transesterification**. For example, a methyl ester could be dissolved in ethanol in the presence of an acid catalyst and converted to an ethyl ester (*Fig. 6*). The reaction is an equilibrium reaction, but if the alcohol to be introduced is used as solvent, it is in large excess and the equilibrium is shifted to the desired ester. Furthermore, if the alcohol to be replaced has a low boiling point, it can be distilled from the reaction as it is substituted, thus shifting the equilibrium to the desired product.

Amides are the least reactive of the acid derivatives and cannot be converted to acid chlorides, acid anhydrides, or esters.

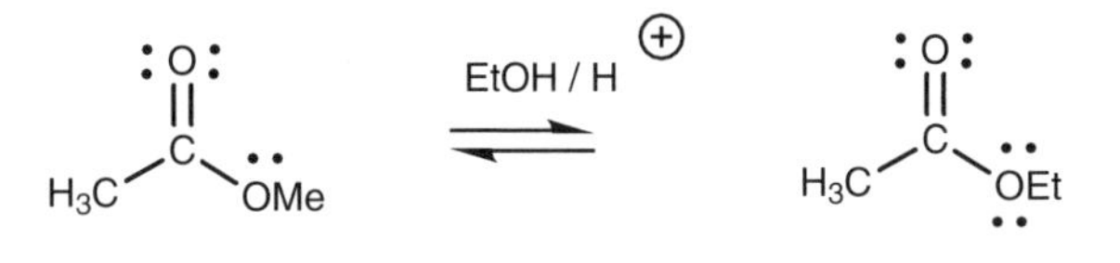

Fig. 6. Transesterification.

Hydrolysis

Reactive acid derivatives (i.e. acid chlorides and acid anhydrides) are hydrolyzed by water to give the constituent carboxylic acids (*Fig. 7*). The reaction is another example of nucleophilic substitution where water acts as the nucleophile. Hydrochloric acid is a byproduct from the hydrolysis of an acid chloride, so pyridine is often added to the reaction mixture to mop it up (*Fig. 3*).

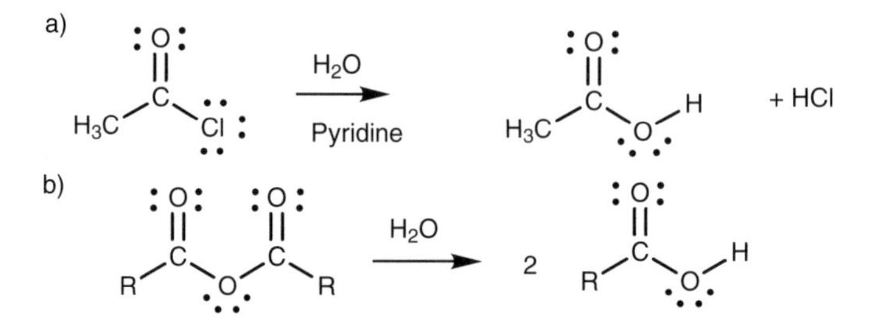

Fig. 7. Hydrolysis of (a) an acid chloride; (b) an acid anhydride.

Esters and amides are less reactive and so the hydrolysis requires more forcing conditions using aqueous sodium hydroxide or aqueous acid with heating (*Fig. 8*).

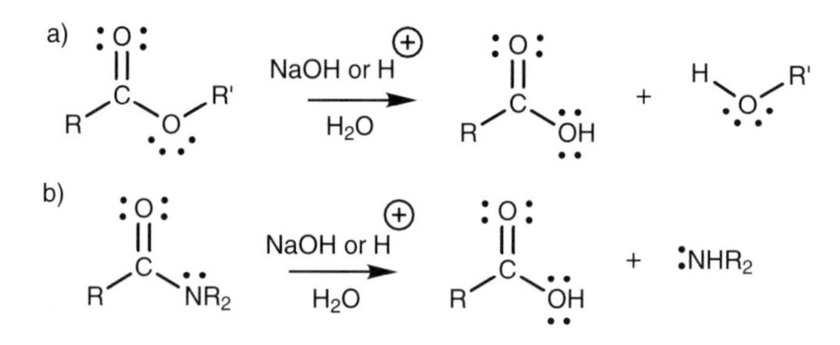

Fig. 8. Hydrolysis of (a) esters; (b) amides.

Under basic conditions, the hydroxide ion acts as the nucleophile by the normal mechanism for nucleophilic substitution (Topic K2). For example, the mechanism of hydrolysis of ethyl acetate is as shown (*Fig. 9*). However, the mechanism does not stop here. The carboxylic acid which is formed reacts with sodium hydroxide to form a water soluble carboxylate ion (*Fig. 10a*). Furthermore, the ethoxide ion which is lost from the molecule is a stronger base than water and undergoes protonation (*Fig. 10b*). The basic hydrolysis of an ester is also known as **saponification** and produces a water soluble carboxylate ion.

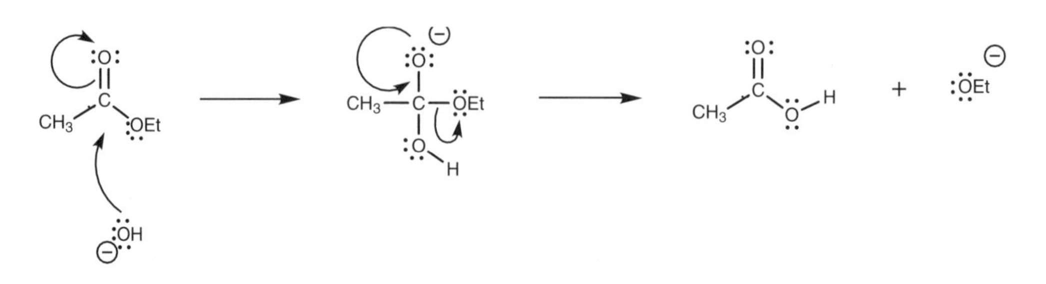

Fig. 9. Mechanism of hydrolysis of ethyl ethanoate.

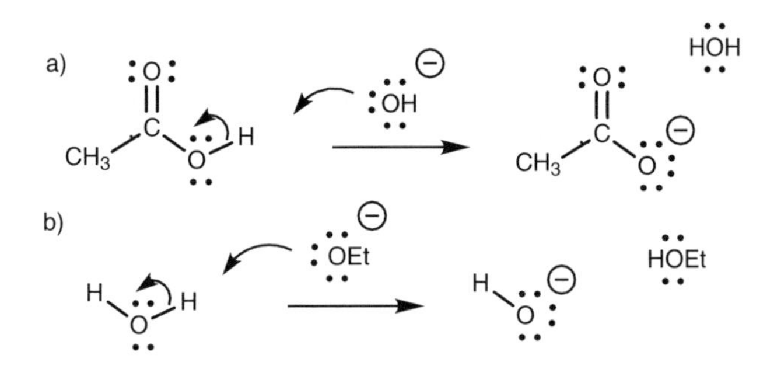

Fig. 10. (a) Ionization of a carboxylic acid; (b) neutralization of the ethoxide ion.

The same mechanism is involved in the basic hydrolysis of an amide and also results in a water soluble carboxylate ion. The leaving group from an amide is initially charged (i.e. $R_2N:^-$). However, this is a strong base and reacts with water to form a free amine plus a hydroxide ion.

In the basic hydrolysis of esters and amides, the formation of a carboxylate ion is irreversible and so serves to drive the reaction to completion.

In order to isolate the carboxylic acid rather than the salt, it is necessary to add acid (e.g. dilute HCl) to the aqueous solution. The acid protonates the carboxylate salt to give the carboxylic acid which (in most cases) is no longer soluble in aqueous solution and precipitates out as a solid or as an oil.

The mechanism for acid–catalyzed hydrolysis (*Fig. 11*) involves water acting as a nucleophile. However, water is a poor nucleophile since it gains an unfavorable

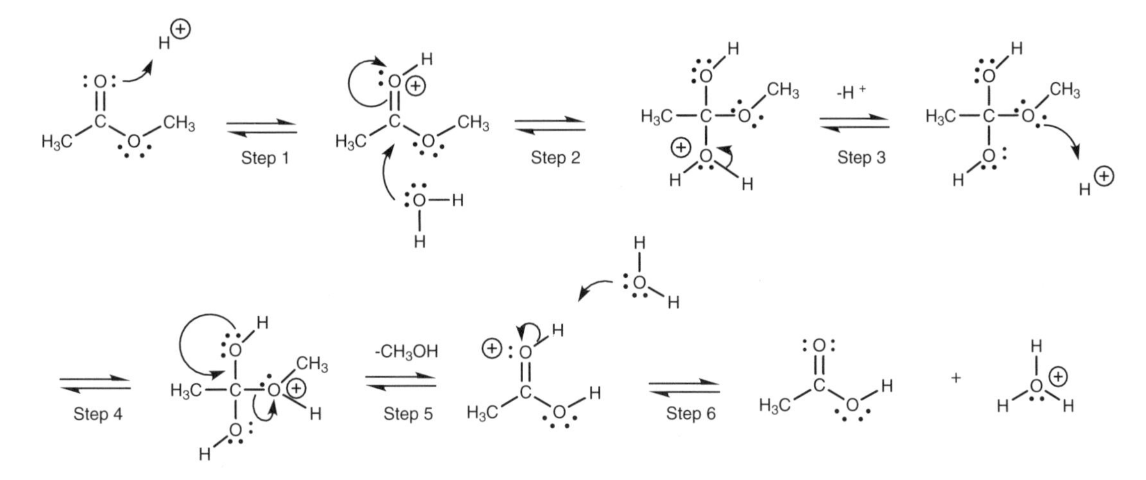

Fig. 11. Mechanism for the acid-catalyzed hydrolysis of an ester.

positive charge when it forms a bond. Therefore, the carbonyl group has to be activated which occurs when the carbonyl oxygen is protonated by the acid catalyst (Step 1). Nucleophilic attack by water is now favored since it neutralizes the unfavorable positive charge on the carbonyl oxygen (Step 2). The intermediate has a positive charge on the oxygen derived from water, but this is neutralized by losing the attached proton such that the oxygen gains the electrons in the O–H bond

(Step 3). Another protonation now takes place (Step 4). This is necessary in order to convert a poor leaving group (the methoxide ion) into a good leaving group (methanol). The π bond can now be reformed (Step 5) with loss of methanol. Finally, water can act as a base to remove the proton from the carbonyl oxygen (Step 6).

The acid-catalyzed hydrolysis of an ester is not as effective as basic hydrolysis since all the steps in the mechanism are reversible and there is no salt formation to pull the reaction through to products. Therefore, it is important to use an excess of water in order to shift the equilibria to the products. In contrast to esters, the hydrolysis of an amide in acid does result in the formation of an ion (*Fig. 12*). The leaving group here is an amine and since amines are basic, they will react with the acid to form a water soluble aminum ion. This is an irreversible step which pulls the equilibrium through to the products.

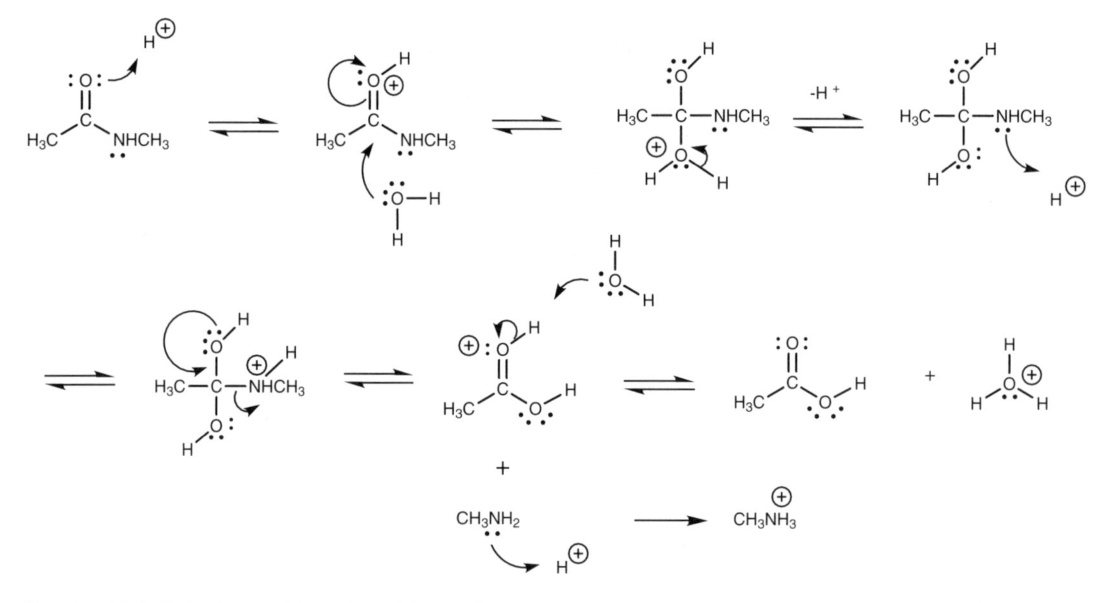

Fig. 12. Hydrolysis of an amide under acidic conditions.

In the acid-catalyzed hydrolysis of an ester, only a catalytic amount of acid is required since the protons used during the reaction mechanism are regenerated. However with an amide, at least one equivalent of acid is required due to the ionization of the amine.

Friedel–Crafts acylation

Acid chlorides can be treated with aromatic rings in the presence of a Lewis acid to give aromatic ketones (*Fig. 13*). The reaction involves formation of an acylium ion from the acid chloride, followed by electrophilic substitution of the aromatic ring (Topic I3).

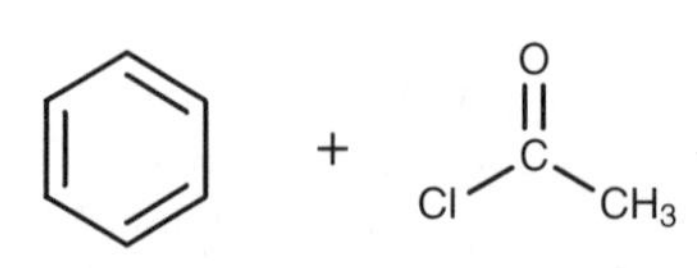

AlCl3

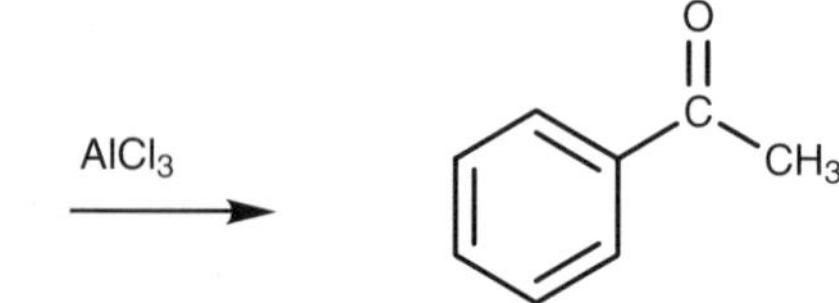

Fig. 13. Friedel–Crafts acylation.

Grignard reaction

Acid chlorides and esters react with two equivalents of a Grignard reagent (Topic L7) to produce a tertiary alcohol where two extra alkyl groups are provided by the Grignard reagent (*Fig. 14*).

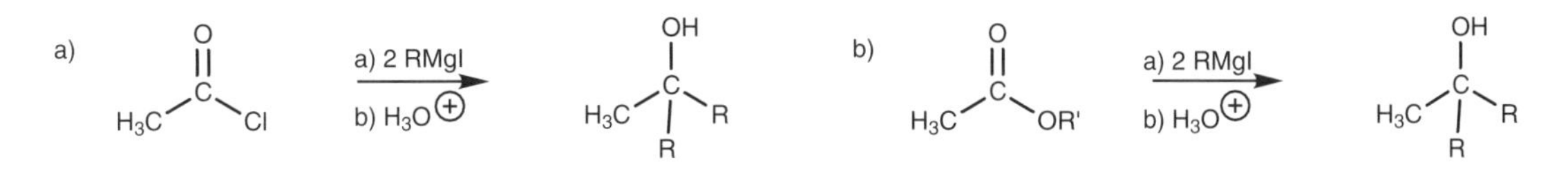

Fig. 14. Grignard reaction with (a) an acid chloride; and (b) an ester to produce a tertiary alcohol.

There are two reactions involved in this process (*Fig. 15*). The acid chloride reacts with the first equivalent of Grignard reagent in a typical nucleophilic substitution to produce an intermediate ketone. However, this ketone is also reactive to Grignard reagents and immediately reacts with a second equivalent of Grignard reagent by the nucleophilic addition mechanism described for aldehydes and ketones (Topic J4).

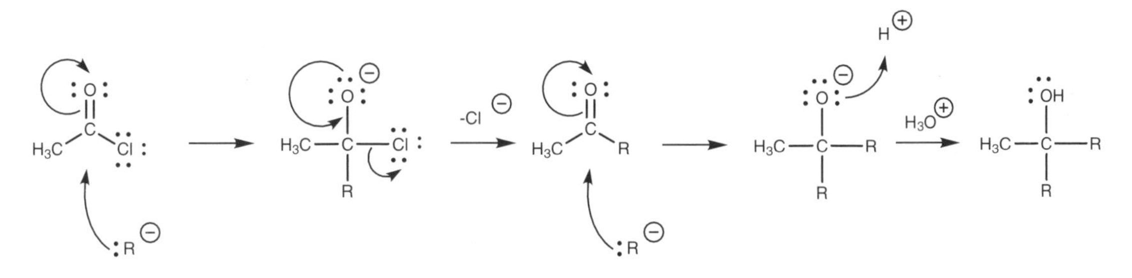

Fig. 15. Mechanism of the Grignard reaction with an acid chloride.

Carboxylic acids react with Grignard reagents in an acid–base reaction resulting in formation of the carboxylate ion and formation of an alkane from the Grignard reagent (*Fig. 16*). This has no synthetic use and it is important to protect carboxylic acids when carrying out Grignard reactions on another part of the molecule so that the Grignard reagent is not wasted.

Fig. 16. Acid–base reaction of a Grignard reagent with a carboxylic acid.

Organolithium reactions

Esters react with two equivalents of an organolithium reagent to give a tertiary alcohol where two of the alkyl groups are derived from the organolithium reagent (*Fig. 17*). The mechanism of the reaction is the same as that described in the Grignard reaction, that is, nucleophilic substitution to a ketone followed by nucleophilic addition. It is necessary to protect any carboxylic acids present when carrying out organolithium reactions since one equivalent of the organolithium reagent would be wasted in an acid–base reaction with the carboxylic acid.

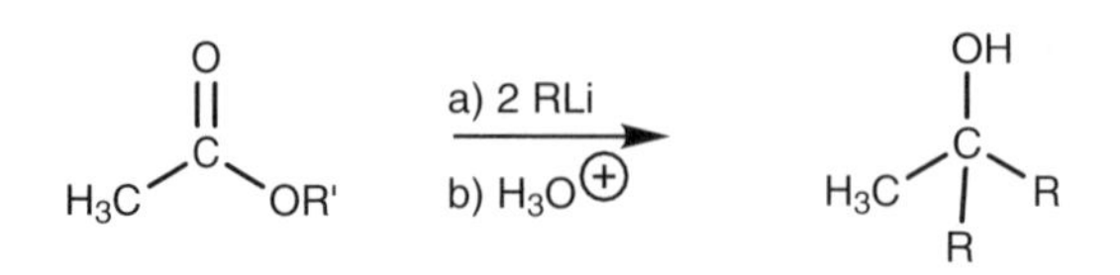

Fig. 17. Reaction of an ester with an organolithium reagent to form a tertiary alcohol.

Organocuprate reactions

Acid chlorides react with diorganocuprate reagents (Topic L7) to form ketones (*Fig. 18*). Like the Grignard reaction, an alkyl group displaces the chloride ion to produce a ketone. However, unlike the Grignard reaction, the reaction stops at the ketone stage. The mechanism is thought to be radical based rather than a nucleophilic substitution. This reaction does not take place with carboxylic acids, acid anhydrides, esters, or amides.

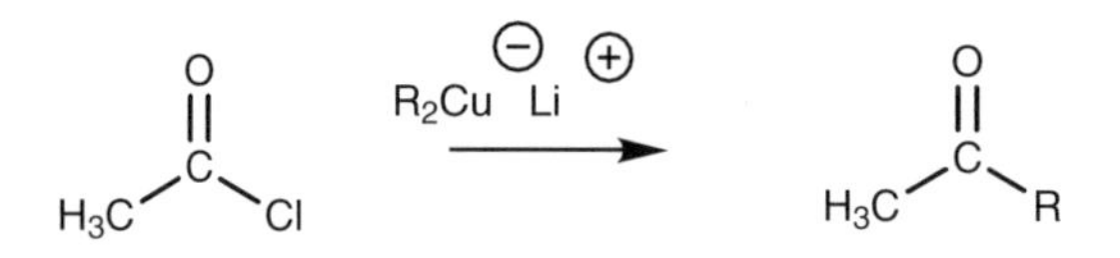

Fig. 18. Reaction of an acid chloride with a diorganocuprate reagent to produce a ketone.

Reduction

Carboxylic acids, acid chlorides, acid anhydrides and esters are reduced to primary alcohols on treatment with lithium aluminum hydride ($LiAlH_4$; *Fig. 19*). The reaction involves nucleophilic substitution by a hydride ion to give an intermediate aldehyde. This cannot be isolated since the aldehyde immediately undergoes a nucleophilic addition reaction with another hydride ion (Topic J4; *Fig. 20*). The detailed mechanism is as shown in *Fig. 21*.

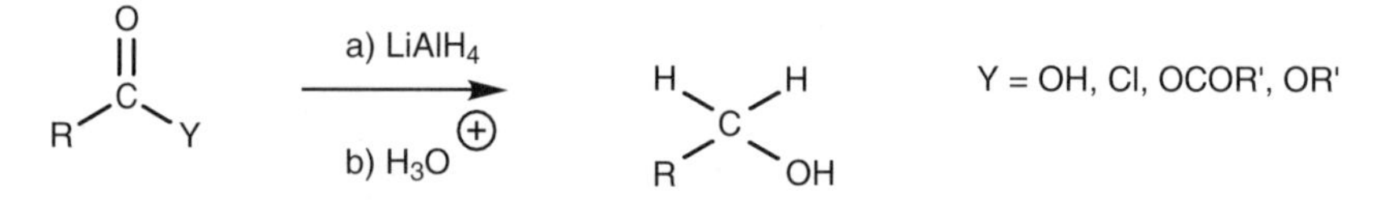

Fig. 19. Reduction of acid chlorides, acid anhydrides, and esters with lithium aluminum hydride.

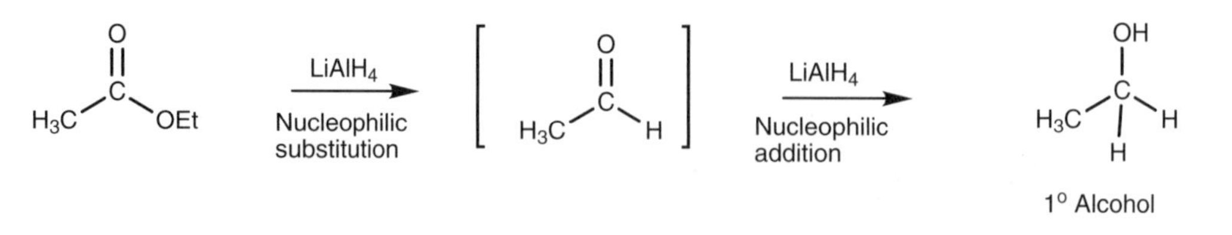

Fig. 20. Intermediate involved in the $LiAlH_4$ reduction of an ester.

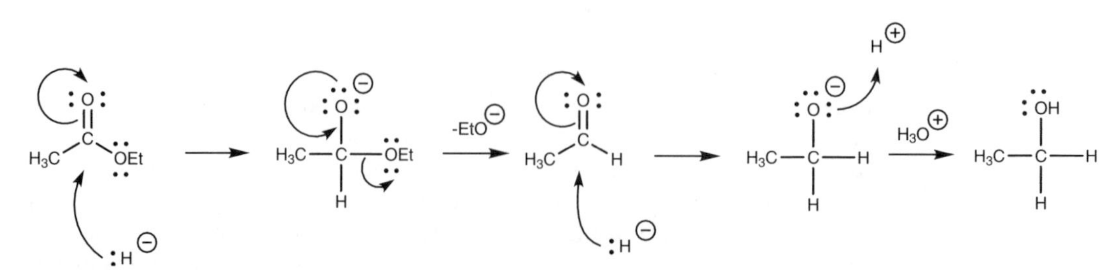

Fig. 21. Mechanism for the $LiAlH_4$ reduction of an ester to a primary alcohol.

Amides differ from carboxylic acids and other acid derivatives in their reaction with $LiAlH_4$. Instead of forming primary alcohols, amides are reduced to amines (*Fig. 22*). The mechanism (*Fig. 23*) involves addition of the hydride ion to form an intermediate which is converted to an organoaluminum intermediate. The difference in this mechanism is the intervention of the nitrogen's lone pair of electrons. These are fed into the electrophilic center to eliminate the oxygen which is then followed by the second hydride addition.

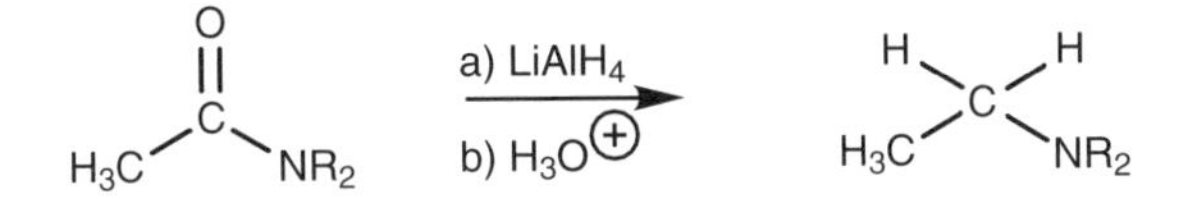

Fig. 22. Reduction of an amide to an amine.

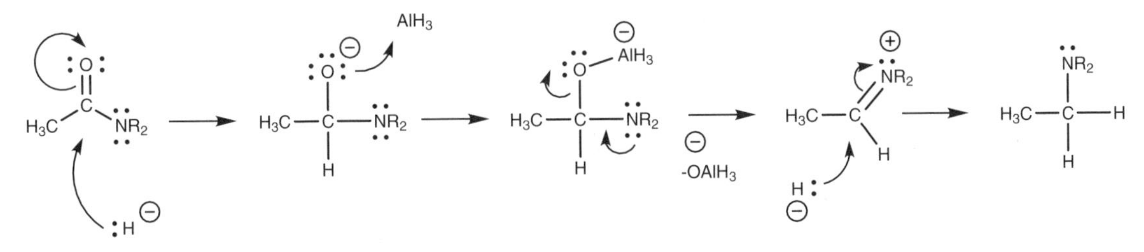

Fig. 23. Mechanism for the $LiAlH_4$ reduction of an amide to an amine.

Although acid chlorides and acid anhydrides are converted to tertiary alcohols with $LiAlH_4$, there is little synthetic advantage in this since the same reaction can be achieved on the more readily available esters and carboxylic acids. However, since acid chlorides are more reactive than carboxylic acids, they can be treated with a milder hydride-reducing agent and this allows the synthesis of aldehydes (*Fig. 24*). The hydride reagent used (lithium tri-*tert*-butoxyaluminum hydride) contains three bulky alkoxy groups which lowers the reactivity of the remaining hydride ion. This means that the reaction stops after nucleophilic substitution with one hydride ion. Another sterically hindered hydride reagent – diisobutylaluminum hydride (DIBAH) – is used to reduce esters to aldehydes (*Fig. 24*). Normally low temperatures are needed to avoid over-reduction.

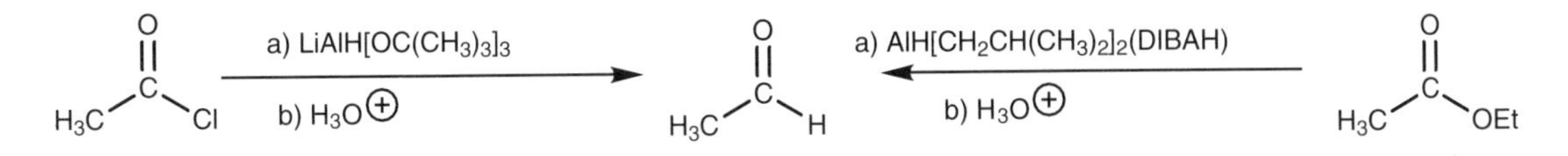

Fig. 24. Reduction of an acid chloride and an ester to an aldehyde.

Borane (B_2H_6) can be used as a reducing agent to convert carboxylic acids to primary alcohols. One advantage of using borane rather than $LiAlH_4$ is the fact that the former does not reduce any nitro groups which might be present. $LiAlH_4$ reduces a nitro group (NO_2) to an amino group (NH_2).

It is worth noting that carboxylic acids and acid derivatives are not reduced by the milder reducing agent – sodium borohydride ($NaBH_4$). This reagent can therefore be used to reduce aldehydes and ketones without affecting any carboxylic acids or acid derivatives which might be present.

Dehydration of primary amides

Primary amides are dehydrated to nitriles using a dehydrating agent such as thionyl chloride ($SOCl_2$), phosphorus pentoxide (P_2O_5), phosphoryl trichloride ($POCl_3$), or acetic anhydride (*Fig. 25*).

The mechanism for the dehydration of an amide with thionyl chloride is shown

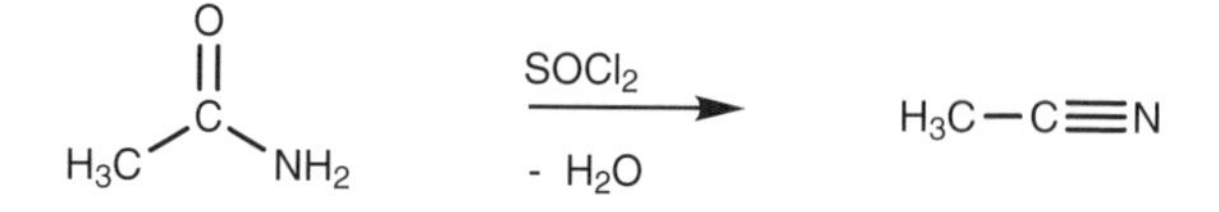

Fig. 25. Conversion of a primary amide to a nitrile.

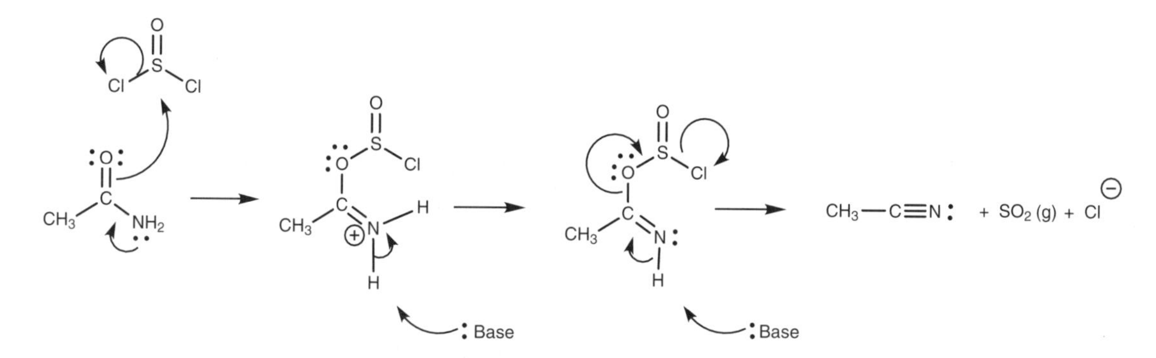

Fig. 26. Mechanism for the dehydration of a primary amide to a nitrile.

in *Fig. 26*. Although the reaction is the equivalent of a dehydration, the mechanism shows that water itself is not eliminated. The reaction is driven by the loss of one sulfur dioxide as a gas.

K7 ENOLATE REACTIONS

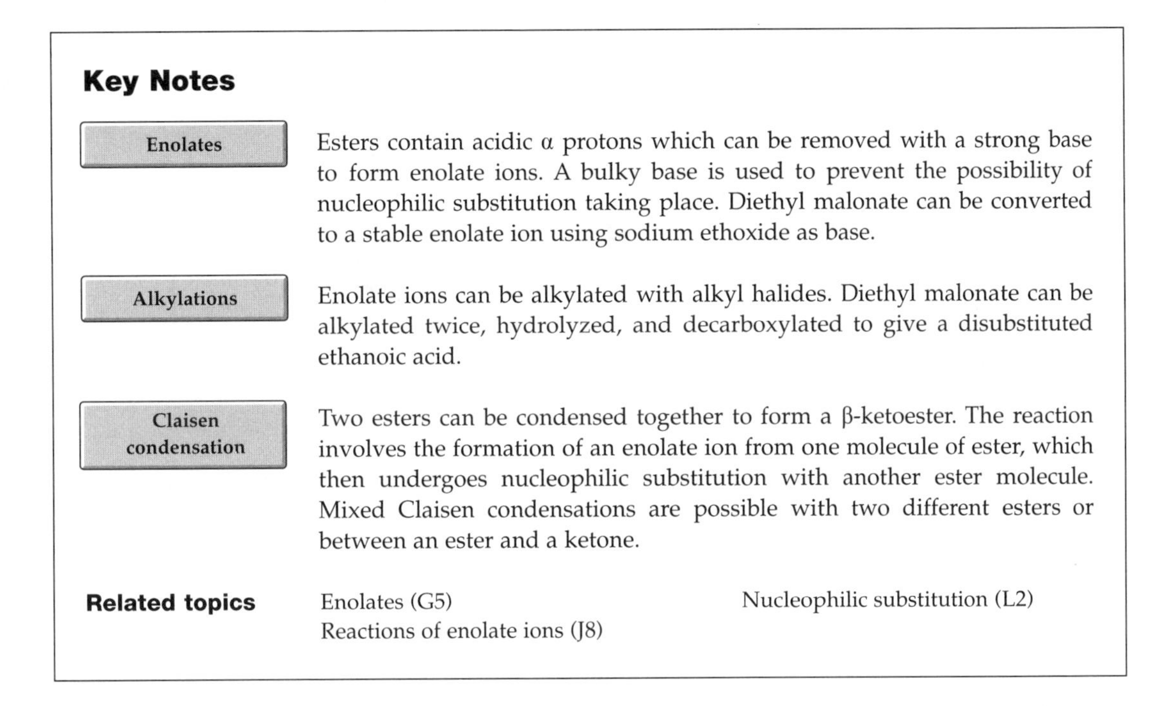

Key Notes

Enolates

Esters contain acidic α protons which can be removed with a strong base to form enolate ions. A bulky base is used to prevent the possibility of nucleophilic substitution taking place. Diethyl malonate can be converted to a stable enolate ion using sodium ethoxide as base.

Alkylations

Enolate ions can be alkylated with alkyl halides. Diethyl malonate can be alkylated twice, hydrolyzed, and decarboxylated to give a disubstituted ethanoic acid.

Claisen condensation

Two esters can be condensed together to form a β-ketoester. The reaction involves the formation of an enolate ion from one molecule of ester, which then undergoes nucleophilic substitution with another ester molecule. Mixed Claisen condensations are possible with two different esters or between an ester and a ketone.

Related topics

Enolates (G5)
Reactions of enolate ions (J8)
Nucleophilic substitution (L2)

Enolates

Enolate ions can be formed from aldehydes and ketones (Topic J8) containing protons on an α-carbon. Enolate ions can also be formed from esters if they have protons on an α-carbon (*Fig. 1*). Such protons are slightly acidic (Topic G5) and can be removed on treatment with a powerful base such as lithium diisopropylamide (LDA). LDA acts as a base rather than as a nucleophile since it is a bulky molecule and this prevents it attacking the carbonyl group in a nucleophilic substitution reaction.

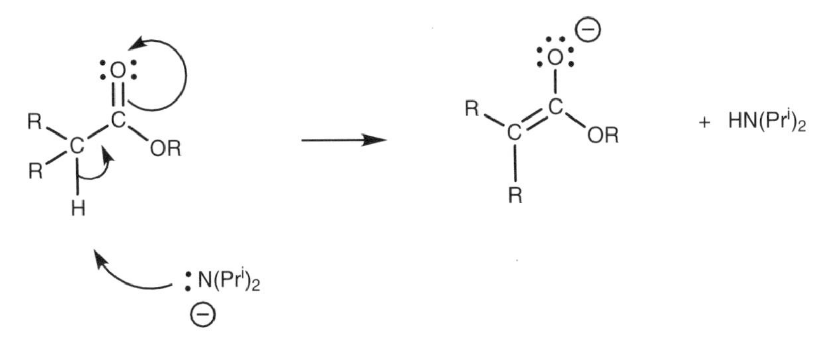

Fig. 1. Enolate ion formation.

Formation of enolate ions is easier if there are two esters flanking the α-carbon since the α-proton will be more acidic. The acidic proton in diethyl malonate can be removed with a weaker base than LDA (e.g. sodium ethoxide; *Fig. 2*). The enolate ion is more stable since the charge can be delocalized over both carbonyl groups.

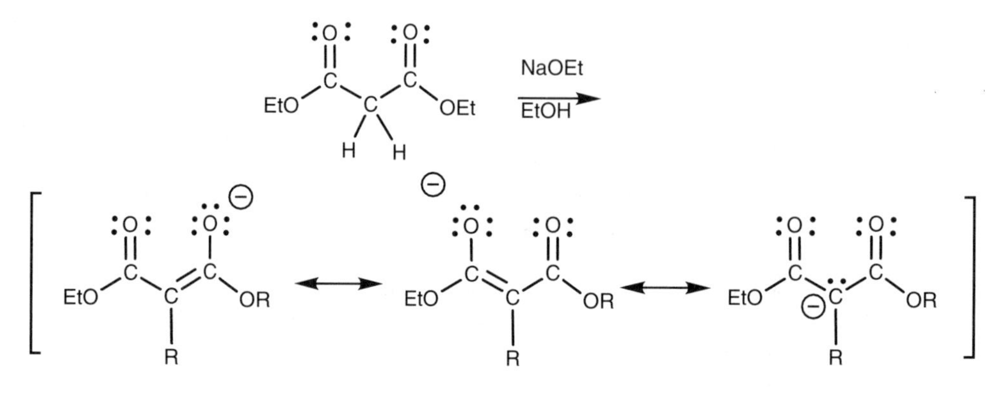

Fig. 2. Formation of an enolate ion from diethyl malonate.

Alkylations

Enolate ions can be alkylated with alkyl halides through the S_N2 nucleophilic substitution of an alkyl halide (Topic L2; *Fig. 3*).

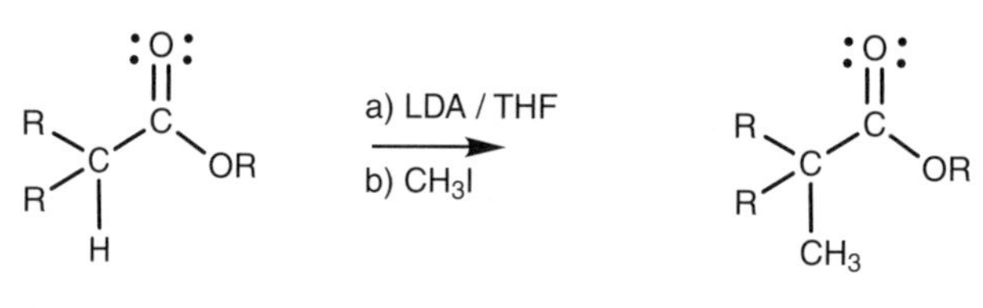

Fig. 3. α-Alkylation of an ester.

Although simple esters can be converted to their enolate ions and alkylated, the use of a molecule such as diethyl malonate is far more effective. This is because the α-protons of diethyl malonate (pK_a 10–12) are more acidic than the α-protons of a simple ester such as ethyl acetate (pK_a 25) and can be removed by a milder base. It is possible to predict the base required to carry out the deprotonation reaction by considering the pK_a value of the conjugate acid for that base. If this pK_a is higher than the pK_a value of the ester, then the deprotonation reaction is possible. For example, the conjugate acid of the ethoxide ion is ethanol (pK_a 16) and so any ester having a pK_a less than 16 will be deprotonated by the ethoxide ion. Therefore, diethyl malonate is deprotonated but not ethyl acetate. A further point worth noting is that the ethoxide ion is strong enough to deprotonate the diethyl malonate quantitatively such that all the diethyl malonate is converted to the enolate ion. This prevents the possibility of any competing Claisen reaction (see below) since that reaction requires the presence of unaltered ester. Diethyl malonate can be converted quantitatively to its enolate with ethoxide ion, alkylated with an alkyl halide, treated with another equivalent of base, then alkylated with a second different alkyl halide (*Fig. 4*). Subsequent hydrolysis and decarboxylation of the diethyl ester results in the formation of the carboxylic acid. The decarboxylation mechanism (*Fig. 5*) is dependent on the presence of the other carbonyl group at the β-position.

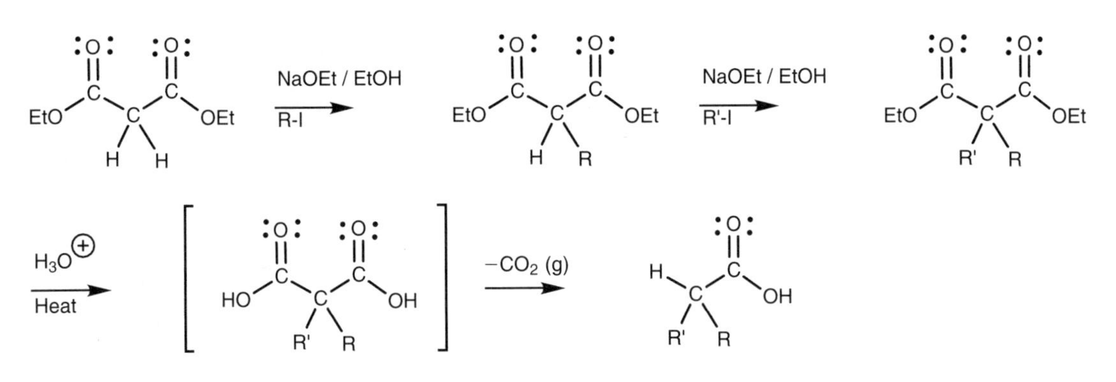

Fig. 4. Alkylations of diethyl malonate.

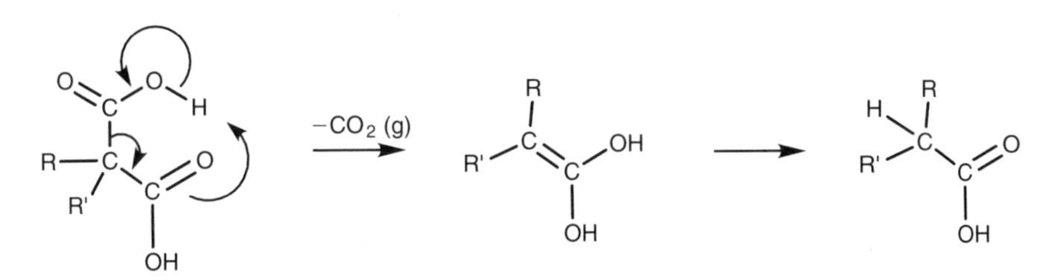

Fig. 5. Decarboxylation mechanism.

The final product can be viewed as a disubstituted ethanoic acid. In theory, this product could also be synthesized from ethyl ethanoate. However, the use of diethyl malonate is superior since the presence of two carbonyl groups allows easier formation of the intermediate enolate ions.

Claisen condensation

The Claisen reaction involves the condensation or linking of two ester molecules to form a β-ketoester (*Fig. 6*). This reaction can be viewed as the ester equivalent of the Aldol reaction (Topic J8). The reaction involves the formation of an enolate ion from one ester molecule, which then undergoes nucleophilic substitution with a second ester molecule (*Fig. 7*, Step 1). The ethoxide ion which is formed in step 2 removes an α-proton from the β-ketoester in step 3 to form a stable enolate ion and this drives the reaction to completion. The final product is isolated by protonating the enolate ion with acid.

Two different esters can be used in the Claisen condensation as long as one of the esters has no α-protons and cannot form an enolate ion (*Fig. 8*). β-Diketones can be synthesized from the mixed Claisen condensation of a ketone with an ester (*Fig. 9*). Again, it is advisable to use an ester which cannot form an enolate ion to prevent competing Claisen condensations.

In both these last two examples, a very strong base is used in the form of LDA such that the enolate ion is formed quantitatively (from ethyl acetate and acetone respectively). This prevents the possibility of self-Claisen condensation and limits the reaction to the crossed Claisen condensation.

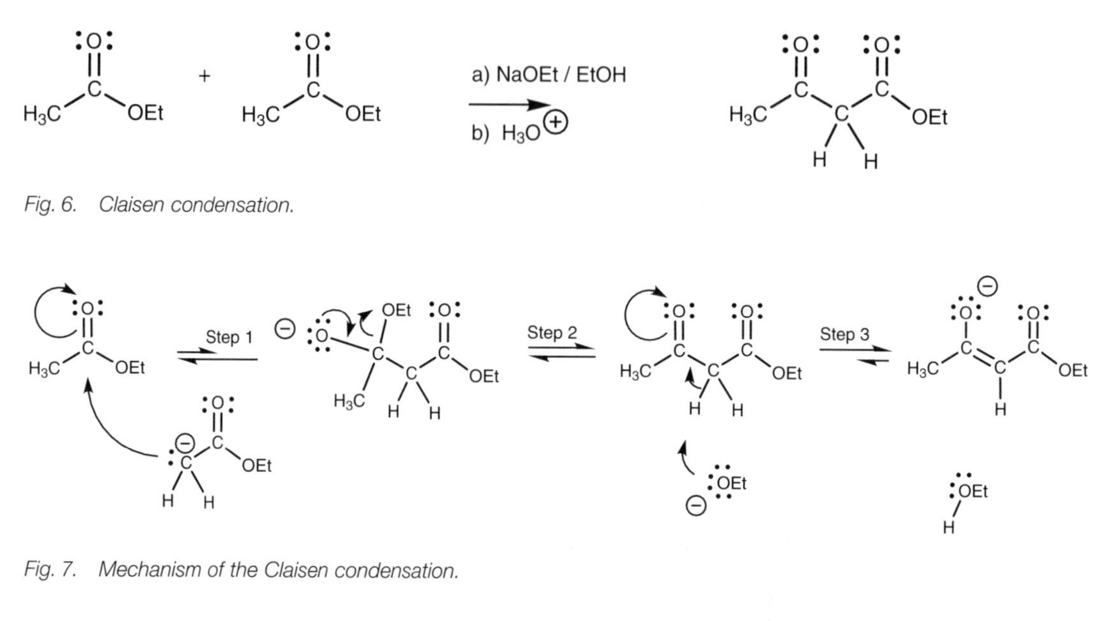

Fig. 6. Claisen condensation.

Fig. 7. Mechanism of the Claisen condensation.

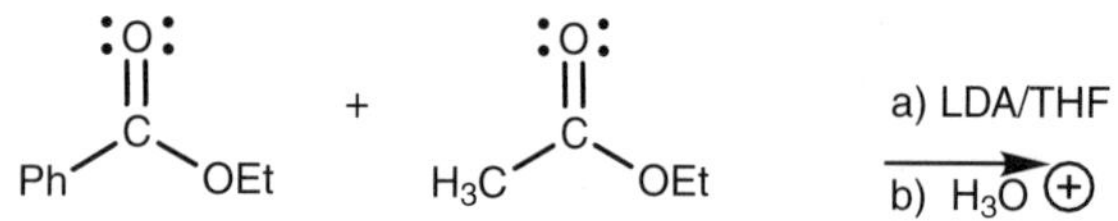

Fig. 8. Claisen condensation of two different esters.

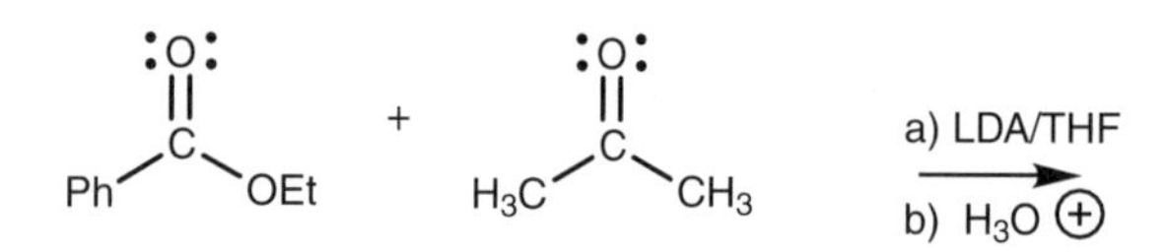

Fig. 9. Claisen condensation of a ketone with an ester.

L1 PREPARATION AND PHYSICAL PROPERTIES

Key Notes

Preparation

Alkenes are converted to alkyl halides by reaction with hydrogen halides. Treatment with halogens results in dihaloalkanes. Tertiary alcohols can be converted to alkyl halides on treatment with hydrogen halides, whereas primary and secondary alcohols are best converted by using thionyl chloride or phosphorus tribromide.

Structure

Alkyl halides consist of an alkyl group linked to a halogen. The carbon linked to the halogen is sp^3 hybridized and tetrahedral. The carbon–halogen bond length increases and the bond strength decreases as the halogen increases in size.

Bonding

The C–halogen bond (C–X) is a polar σ bond where the halogen is slightly negative and the carbon is slightly positive. Intermolecular bonding is by weak van der Waals interactions.

Properties

Alkyl halides have a dipole moment. They are poorly soluble in water, but dissolve in organic solvents. They react as electrophiles at the carbon center.

Reactions

Alkyl halides undergo nucleophilic substitution reactions and elimination reactions.

Related topics

sp^3 Hybridization (A3)
Intermolecular bonding (C3)
Properties and reactions (C4)
Electrophilic addition to symmetrical alkenes (H3)
Reactions of alcohols (M4)

Preparation

Alkenes can be treated with hydrogen halides (HCl, HBr, and HI) or halogens (Cl_2 and Br_2) to give alkyl halides and dihaloalkanes respectively (Topic H3). An extremely useful method of preparing alkyl halides is to treat an alcohol with a hydrogen halide (HX = HCl, HBr, or HI). The reaction works best for tertiary alcohols (Topic M4). Primary and secondary alcohols can be converted to alkyl halides more effectively by treating them with thionyl chloride ($SOCl_2$) or phosphorus tribromide (PBr_3). The conditions are less acidic and less likely to cause acid-catalyzed rearrangements.

Structure

Alkyl halides consist of an alkyl group linked to a halogen atom (F, Cl, Br, or I) by a single (σ) bond. The carbon atom linked to the halogen atom is sp^3 hybridized and has a tetrahedral geometry with bond angles of approximately 109°. The carbon–halogen bond length increases with the size of the halogen atom and this is associated with a decrease in bond strength. For example, C–F bonds are shorter and stronger than C–Cl bonds.

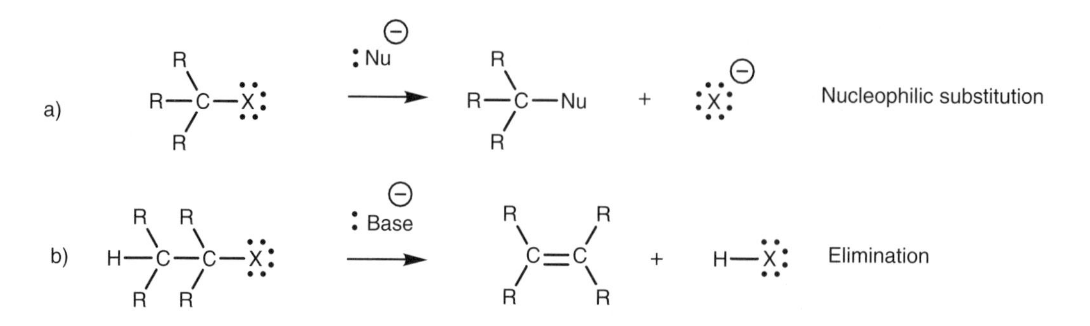

Fig. 1. Reactions of alkyl halides.

Bonding

The carbon–halogen bond (referred to as C–X from here on) is a σ bond. The bond is polar since the halogen atom is more electronegative than carbon, resulting in the halogen being slightly negative and the carbon being slightly positive. Intermolecular hydrogen bonding or ionic bonding is not possible between alkyl halide molecules and the major intermolecular bonding force consists of weak van der Waals interactions.

Properties

The polar C–X bond means that alkyl halides have a substantial dipole moment. Alkyl halides are poorly soluble in water, but are soluble in organic solvents. They have boiling points which are similar to alkanes of comparable molecular weight. The polarity also means that the carbon is an electrophilic center and the halogen is a nucleophilic center. Halogens are extremely weak nucleophilic centers and therefore, alkyl halides are more likely to react as electrophiles at the carbon center.

Reactions

The major reactions undergone by alkyl halides are (a) nucleophilic substitution where an attacking nucleophile replaces the halogen (*Fig. 1a*), and (b) elimination where the alkyl halide loses HX and is converted to an alkene (*Fig. 1b*).

L2 NUCLEOPHILIC SUBSTITUTION

Key Notes

Definition

Nucleophilic substitution of an alkyl halide involves the substitution of the halogen atom with a different nucleophile. The halogen is lost as a halide ion. There are two types of mechanism for alkyl halides – S_N1 and S_N2.

S_N2 Mechanism

The S_N2 mechanism is a concerted process where the incoming nucleophile forms a bond to the reaction center at the same time as the C–X bond is broken. The transition state involves the incoming nucleophile approaching from one side of the molecule and the outgoing halide departing from the other side. As a result, the reaction center is inverted during the process. The reaction is second order since the rate is dependent both on the alkyl halide and the incoming nucleophile. Primary and secondary alkyl halides can undergo the S_N2 mechanism, but tertiary alkyl halides react only very slowly.

S_N1 Mechanism

The S_N1 mechanism is a two-stage mechanism where the first stage is the rate determining step. In the first stage, the C–X bond is broken and the halogen is lost as a halide ion. The remaining alkyl portion becomes a planar carbocation. Since the first stage is the rate-determining step, the rate is dependent on the concentration of the alkyl halide alone. In the second stage of the mechanism, the incoming nucleophile can bond to either side of the carbocation to regenerate an sp^3 hybridized, tetrahedral center. The reaction is not stereospecific. Asymmetric alkyl halides will be fully or partially racemized during the reaction.

Related topics

sp^2 Hybridization (A4)
Configurational isomers – optical isomers (D3)
Charged species (E2)
Organic structures (E4)
Nucleophilic substitution (K2)
Elimination versus substitution (L5)
Reactions of alcohols (M4)

Definition

The presence of a strongly electrophilic carbon center makes alkyl halides susceptible to nucleophilic attack whereby a nucleophile displaces the halogen as a nucleophilic halide ion (*Fig. 1*). The reaction is known as nucleophilic substitution and there are two types of mechanism – the S_N1 and S_N2 mechanisms. Carboxylic acids and carboxylic acid derivatives also undergo nucleophilic substitutions (Topic K2), but the mechanisms are entirely different.

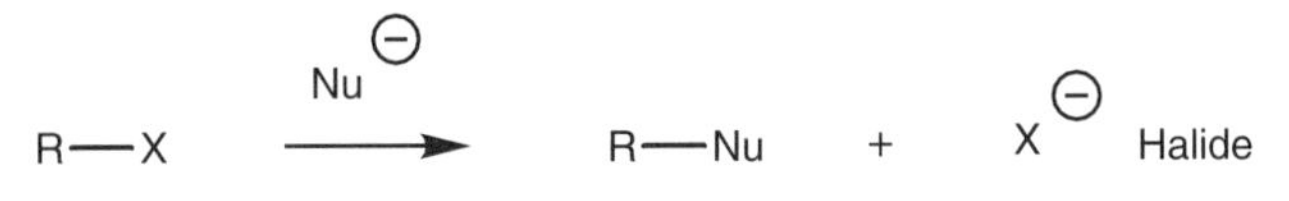

Fig. 1. Nucleophilic substitution.

S_N2 Mechanism

The reaction between methyl iodide and a hydroxide ion is an example of the S_N2 mechanism (*Fig. 2*). The hydroxide ion is a nucleophile and uses one of its lone pair of electrons to form a new bond to the electrophilic carbon of the alkyl halide. At the same time, the C–I bond breaks. Both electrons in that bond move onto the iodine to give it a fourth lone pair of electrons and a negative charge. Since iodine is electronegative, it can stabilize this charge, so the overall process is favored.

In the transition state for this process (*Fig. 3*), the new bond from the incoming nucleophile is partially formed and the C–X bond is partially broken. The reaction center itself (CH_3) is planar. This transition state helps to explain several other features of the S_N2 mechanism. First of all, both the alkyl halide and the nucleophile are required to form the transition state which means that the reaction rate is dependent on both components. Secondly, it can be seen that the hydroxide ion approaches iodomethane from one side while the iodide leaves from the opposite side. The hydroxide and the iodide ions are negatively charged and will repel each other, so it makes sense that they are as far apart as possible in the transition state. In addition, the hydroxide ion has to gain access to the reaction center – the electrophilic carbon. There is more room to attack from the 'rear' since the large iodine atom blocks approach from the other side. Lastly from an orbital point of view, it is proposed that the orbital from the incoming nucleophile starts to overlap with the empty antibonding orbital of the C–X bond (*Fig. 4*). As this interaction increases, the bonding interaction between the carbon and the halogen decreases until a transition state is reached where the incoming and outgoing nucleophiles

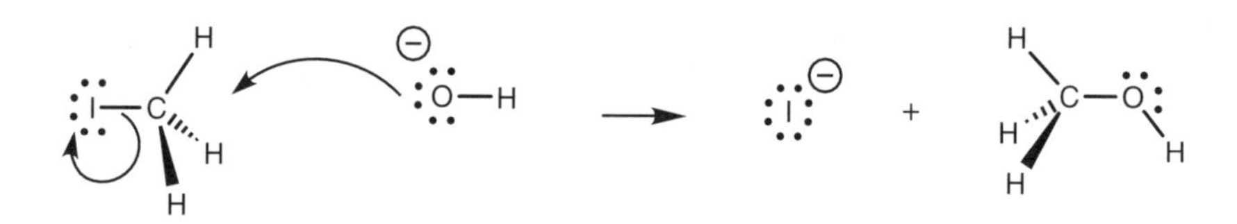

Fig. 2. S_N2 Mechanism for nucleophilic substitution.

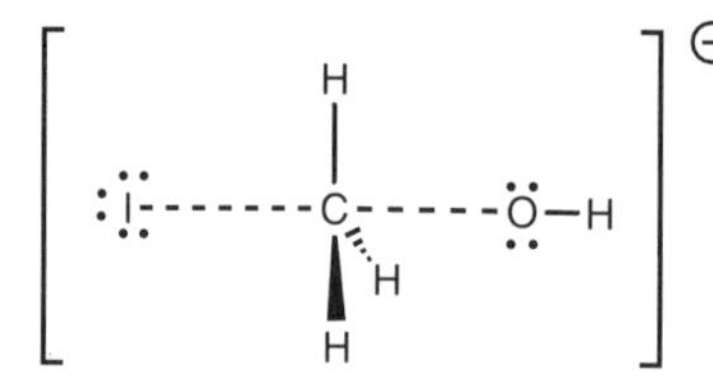

Fig. 3. Transition state for S_N2 nucleophilic substitution.

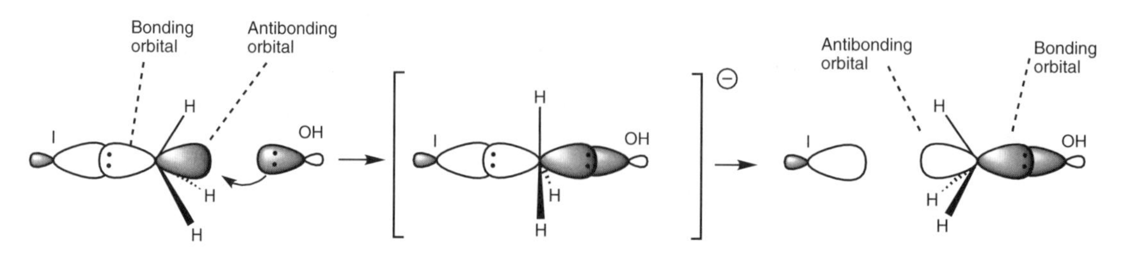

Fig. 4. Orbital interactions in the S_N2 mechanism.

are both partially bonded. The orbital geometry requires the nucleophiles to be on opposite sides of the molecule.

A third interesting feature about this mechanism concerns the three substituents on the carbon. Both the iodide and the alcohol product are tetrahedral compounds with the three hydrogens forming an 'umbrella' shape with the carbon (*Fig. 5*). However, the 'umbrella' is pointing in a different direction in the alcohol product compared to the alkyl halide. This means that the 'umbrella' has been turned inside out during the mechanism. In other words, the carbon center has been 'inverted'. The transition state is the halfway house in this inversion.

There is no way of telling whether inversion has taken place in a molecule such as iodomethane, but proof of this inversion can be obtained by looking at the nucleophilic substitution of asymmetric alkyl halides with the hydroxide ion (*Fig. 6*). Measuring the optical activity of both the alkyl halide and the alcohol allows the configuration of each enantiomer to be identified. This in turn demonstrates that inversion of the asymmetric center takes place. This inversion is known as the 'Walden Inversion' and the mechanism is known as the S_N2 mechanism. The S_N stands for 'substitution nucleophilic'. The 2 signifies that the rate of reaction is **second order** or **bimolecular** and depends on both the concentration of the nucleophile and the concentration of the alkyl halide. The S_N2 mechanism is possible for the nucleophilic substitutions of primary and secondary alkyl halides, but is difficult for tertiary alkyl halides. We can draw a general mechanism (*Fig. 7*) to account for a range of alkyl halides and charged nucleophiles. The mechanism is much the same with nucleophiles such as ammonia or amines – the only difference being that a salt is formed and an extra step is required in order to gain the free

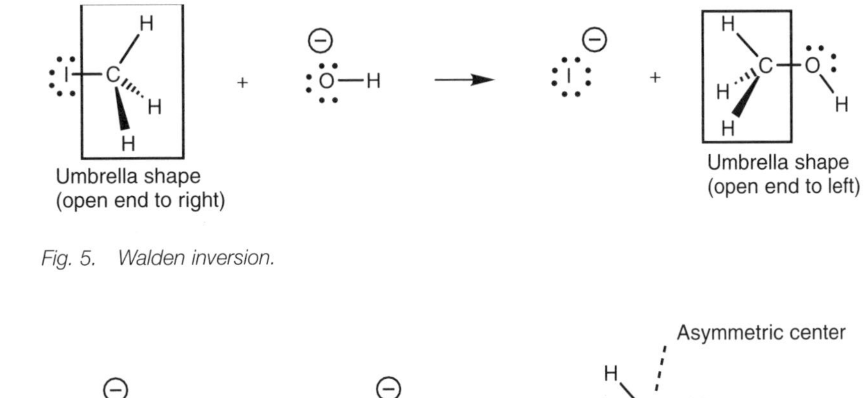

Fig. 5. Walden inversion.

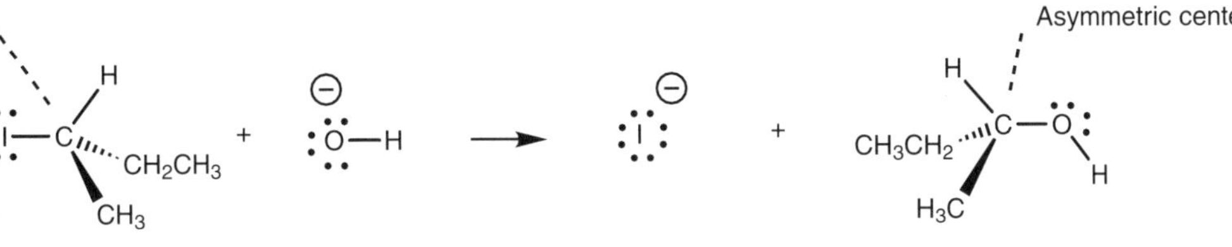

Fig. 6. Walden inversion of an asymmetric center.

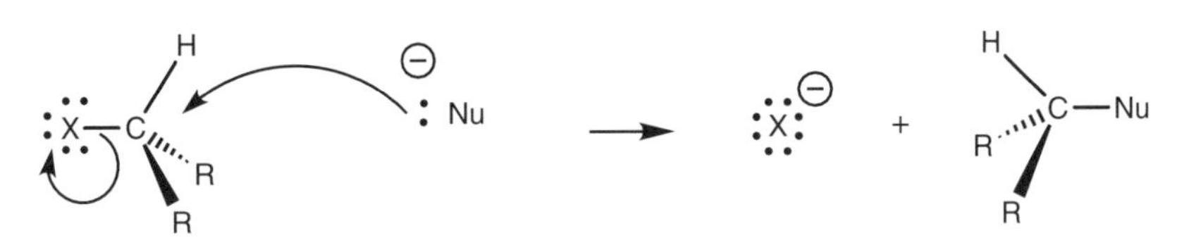

Fig. 7. General mechanism for the S_N2 nucleophilic substitution of alkyl halides.

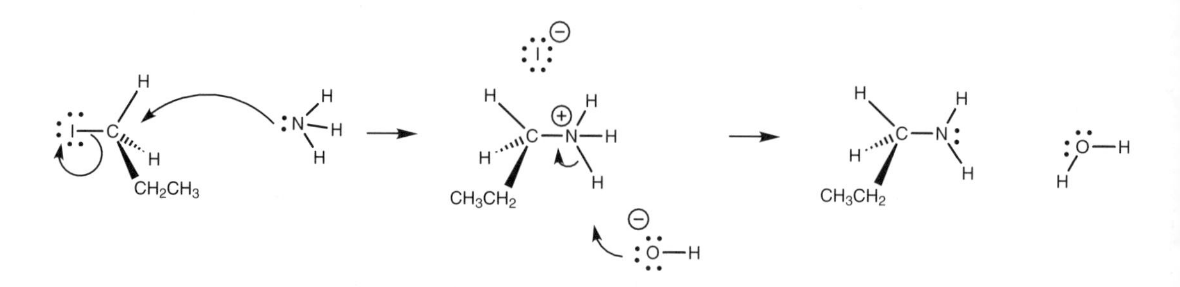

Fig. 8. S_N2 mechanism for the reaction of 1-iodopropane with ammonia.

amine. As an example, we shall consider the reaction between ammonia and 1-iodopropane (*Fig. 8*). Ammonia's nitrogen atom is the nucleophilic center for this reaction and uses its lone pair of electrons to form a bond to the alkyl halide. As a result, the nitrogen will effectively lose an electron and will gain a positive charge. The C–I bond is broken as previously described and an iodide ion is formed as a leaving group, which then acts as a counterion to the alkylammonium salt. The free amine can be obtained by reaction with sodium hydroxide. This neutralizes the amine to the free base which becomes insoluble in water and precipitates as a solid or as an oil.

The reaction of ammonia with an alkyl halide is a nucleophilic substitution as far as the alkyl halide is concerned. However, the same reaction can be viewed as an alkylation from the ammonia's point of view. This is because the ammonia has gained an alkyl group from the reaction.

Primary alkyl halides undergo the S_N2 reaction faster than secondary alkyl halides. Tertiary alkyl halides react extremely slowly if at all.

S_N1 Mechanism

When an alkyl halide is dissolved in a protic solvent such as ethanol or water, it is exposed to a nonbasic nucleophile (i.e. the solvent molecule). Under these conditions, the order of reactivity to nucleophilic substitution changes dramatically from that observed in the S_N2 reaction, such that tertiary alkyl halides are more reactive then secondary alkyl halides, with primary alkyl halides not reacting at all. Clearly a different mechanism must be involved. As an example, we shall consider the reaction of 2-iodo-2-methylpropane with water (*Fig. 9*). Here, the rate of reaction depends on the concentration of the alkyl halide alone and the concentration of the attacking nucleophile has no effect. Clearly, the nucleophile has to be present if the reaction is to take place, but it does not matter whether there is one equivalent of the nucleophile or an excess. Since the reaction rate only depends on the alkyl halide, the mechanism is known as the S_N1 reaction, where S_N stands for substitution nucleophilic and the 1 shows that the reaction is **first order** or **unimolecular**, that is, only one of the reactants affects the reaction rate.

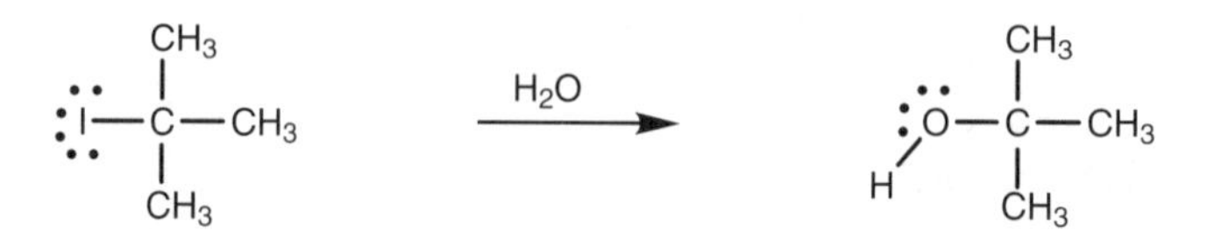

Fig. 9. Reaction of 2-iodo-2-methylpropane with water.

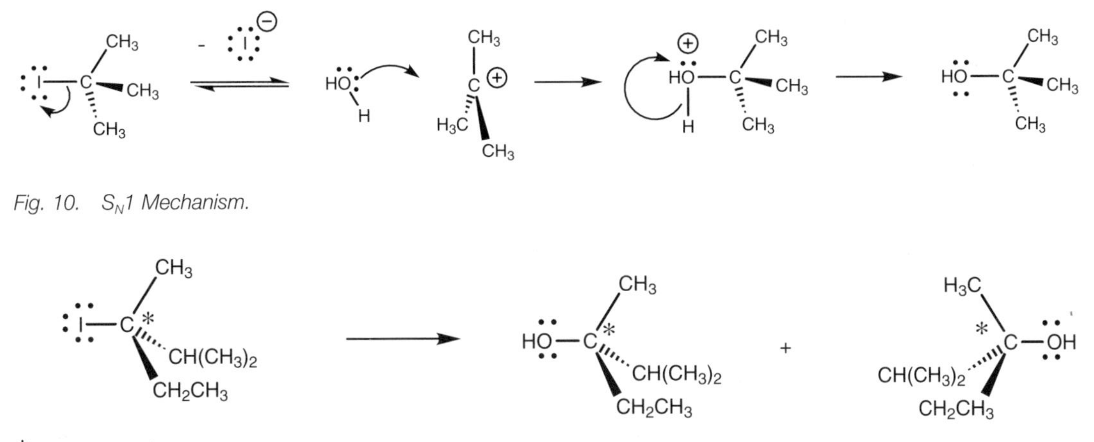

Fig. 10. S_N1 Mechanism.

* = Asymmetric center

Fig. 11. Racemization of an asymmetric center during S_N1 nucleophilic substitution.

There are two steps in the S_N1 mechanism (*Fig. 10*). The first step is the rate-determining step and involves loss of the halide ion. The C–I bond breaks with both electrons in the bond moving onto the iodine atom to give it a fourth lone pair of electrons and a negative charge. The alkyl portion becomes a planar carbocation where all three alkyl groups are as far apart from each other as possible. The central carbon atom is now sp^2 hybridized with an empty $2p_y$ orbital. In the second step, water acts as a nucleophile and reacts with the carbocation to form an alcohol.

In the mechanism shown, we have shown the water molecule coming in from the left hand side, but since the carbocation is planar, the water can attack equally well from the right hand side. Since the incoming nucleophile can attack from either side of the carbocation, there is no overall inversion of the carbon center. This is significant when the reaction is carried out on chiral molecules. For example, if a chiral alkyl halide reacts with water by the S_N1 mechanism, both enantiomeric alcohols would be formed resulting in a racemate (Topic D3; *Fig. 11*). However, it has to be stated that **total** racemization does not usually occur in S_N1 reactions. This is because the halide ion (departing from one side of the molecule) is still in the vicinity when the attacking nucleophile makes its approach. As a result the departing halide ion can hinder the approach of the attacking nucleophile from that particular side. The term stereo**specific** indicates that the mechanism results in one specific stereochemical outcome (e.g. the S_N2 mechanism). This is distinct from a reaction which is stereo**selective** where the mechanism can lead to more than one stereochemical outcome, but where there is a preference for one outcome over another. Many S_N1 reactions will show a slight stereoselectivity.

L3 FACTORS AFFECTING S_N2 VERSUS S_N1 REACTIONS

Key Notes

S_N1 versus S_N2

The nature of the nucleophile, the solvent, and the alkyl halide determine whether nucleophilic substitution takes place by the S_N1 or the S_N2 mechanism. With polar aprotic solvents, primary alkyl halides react faster than secondary halides by the S_N2 mechanism, whereas tertiary alkyl halides hardly react at all. With polar protic solvents and nonbasic nucleophiles, tertiary alkyl halides react faster than secondary alkyl halides by the S_N1 mechanism, and primary halides do not react. The reactivity of primary, secondary, and tertiary alkyl halides is controlled by electronic and steric factors.

Solvent

Polar, aprotic solvents are used for S_N2 reactions since they solvate cations but not anions. As a result, nucleophiles are 'naked' and more reactive. Protic solvent such as water or alcohol are used in S_N1 reactions since they solvate and stabilize the intermediate carbocation. The nucleophile is also solvated, but this has no effect on the reaction rate since the rate is dependent on the concentration of the alkyl halide.

Nucleophilicity

The rate of the S_N2 reaction increases with the nucleophilic strength of the incoming nucleophile. The rate of the S_N1 reaction is unaffected by the nature of the nucleophile.

Leaving group

The reaction rates of both the S_N1 and the S_N2 reaction is increased if the leaving group is a stable ion and a weak base. Iodide is a better leaving group than bromide and bromide is a better leaving group than chloride. Alkyl fluorides do not undergo nucleophilic substitution.

Alkyl halides – S_N2

Tertiary alkyl groups are less likely to react by the S_N2 mechanism than primary or secondary alkyl halides since the presence of three alkyl groups linked to the reaction center lowers the electrophilicity of the alkyl halide by inductive effects. Tertiary alkyl halides have three bulky alkyl groups attached to the reaction center which act as steric shields and hinder the approach of nucleophiles. Primary alkyl halides only have one alkyl group attached to this center and so access is easier.

Alkyl halides – S_N1

Formation of a planar carbocation in the first stage of the S_N1 mechanism is favored for tertiary alkyl halides since it relieves the steric strain in the crowded tetrahedral alkyl halide. The carbocation is also more accessible to an incoming nucleophile. The formation of the carbocation is helped by electronic factors involving the inductive and hyperconjugation effects of the three neighboring alkyl groups. Such inductive and hyperconjugation effects are greater in carbocations formed from tertiary alkyl halides than from those formed from primary or secondary alkyl halides.

Determining the mechanism

Measuring how the reaction rate is affected by the concentration of the alkyl halide and the nucleophile determines whether a nucleophilic substitution is S_N2 or S_N1. Measuring the optical activity of products from the nucleophilic substitution of asymmetric alkyl halides indicates the type of mechanism involved. A pure enantiomeric product indicates an S_N2 reaction. A partially or fully racemized product indicates an S_N1 reaction.

Related topics

Base strength (G3)
Carbocation stabilization (H5)
Nucleophilic substitution (L2)

S_N1 versus S_N2

There are two different mechanisms involved in the nucleophilic substitution of alkyl halides. When polar aprotic solvents are used, the S_N2 mechanism is preferred. Primary alkyl halides react more quickly than secondary alkyl halides, with tertiary alkyl halides hardly reacting at all. Under protic solvent conditions with nonbasic nucleophiles (e.g. dissolving the alkyl halide in water or alcohol), the S_N1 mechanism is preferred and the order of reactivity is reversed. Tertiary alkyl halides are more reactive than secondary alkyl halides, and primary alkyl halides do not react at all.

There are several factors which determine whether substitution will be S_N1 or S_N2 and which also control the rate at which these reactions take place. These include the nature of the nucleophile and the type of solvent used. The reactivity of primary, secondary, and tertiary alkyl halides is controlled by electronic and steric factors.

Solvent

The S_N2 reaction works best in polar aprotic solvents (i.e. solvents with a high dipole moment, but with no O–H or N–H groups). These include solvents such as acetonitrile (CH_3CN) or dimethylformamide (DMF). These solvents are polar enough to dissolve the ionic reagents required for nucleophilic substitution, but they do so by solvating the metal cation rather than the anion. Anions are solvated by hydrogen bonding and since the solvent is incapable of hydrogen bonding, the anions remain unsolvated. Such 'naked' anions retain their nucleophilicity and react more strongly with electrophiles.

Polar, protic solvents such as water or alcohols can also dissolve ionic reagents but they solvate both the metal cation and the anion. As a result, the anion is 'caged' in by solvent molecules. This stabilizes the anion, makes it less nucleophilic and makes it less likely to react by the S_N2 mechanism. As a result, the S_N1 mechanism becomes more important.

The S_N1 mechanism is particularly favored when the polar protic solvent is also a nonbasic nucleophile. Therefore, it is most likely to occur when an alkyl halide is dissolved in water or alcohol. Protic solvents are bad for the S_N2 mechanism since they solvate the nucleophile, but they are good for the S_N1 mechanism. This is because polar protic solvents can solvate and stabilize the carbocation intermediate. If the carbocation is stabilized, the transition state leading to it will also be stabilized and this determines whether the S_N1 reaction is favored or not. Protic solvents will also solvate the nucleophile by hydrogen bonding, but unlike the S_N2 reaction, this does not affect the reaction rate since the rate of reaction is independent of the nucleophile.

Nonpolar solvents are of no use in either the S_N1 or the S_N2 reaction since they cannot dissolve the ionic reagents required for nucleophilic substitution.

Nucleophilicity

The relative nucleophilic strengths of incoming nucleophiles will affect the rate of the S_N2 reaction with stronger nucleophiles reacting faster. A charged nucleophile is stronger than the corresponding uncharged nucleophile (e.g. alkoxide ions are stronger nucleophiles than alcohols). Nucleophilicity is also related to base strength when the nucleophilic atom is the same (e.g. $RO^- > HO^- > RCO_2^- > ROH > H_2O$) (Topic G3). In polar aprotic solvents, the order of nucleophilic strength for the halides is $F^- > Cl^- > Br^- > I^-$.

Since the rate of the S_N1 reaction is independent of the incoming nucleophile, the nucleophilicity of the incoming nucleophile is unimportant.

Leaving group

The nature of the leaving group is important to both the S_N1 and S_N2 reactions – the better the leaving group, the faster the reaction. In the transition states of both reactions, the leaving group has gained a partial negative charge and the better that can be stabilized, the more stable the transition state and the faster the reaction. Therefore, the best leaving groups are the ones which form the most stable anions. This is also related to basicity in the sense that the more stable the anion, the weaker the base. Iodide and bromide ions are stable ions and weak bases, and prove to be good leaving groups. The chloride ion is less stable, more basic and a poorer leaving group. The fluoride ion is a very poor leaving group and as a result alkyl fluorides do not undergo nucleophilic substitution. The need for a stable leaving group explains why alcohols, ethers, and amines do not undergo nucleophilic substitutions since they would involve the loss of a strong base (e.g. RO^- or R_2N^-).

Alkyl halides – S_N2

There are two factors which affect the rate at which alkyl halides undergo the S_N2 reaction – electronic and steric. In order to illustrate why different alkyl halides react at different rates in the S_N2 reaction, we shall compare a primary, secondary, and tertiary alkyl halide (*Fig. 1*).

Alkyl groups have an inductive, electron-donating effect which tends to lower the electrophilicity of the neighboring carbon center. Lowering the electrophilic strength means that the reaction center will be less reactive to nucleophiles. Therefore, tertiary alkyl halides will be less likely to react with nucleophiles than primary alkyl halides, since the inductive effect of three alkyl groups is greater than one alkyl group.

Steric factors also play a role in making the S_N2 mechanism difficult for tertiary halides. An alkyl group is a bulky group compared to a hydrogen atom, and can therefore act like a shield against any incoming nucleophile (*Fig. 2*). A tertiary alkyl halide has three alkyl shields compared to the one alkyl shield of a primary alkyl halide. Therefore, a nucleophile is more likely to be deflected when it approaches a tertiary alkyl halide and fails to reach the electrophilic center.

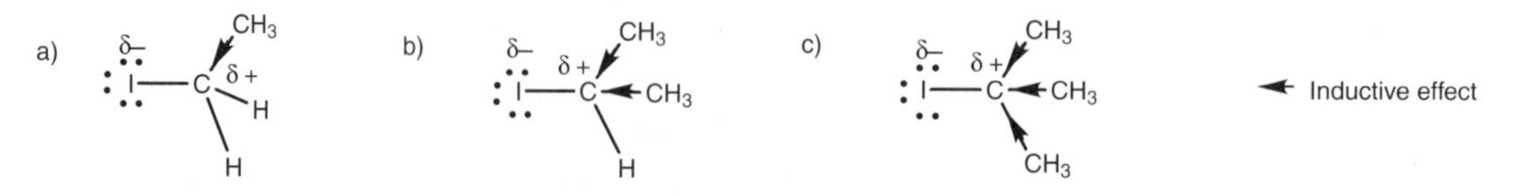

Fig. 1. (a) Iodoethane; (b) 2-iodopropane; (c) 2-iodo-2-methylpropane.

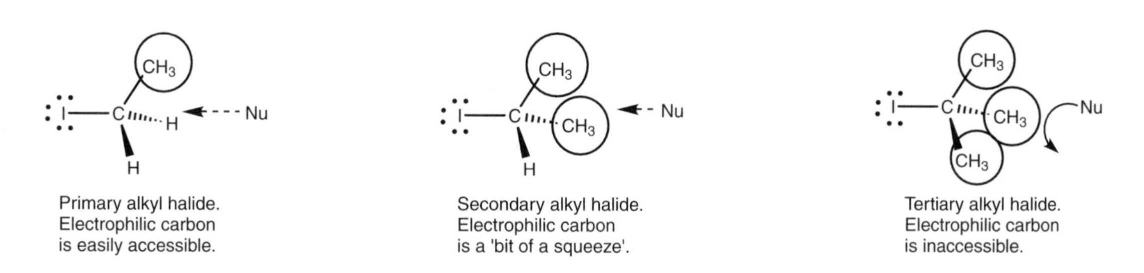

Fig. 2. Steric factors affecting nucleophilic substitution.

Alkyl halides – S_N1

Steric and electronic factors also play a role in the rate of the S_N1 reaction. Since the steric bulk of three alkyl substituents makes it very difficult for a nucleophile to reach the electrophilic carbon center of tertiary alkyl halides, these structures undergo nucleophilic substitution by the S_N1 mechanism instead. In this mechanism, the steric problem is relieved because loss of the halide ion creates a planar carbocation where the alkyl groups are much further apart and where the carbon center is more accessible. Formation of the carbocation also relieves steric strain between the substituents.

Electronic factors also help in the formation of the carbocation since the positive charge can be stabilized by the inductive and hyperconjugative effects of the three alkyl groups (cf. Topic H5; *Fig. 3*).

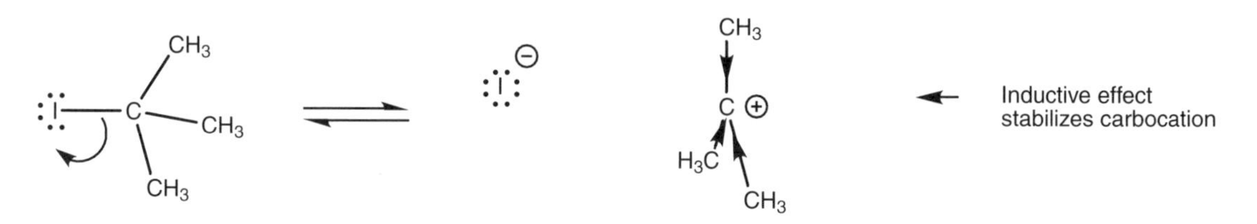

Fig. 3. Inductive effects stabilizing a carbocation.

Both the inductive and hyperconjugation effects are greater when there are three alkyl groups connected to the carbocation center than when there are only one or two. Therefore, tertiary alkyl halides are far more likely to produce a stable carbocation intermediate than primary or secondary alkyl halides. It is important to realize that the reaction rate is determined by how well the **transition state** of the rate determining step is stabilized. In a situation like this where a high energy intermediate is formed (i.e. the carbocation), the transition state leading to it will be closer in character to the intermediate than the starting material. Therefore, any factor which stabilizes the intermediate carbocation also stabilizes the transition state and consequently increases the reaction rate.

Determining the mechanism

It is generally fair to say that the nucleophilic substitution of primary alkyl halides will take place via the S_N2 mechanism, whereas nucleophilic substitution of tertiary alkyl halides will take place by the S_N1 mechanism. In general, secondary alkyl halides are more likely to react by the S_N2 mechanism, but it is not possible to predict this with certainty. The only way to find out for certain is to try out the reaction and see whether the reaction rate depends on the concentration of both reactants (S_N2) or whether it depends on the concentration of the alkyl halide alone (S_N1).

If the alkyl halide is chiral the optical rotation of the product could be measured to see whether it is a pure enantiomer or not. If it is, the mechanism is S_N2. If not, it is S_N1.

L4 ELIMINATION

Key Notes

Definition

Alkyl halides undergo elimination reactions with nucleophiles or bases, where hydrogen halide is lost from the molecule to produce an alkene. There are two commonly occurring mechanisms. The E2 mechanism is the most effective for the synthesis of alkenes from alkyl halides and can be used on primary, secondary, and tertiary alkyl halides. The E1 reaction is not particularly useful from a synthetic point of view and occurs in competition with the S_N1 reaction. Tertiary alkyl halides and some secondary alkyl halides can react by this mechanism, but not primary alkyl halides.

Susceptible β-protons

Elimination is possible if the alkyl halide contains a susceptible β-proton which can be lost during the elimination reaction. β-Protons are situated on the carbon linked to the carbon of the C–X group.

E2 Mechanism

The E2 mechanism is a concerted, one-stage process involving both alkyl halide and nucleophile. The reaction is second order and depends on the concentration of both reactants.

E1 Mechanism

The E1 mechanism is a two-stage process involving loss of the halide ion to form a carbocation, followed by loss of the susceptible proton to form the alkene. The rate determining step is the first stage involving loss of the halide ion. As a result, the reaction is first order, depending on the concentration of the alkyl halide alone. The carbocation intermediate is stabilized by substituent alkyl groups.

E2 versus E1

The E2 reaction is more useful than the E1 reaction in synthesizing alkenes. The use of a strong base in a protic solvent favors the E2 elimination over the E1 elimination. The E1 reaction occurs when tertiary alkyl halides are dissolved in protic solvents.

Related topics

Electrophilic addition to symmetrical alkenes (H3)

Reactions of alcohols (M4)

Definition

Alkyl halides which have a proton attached to a neighboring β-carbon atom can undergo an elimination reaction to produce an alkene plus a hydrogen halide (*Fig. 1*). In essence, this reaction is the reverse of the electrophilic addition of a hydrogen halide to an alkene (Topic H3). There are two mechanisms by which this elimination can take place – the E2 mechanism and the E1 mechanism.

The E2 reaction is the most effective for the synthesis of alkenes from alkyl halides and can be used on primary, secondary, and tertiary alkyl halides. The E1 reaction is not particularly useful from a synthetic point of view and occurs in

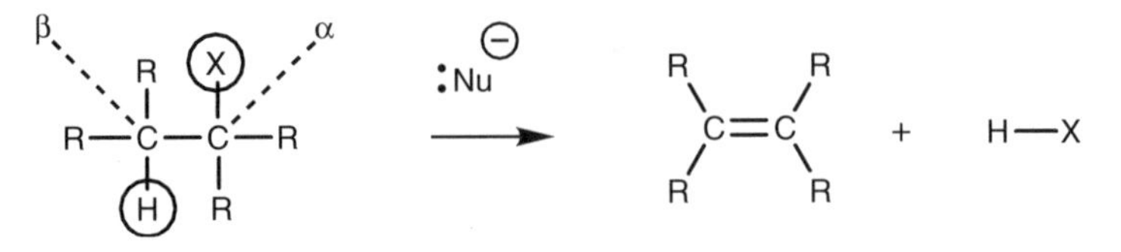

Fig. 1. Elimination of an alkyl halide.

competition with the S_N1 reaction of tertiary alkyl halides. Primary and secondary alkyl halides do not usually react by this mechanism.

Susceptible β-protons

An alkyl halide can undergo an elimination reaction if it has a susceptible proton situated on a β-carbon, that is, the carbon next to the C–X group. This proton is lost during the elimination reaction along with the halide ion. In some respects, there is a similarity here between alkyl halides and carbonyl compounds (*Fig. 2*). Alkyl halides can have susceptible protons at the β-position whilst carbonyl compounds can have acidic protons at their α-position. If we compare both structures, we can see that the acidic/susceptible proton is attached to a carbon neighboring an electrophilic carbon.

E2 Mechanism

The E2 mechanism is a concerted mechanism involving both the alkyl halide and the nucleophile. As a result, the reaction rate depends on the concentration of both reagents and is defined as second order (E2 = Elimination second order). To illustrate the mechanism, we shall look at the reaction of 2-bromopropane with a hydroxide ion (*Fig. 3*).

The mechanism (*Fig. 4*) involves the hydroxide ion forming a bond to the susceptible proton. As the hydroxide ion forms its bond, the C–H bond breaks. Both electrons in that bond could move onto the carbon, but there is a neighboring electrophilic carbon which attracts the electrons and so the electrons move in to form a π bond between the two carbons. At the same time as this π bond is formed, the C–Br bond breaks and both electrons end up on the bromine atom which is lost as a bromide ion.

The E2 elimination is stereospecific, with elimination occurring in an antiperiplanar geometry. The diagrams (*Fig. 5*) show that the four atoms involved in the reaction are in a plane with the H and Br on opposite sides of the molecule.

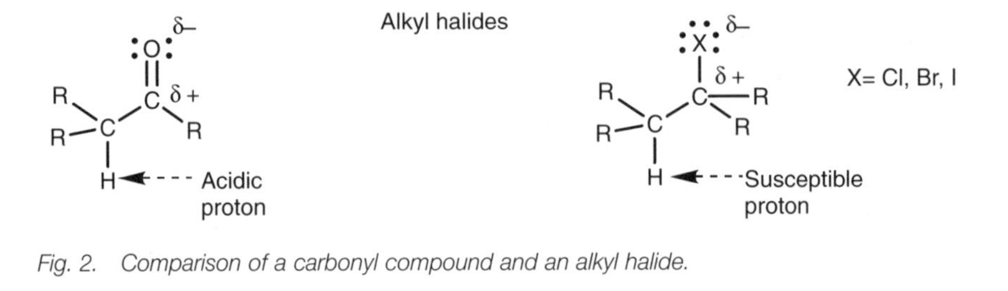

Fig. 2. Comparison of a carbonyl compound and an alkyl halide.

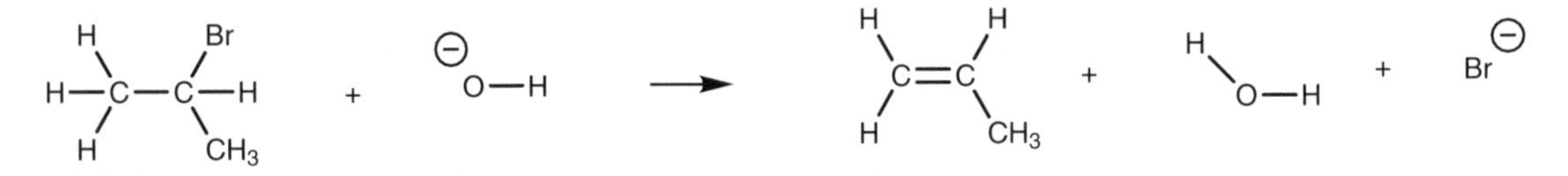

Fig. 3. Reaction of 2-bromopropane with the hydroxide ion.

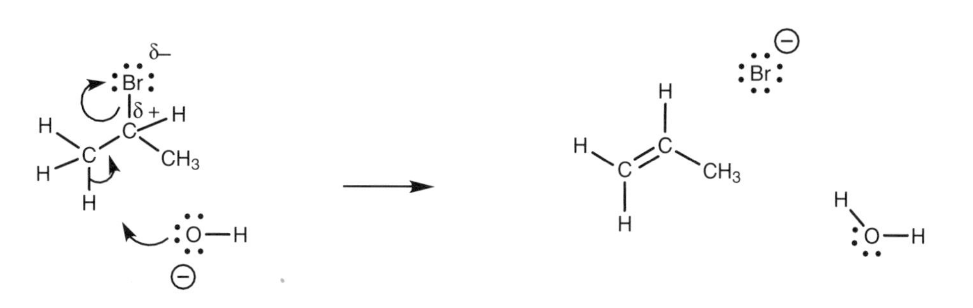

Fig. 4. E2 Elimination mechanism.

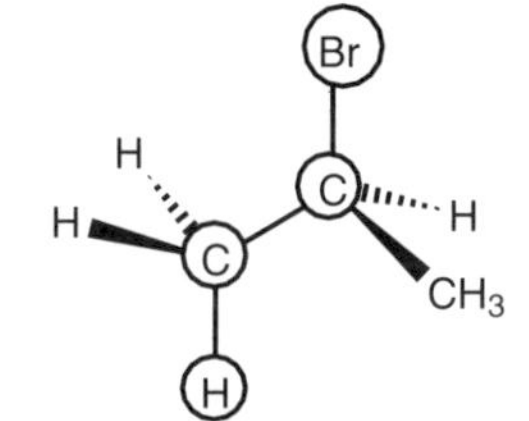

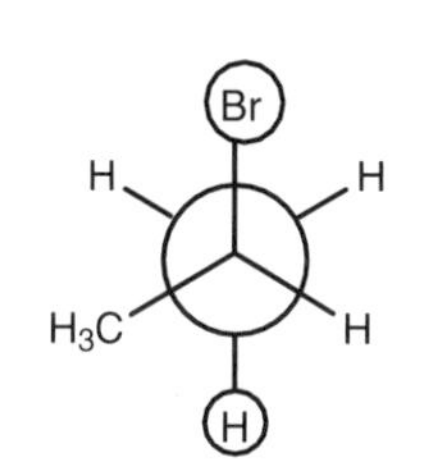

Fig. 5. Relative geometry of the atoms involved in the E2 elimination mechanism.

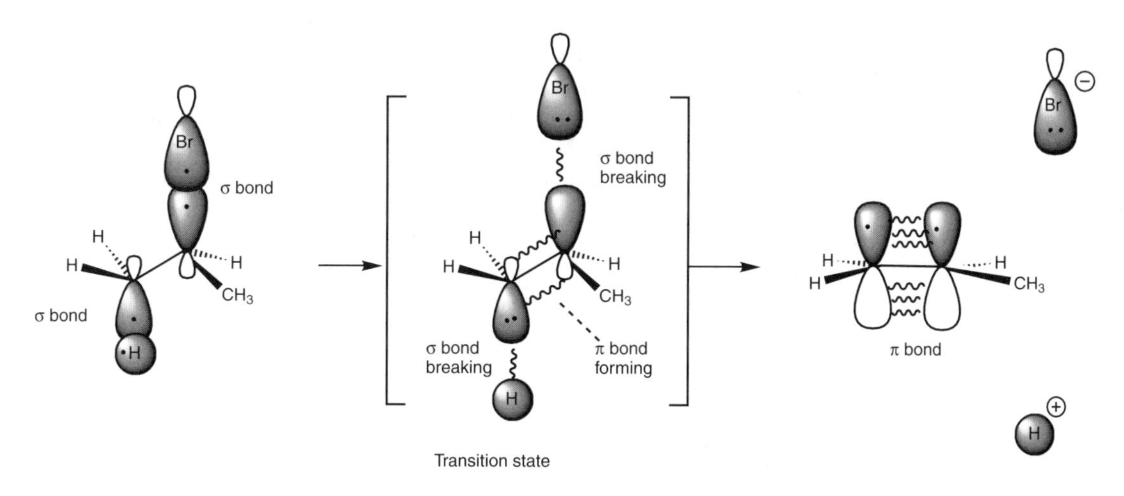

Fig. 6. Orbital diagram of the E2 elimination process.

The reason for this stereospecificity can be explained using orbital diagrams (*Fig. 6*). In the transition state of this reaction, the C–H and C–Br σ bonds are in the process of breaking. As they do so, the sp^3 hybridized orbitals which were used for these σ bonds are changing into *p* orbitals which start to interact with each other to form the eventual π bond. For all this to happen in the one transition state, an antiperiplanar arrangement is essential.

E1 Mechanism

The E1 mechanism usually occurs when an alkyl halide is dissolved in a protic solvent where the solvent can act as a nonbasic nucleophile. These are the same conditions for the S_N1 reaction and so both these reactions usually occur at the

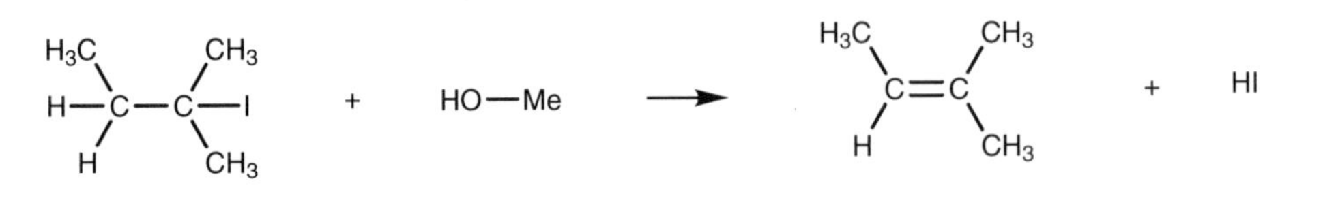

Fig. 7. Elimination reaction of 2-iodo-2-methyl-butane.

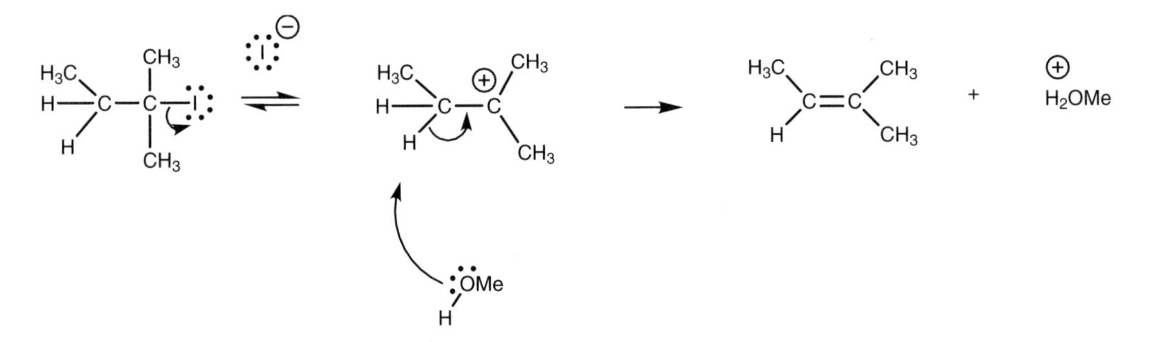

Fig. 8. The E1 mechanism.

same time resulting in a mixture of products. As an example of the E1 mechanism, we shall look at the reaction of 2-iodo-2-methyl-butane with methanol (*Fig. 7*).

There are two stages to this mechanism (*Fig. 8*). The first stage is exactly the same process described for the S_N1 mechanism and that is cleavage of the C–X bond to form a planar carbocation intermediate where the positive charge is stabilized by the three alkyl groups surrounding it. In the second stage, the methanol forms a bond to the susceptible proton on the β-carbon. The C–H bond breaks and both electrons are used to form a π bond to the neighboring carbocation. The first step of the reaction mechanism is the rate-determining step and since this is only dependent on the concentration of the alkyl halide, the reaction is first order (El = elimination first order). There is no stereospecificity involved in this reaction and a mixture of isomers may be obtained with the more stable (more substituted) alkene being favored.

E2 versus E1

The E2 elimination occurs with a strong base (such as a hydroxide or ethoxide ion) in a protic solvent (such as ethanol or water). The E2 reaction is more common than the E1 elimination and more useful. All types of alkyl halide can undergo the E2 elimination and the method is useful for preparing alkenes.

The conditions which suit E1 are the same which suit the S_N1 reaction (i.e. a protic solvent and a nonbasic nucleophile). Therefore, the E1 reaction normally only occurs with tertiary alkyl halides and will be in competition with the S_N1 reaction.

L5 ELIMINATION VERSUS SUBSTITUTION

Key Notes

Introduction

The ratio of substitution and elimination products formed from alkyl halides depends on the reaction conditions as well as the nature of the nucleophile and the alkyl halide.

Primary alkyl halides

Primary alkyl halides will usually undergo S_N2 substitution reactions in preference to E2 elimination reactions. However, the E2 elimination reaction is favored if a strong bulky base is used with heating.

Secondary alkyl halides

Substitution by the S_N2 mechanism is favored over the E2 elimination if the nucleophile is a weak base and the solvent is polar and aprotic. E2 elimination is favored over the S_N2 reaction if a strong base is used in a protic solvent. Elimination is further favored by heating. S_N1 and E1 reactions may be possible when dissolving secondary alkyl halides in protic solvents.

Tertiary alkyl halides

E2 elimination occurs virtually exclusively if a tertiary alkyl halide is treated with a strong base in a protic solvent. Heating a tertiary alkyl halide in a protic solvent is likely to produce a mixture of S_N1 substitution and E1 elimination products, with the former being favored.

Related topics

Base strength (G3)
Nucleophilic substitution (L2)
Factors affecting S_N2 and S_N1 reactions (L3)
Elimination (L4)

Introduction

Alkyl halides can undergo both elimination and substitution reactions and it is not unusual to find both substitution and elimination products present. The ratio of the products will depend on the reaction conditions, the nature of the nucleophile and the nature of the alkyl halide.

Primary alkyl halides

Primary alkyl halides undergo the S_N2 reaction with a large range of nucleophiles (e.g. RS^-, I^-, CN^-, NH_3, or Br^-) in polar aprotic solvents such as hexamethylphosphoramide (HMPA; $[(CH_3)_2N]_3PO$). However, there is always the possibility of some E2 elimination occurring as well. Nevertheless, substitution is usually favored over elimination, even when using strong bases such as HO^- or EtO^-. If E2 elimination of a primary halide is desired, it is best to use a strong bulky base such as *tert*-butoxide $[(CH_3)_3C\text{–}O^-]$. With a bulky base, the elimination product is favored over the substitution product since the bulky base experiences more steric hindrance in its approach to the electrophilic carbon than it does to the acidic β-proton.

Thus, treatment of a primary halide (*Fig. 1*) with an ethoxide ion is likely to give

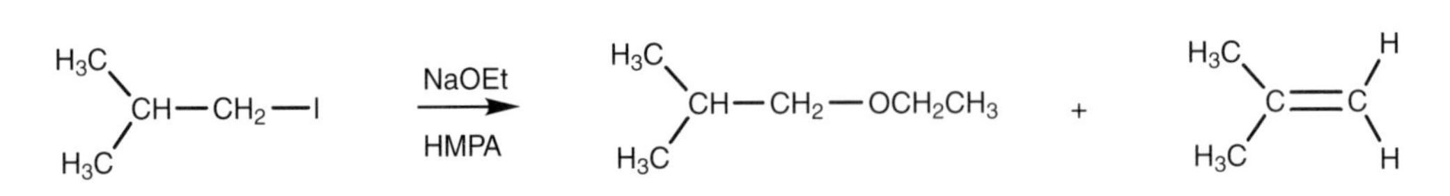

Fig. 1. Reaction of 1-iodo-2-methylpropane with sodium ethoxide.

a mixture of an ether arising from S_N2 substitution along with an alkene arising from E2 elimination, with the ether being favored. By using sodium *tert*-butoxide instead, the preferences would be reversed.

Increasing the temperature of the reaction shifts the balance from the S_N2 reaction to the elimination reaction. This is because the elimination reaction has a higher activation energy due to more bonds being broken. The S_N1 and E1 reactions do not occur for primary alkyl halides.

Secondary alkyl halides

Secondary alkyl halides can undergo both S_N2 and E2 reactions to give a mixture of products. However, the substitution product predominates if a polar aprotic solvent is used and the nucleophile is a weak base. Elimination will predominate if a strong base is used as the nucleophile in a polar, protic solvent. In this case, bulky bases are not so crucial and the use of ethoxide in ethanol will give more elimination product than substitution product. Increasing the temperature of the reaction favors E2 elimination over S_N2 substitution as explained above.

If weakly basic or nonbasic nucleophiles are used in protic solvents, elimination and substitution may occur by the S_N1 and E1 mechanisms to give mixtures.

Tertiary alkyl halides

Tertiary alkyl halides are essentially unreactive to strong nucleophiles in polar, aprotic solvents – the conditions for the S_N2 reaction. Tertiary alkyl halides can undergo E2 reactions when treated with a strong base in a protic solvent and will do so in good yield since the S_N2 reaction is so highly disfavored. Under nonbasic conditions in a protic solvent, E1 elimination and S_N1 substitution both take place.

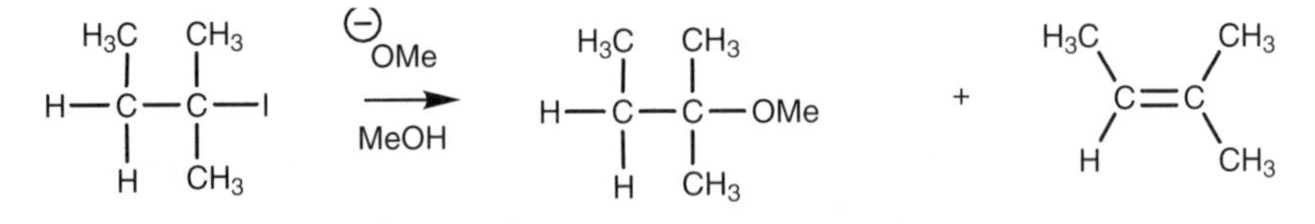

Fig. 2. Reaction of 1-iodo-2-methylbutane with methoxide ion.

A tertiary alkyl halide treated with sodium methoxide could give an ether or an alkene (*Fig.* 2). A protic solvent is used here and this favors both the S_N1 and E1 mechanisms. However, a strong base is also being used and this favors the E2 mechanism. Therefore, the alkene would be expected to be the major product with only a very small amount of substitution product arising from the S_N2 reaction. Heating the same alkyl halide in methanol alone means that the reaction is being carried out in a protic solvent with a nonbasic nucleophile (MeOH). These conditions would result in a mixture of substitution and elimination products arising from the S_N1 and E1 mechanisms. The substitution product would be favored over the elimination product.

L6 Reactions of alkyl halides

Key Notes

Nucleophilic substitution

Primary and secondary alkyl halides can be converted to a large variety of different functional groups such as alcohols, ethers, thiols, thioethers, esters, amines, azides, alkyl iodides, and alkyl chlorides. Larger carbon skeletons can be created by the reaction of alkyl halides with nitrile, acetylide, and enolate ions. Substitution reactions of tertiary alkyl halides are more difficult and less likely to give good yields.

Eliminations

Elimination reactions of alkyl halides allow the synthesis of alkenes and are best carried out using a strong base in a protic solvent with heating. Primary, secondary, and tertiary alkyl halides can all be used. However, the use of a bulky base is advisable for primary alkyl halides. The most substituted alkene is favored if there are several possible alkene products.

Related topics Nucleophilic substitution (L2) Elimination (L4)

Nucleophilic substitution

The nucleophilic substitution of alkyl halides is one of the most powerful methods of obtaining a wide variety of different functional groups. Therefore, it is possible to convert a variety of primary and secondary alkyl halides to alcohols, ethers, thiols, thioethers, esters, amines, and azides (*Fig. 1*). Alkyl iodides and alkyl chlorides can also be synthesized from other alkyl halides.

It is also possible to construct larger carbon skeletons using alkyl halides. A simple example is the reaction of an alkyl halide with a cyanide ion (*Fig. 2*). This is an

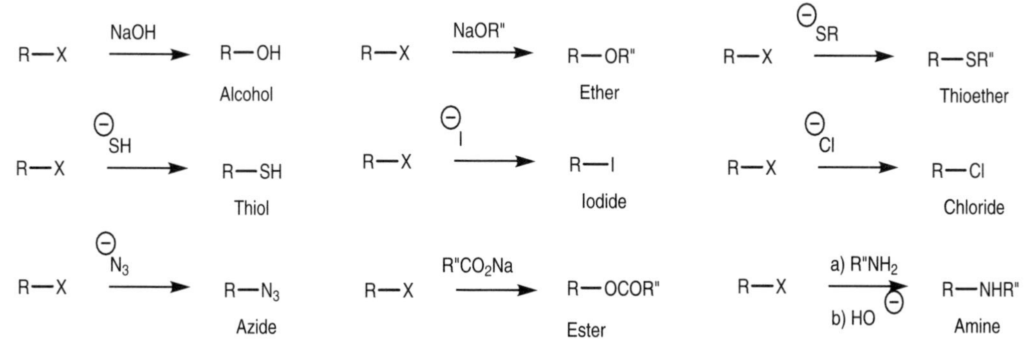

Fig. 1. Nucleophilic substitutions of alkyl halides.

$$\text{R—X} \xrightarrow{\text{NaCN}} \underset{\text{Nitrile}}{\text{R—CN}} \xrightarrow{H_3O^{+}} \text{R—}CO_2H$$

Fig. 2. Constructing larger carbon skeletons from an alkyl halide.

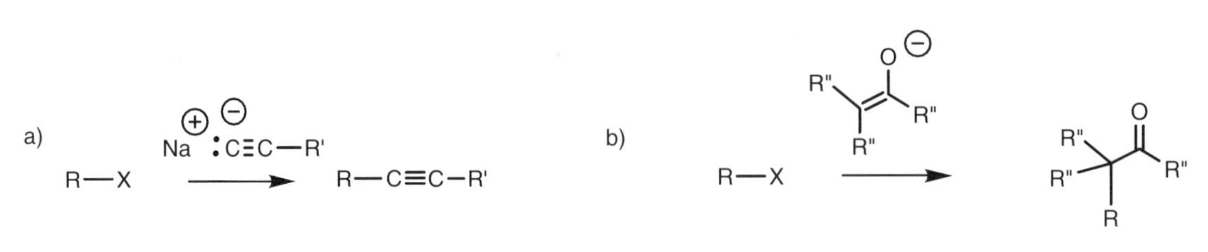

Fig. 3. (a) Synthesis of a disubstituted alkyne; (b) alkylation of an enolate ion.

important reaction since the nitrile product can be hydrolyzed to give a carboxylic acid (Topic O3).

The reaction of an acetylide ion with a primary alkyl halide allows the synthesis of disubstituted alkynes (*Fig. 3a*). The enolate ions of esters or ketones can also be alkylated with alkyl halides to create larger carbon skeletons (*Fig. 3b*; Topics K7 and J9). The most successful nucleophilic substitutions are with primary alkyl halides. With secondary and tertiary alkyl halides, the elimination reaction may compete, especially when the nucleophile is a strong base. The substitution of tertiary alkyl halides is best done in a protic solvent with weakly basic nucleophiles. However, yields may be poor.

Eliminations

Elimination reactions of alkyl halides (**dehydrohalogenations**) are a useful method of synthesizing alkenes. For best results, a strong base (e.g. NaOEt) should be used in a protic solvent (EtOH) with a secondary or tertiary alkyl halide. The reaction proceeds by an E2 mechanism. Heating increases the chances of elimination over substitution.

For primary alkyl halides, a strong, bulky base (e.g. $NaOBu^t$) should be used. The bulk hinders the possibility of the S_N2 substitution and encourages elimination by the E2 mechanism. The advantage of the E2 mechanism is that it is higher yielding than the E1 mechanism and is also stereospecific. The geometry of the product obtained is determined by the antiperiplanar geometry of the transition state. For example, the elimination in *Fig. 4* gives the (*E*)-isomer and none of the (*Z*)-isomer.

If the elimination occurs by the E1 mechanism, the reaction is more likely to

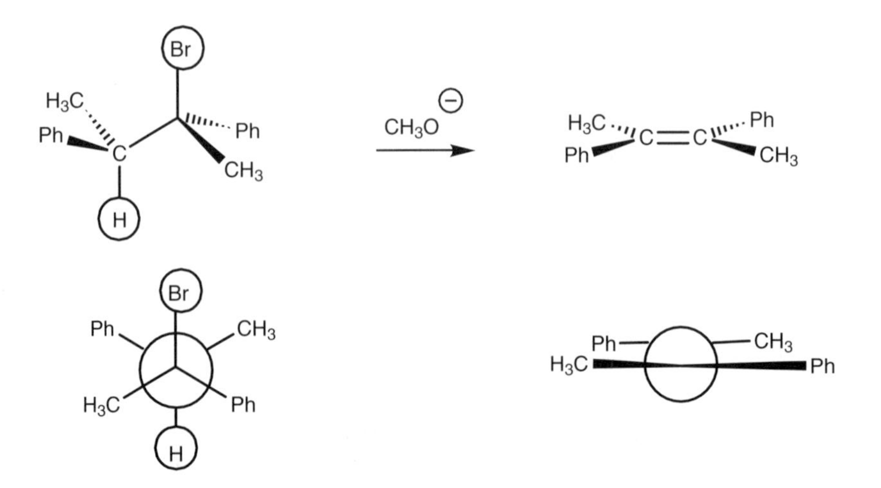

Fig. 4. Stereochemistry of the E2 elimination reaction.

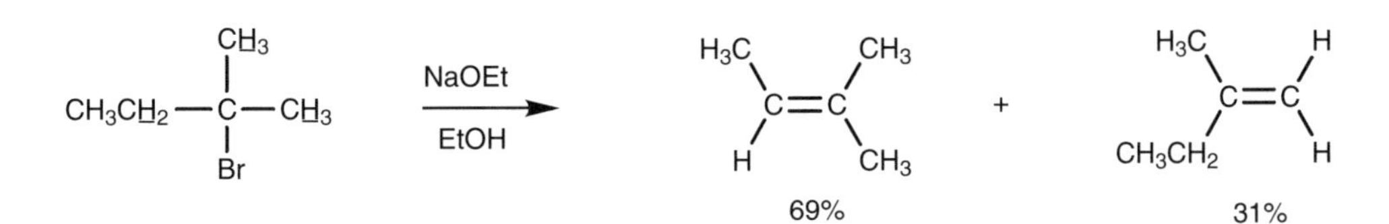

Fig. 5. Example of Zaitsev's rule.

compete with the S_N1 reaction and a mixture of substitution and elimination products is likely.

The E2 elimination requires the presence of a β-proton. If there are several options available, a mixture of alkenes will be obtained, but the favored alkene will be the most substituted (and most stable) one (**Zaitsev's rule**; *Fig. 5*).

The transition state for the reaction resembles the product more than the reactant and so the factors which stabilize the product also stabilize the transition state and make that particular route more likely (see Topic L4). However, the opposite preference is found when potassium *tert*-butoxide is used as base, and the less substituted alkene is favored.

L7 ORGANOMETALLIC REACTIONS

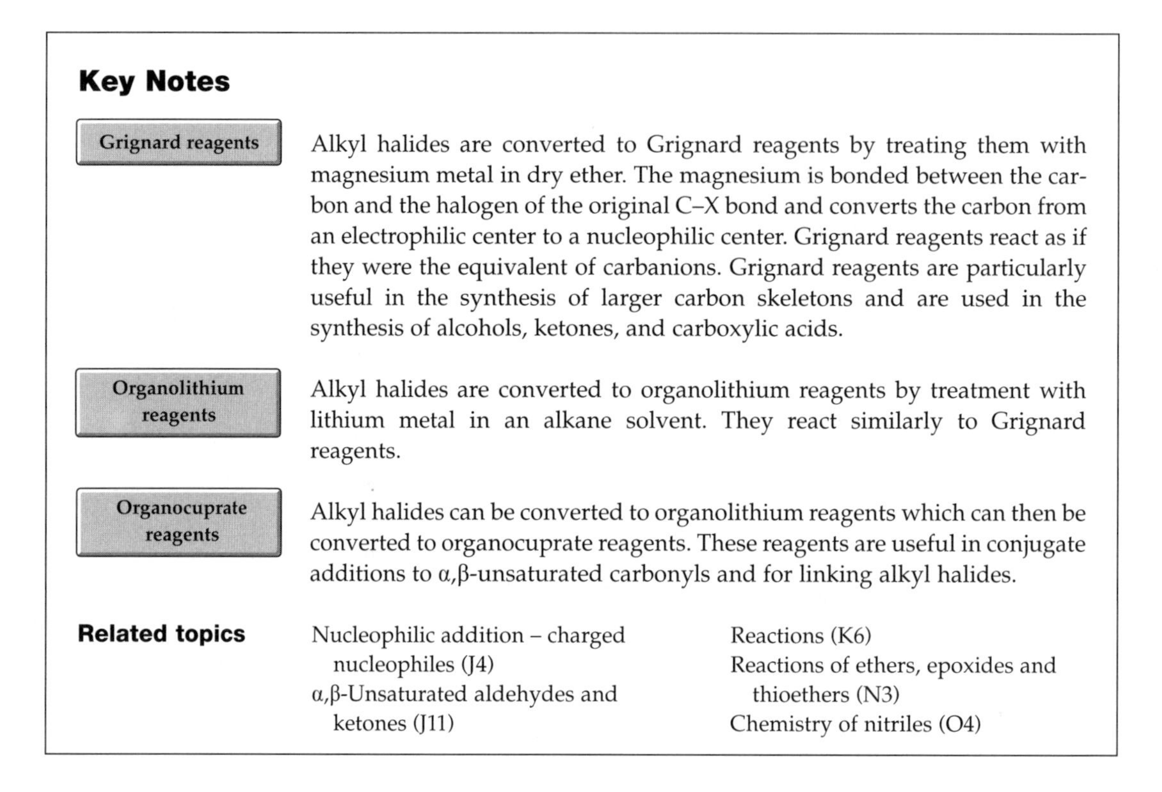

Key Notes

Grignard reagents

Alkyl halides are converted to Grignard reagents by treating them with magnesium metal in dry ether. The magnesium is bonded between the carbon and the halogen of the original C–X bond and converts the carbon from an electrophilic center to a nucleophilic center. Grignard reagents react as if they were the equivalent of carbanions. Grignard reagents are particularly useful in the synthesis of larger carbon skeletons and are used in the synthesis of alcohols, ketones, and carboxylic acids.

Organolithium reagents

Alkyl halides are converted to organolithium reagents by treatment with lithium metal in an alkane solvent. They react similarly to Grignard reagents.

Organocuprate reagents

Alkyl halides can be converted to organolithium reagents which can then be converted to organocuprate reagents. These reagents are useful in conjugate additions to α,β-unsaturated carbonyls and for linking alkyl halides.

Related topics

Nucleophilic addition – charged nucleophiles (J4)
α,β-Unsaturated aldehydes and ketones (J11)
Reactions (K6)
Reactions of ethers, epoxides and thioethers (N3)
Chemistry of nitriles (O4)

Grignard reagents

Alkyl halides of all types (1°, 2° and 3°) react with magnesium in dry ether to form Grignard reagents, where the magnesium is 'inserted' between the halogen and the alkyl chain (*Fig. 1*).

These reagents are extremely useful in organic synthesis and can be used in a wide variety of reactions. Their reactivity reflects the polarity of the atoms present. Since magnesium is a metal it is electropositive, which means that the electrons in the C–Mg bond spend more of their time closer to the carbon making it slightly negative and a nucleophilic center. This reverses the character of this carbon since it is an electrophilic center in the original alkyl halide (Topic E4). In essence, a Grignard reagent can be viewed as providing the equivalent of a carbanion. The carbanion is not a distinct species, but the reactions which take place can be explained as if it was.

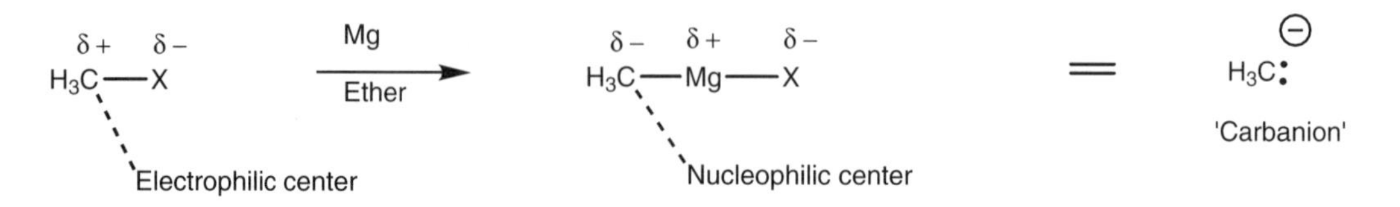

Fig. 1. Formation of a Grignard reagent (X=Cl, Br, I).

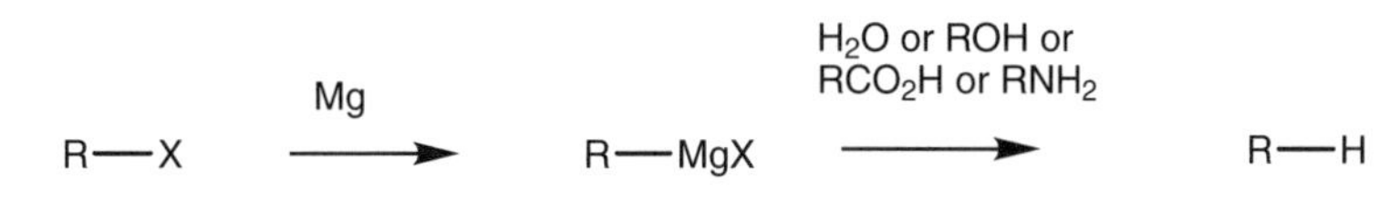

Fig. 2. Conversion of a Grignard reagent to an alkane.

A Grignard reagent can react as a base with water to form an alkane. This is one way of converting an alkyl halide to an alkane. The same acid–base reaction can take place with a variety of proton donors (Brønsted acids) including functional groups such as alcohols, carboxylic acids, and amines (*Fig. 2*). This can prove a disadvantage if the Grignard reagent is intended to react at some other site on the target molecule. In such cases, functional groups containing an X–H bond (where X= a heteroatom) would have to be protected before the Grignard reaction is carried out.

The reactions of Grignard reagents have been covered elsewhere and include reaction with aldehydes and ketones to give alcohols (Topic J4), reaction with acid chlorides and esters to give tertiary alcohols (Topic K6), reaction with carbon dioxide to give carboxylic acids (Topic K4), reaction with nitriles to give ketones (Topic O4), and reaction with epoxides to give alcohols (Topic N3).

Organolithium reagents

Alkyl halides can be converted to organolithium reagents using lithium metal in an alkane solvent (*Fig. 3*).

Organolithium reagents react similarly to Grignard reagents. For example, reaction with aldehydes and ketones results in nucleophilic addition to give secondary and tertiary alcohols respectively (Topic K6).

$$R\text{—}X \xrightarrow[\text{Pentane}]{\text{Li}} R\text{—}Li$$

Fig. 3. Formation of organolithium reagents.

$$2\ Li\text{—}CH_3 \xrightarrow{\text{CuI}} Li^+\ (Me_2Cu^-)$$

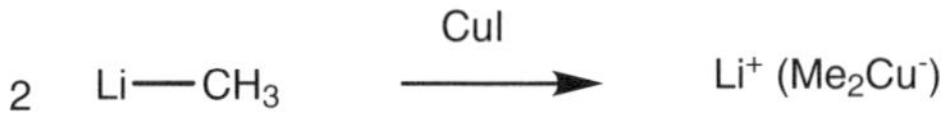

Fig. 4. Formation of organocuprate reagents.

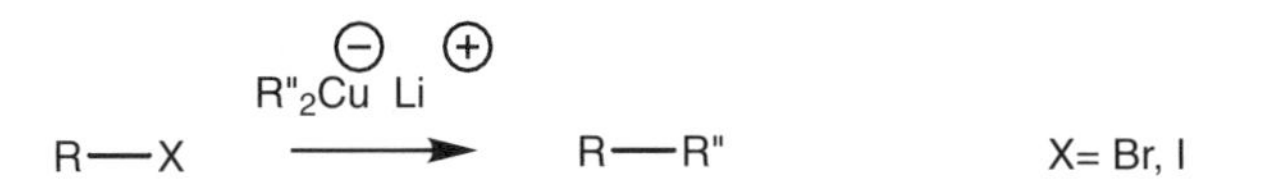

Fig. 5. Reaction of an organocuprate reagent with an alkyl halide.

Organocuprate reagents

Organocuprate reagents (another source of carbanion equivalents) are prepared by the reaction of one equivalent of cuprous iodide with two equivalents of an organolithium reagent (*Fig. 4*).

These reagents are useful in the 1,4-addition of alkyl groups to α,β-unsaturated carbonyl systems (Topic J11) and can also be reacted with alkyl halides to produce larger alkanes (*Fig. 5*). The mechanism is thought to be radical based.

M1 PREPARATION OF ALCOHOLS

Key Notes

Functional group transformation

Functional groups such as alkyl halides, carboxylic acids, esters, alkenes, aldehydes, ketones, and ethers can be transformed into alcohols.

C–C bond formation

Alcohols can be formed from epoxides, aldehydes, ketones, esters, and acid chlorides as a consequence of C–C bond formation with Grignard or organolithium reagents.

Related topics

Electrophilic addition to symmetrical alkenes (H3)
Hydroboration of alkenes (H7)
Nucleophilic addition – charged nucleophiles (J4)
Reactions (K6)
Reactions of alkyl halides (L6)
Reactions of ethers, epoxides and thioethers (N3)

Functional group transformation

Alcohols can be synthesized by nucleophilic substitution of alkyl halides (Topic L6), hydrolysis of esters (Topic K6), reduction of carboxylic acids or esters (Topic K6), reduction of aldehydes or ketones (Topic J4), electrophilic addition of alkenes (Topic H3), hydroboration of alkenes (Topic H7), or substitution of ethers (Topic N3).

C–C bond formation

Alcohols can also be formed from epoxides (Topic N3), aldehydes (Topics J4 & H10), ketones (Topics J4 and H10), esters (Topic K6), and acid chlorides (Topic K6) as a consequence of C–C bond formation. These reactions involve the addition of carbanion equivalents through the use of Grignard or organolithium reagents and are described in detail elsewhere in the text.

M2 Preparation of phenols

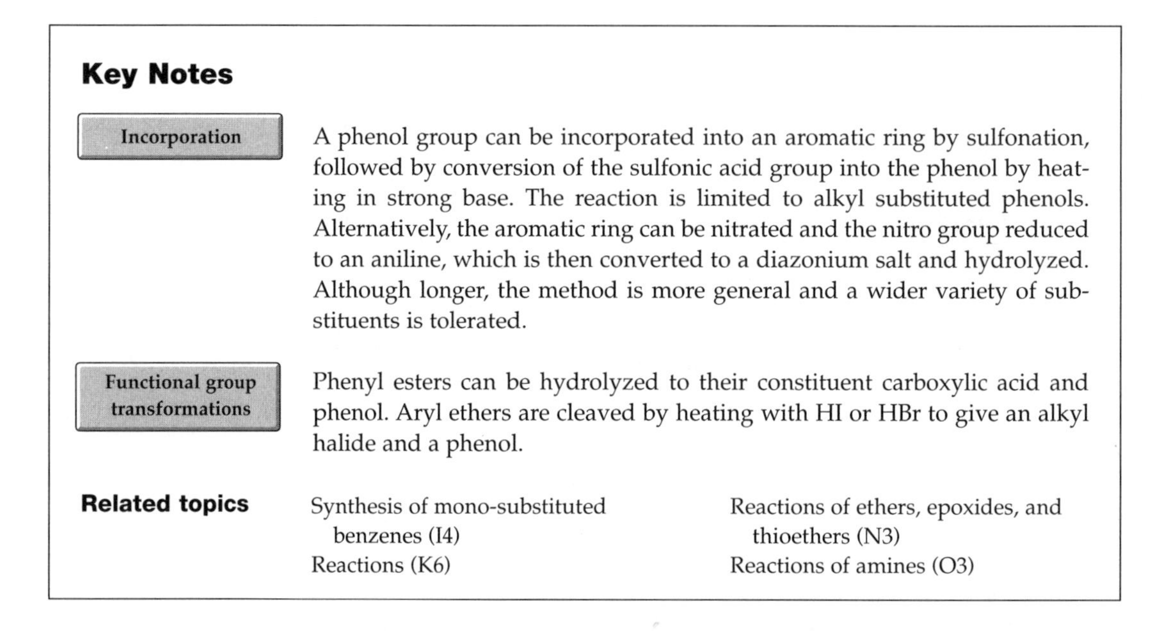

Key Notes

Incorporation

A phenol group can be incorporated into an aromatic ring by sulfonation, followed by conversion of the sulfonic acid group into the phenol by heating in strong base. The reaction is limited to alkyl substituted phenols. Alternatively, the aromatic ring can be nitrated and the nitro group reduced to an aniline, which is then converted to a diazonium salt and hydrolyzed. Although longer, the method is more general and a wider variety of substituents is tolerated.

Functional group transformations

Phenyl esters can be hydrolyzed to their constituent carboxylic acid and phenol. Aryl ethers are cleaved by heating with HI or HBr to give an alkyl halide and a phenol.

Related topics

Synthesis of mono-substituted benzenes (I4)
Reactions (K6)
Reactions of ethers, epoxides, and thioethers (N3)
Reactions of amines (O3)

Incorporation

Phenol groups can be incorporated into an aromatic ring by sulfonation of the aromatic ring (Topic I4) followed by melting the product with sodium hydroxide to convert the sulfonic acid group to a phenol (*Fig. 1*). The reaction conditions are harsh and only alkyl-substituted phenols can be prepared by this method.

A more general method of synthesizing phenols is to hydrolyze a diazonium salt (Topic O3), prepared from an aniline group (NH_2) (Topic I4; *Fig. 2*).

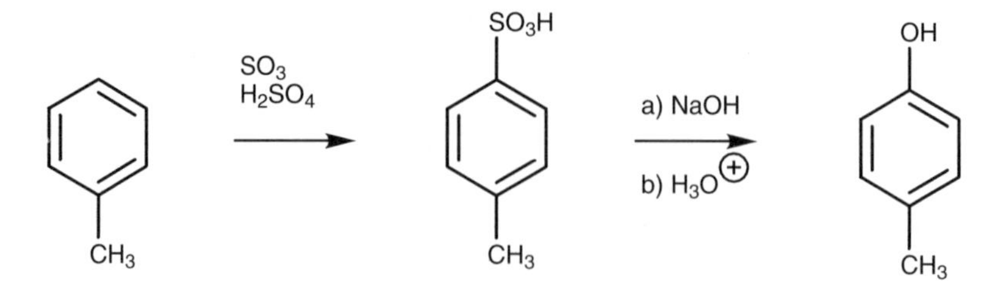

Fig. 1. Synthesis of a phenol via sulfonation.

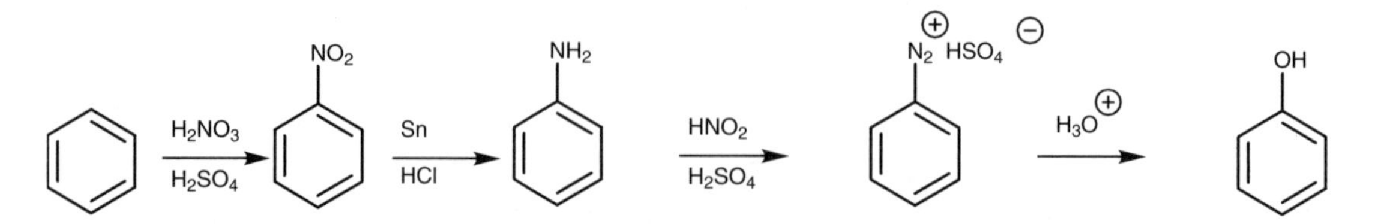

Fig. 2. Synthesis of a phenol via a diazonium salt.

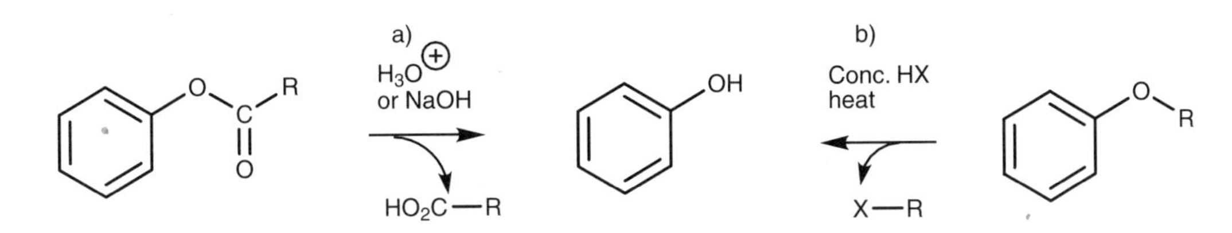

Fig. 3. Functional group transformations to a phenol.

Functional group transformation

Various functional groups can be converted to phenols. Sulfonic acids and amino groups have already been mentioned. Phenyl esters can be hydrolyzed as described in Topic K6 (*Fig. 3a*). Aryl ethers can be cleaved as described in Topic N3 (*Fig. 3b*). The bond between the alkyl group and oxygen is specifically cleaved since the Ar–OH bond is too strong to be cleaved.

M3 PROPERTIES OF ALCOHOLS AND PHENOLS

Key Notes

Alcohols

The carbon and oxygen atoms of the alcohol group are sp^3 hybridized such that the C–O–H bond angle is approximately 109°. Hydrogen bonding means that alcohols have higher boiling points than comparable alkanes. Alcohols of low molecular weight are soluble in water and can act as weak acids and weak bases. Alcohols are polar. The oxygen atom is a nucleophilic center while the neighboring carbon and hydrogen are weak electrophilic centers. Alcohols will not react with nucleophiles, but will react with strong bases in an acid–base reaction to form an alkoxide ion. An alcohol's C–O bond can be split if the hydroxyl group is converted into a better leaving group.

Phenols

Phenols have an OH group directly linked to an aromatic ring. The oxygen is sp^3 hybridized and the aryl carbon is sp^2 hybridized. Phenols are polar compounds which are capable of intermolecular hydrogen bonding such that phenols have higher boiling points than nonphenolic aromatic structures of comparable molecular weight. Hydrogen bonding also permits moderate water solubility and phenols act as weak acids in aqueous solution. Phenols are stronger acids than alcohols but weaker acids than carboxylic acids. They are soluble as their phenoxide salts in sodium hydroxide solution, but insoluble in sodium hydrogen carbonate solution.

Related topics

sp^3 Hybridization (A3)
sp^2 Hybridization (A4)
Intermolecular bonding (C3)
Properties and reactions (C4)
Organic structures (E4)
Acid strength (G2)
Base strength (G3)
Reactions of alcohols (M4)

Alcohols

The alcohol functional group (R_3C–OH) has the same geometry as water, with a C–O–H bond angle of approximately 109°. Both the carbon and the oxygen are sp^3 hybridized. The presence of the O–H group means that intermolecular hydrogen bonding is possible which accounts for the higher boiling points of alcohols compared with alkanes of similar molecular weight. Hydrogen bonding also means that alcohols are more soluble in protic solvents than alkanes of similar molecular weight. In fact, the smaller alcohols (methanol, ethanol, propanol, and *tert*-butanol) are completely miscible in water. With larger alcohols, the hydrophobic character of the bigger alkyl chain takes precedence over the polar alcohol group and so larger alcohols are insoluble in water.

The O–H and C–O bonds are both polarized due to the electronegative oxygen, such that oxygen is slightly negative and the carbon and hydrogen atoms are slightly positive. This means that the oxygen serves as a nucleophilic center while

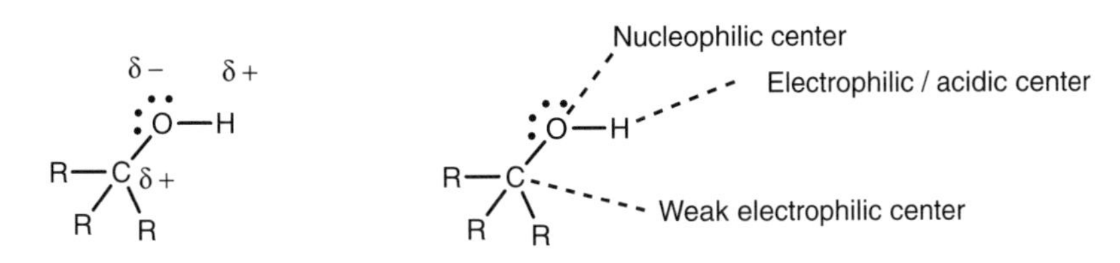

Fig. 1. Bond polarization and nucleophilic and electrophilic centers.

the hydrogen and the carbon atoms serve as weak electrophilic centers (Topic E4; *Fig. 1*).

Due to the presence of the nucleophilic oxygen and electrophilic proton, alcohols can act both as weak acids and as weak bases when dissolved in water (*Fig. 2*). However, the equilibria in both cases is virtually completely weighted to the unionized form.

Alcohols will commonly react with stronger electrophiles than water. However, they are less likely to react with nucleophiles unless the latter are also strong bases, in which case the acidic proton is abstracted to form an **alkoxide** ion (RO^-; *Fig. 3*). Alkoxide ions are extremely useful reagents in organic synthesis. However, they cannot be used if water is the solvent since the alkoxide ion would act as a base and abstract a proton from water to regenerate the alcohol. Therefore, an alcohol would have to be used as solvent instead of water.

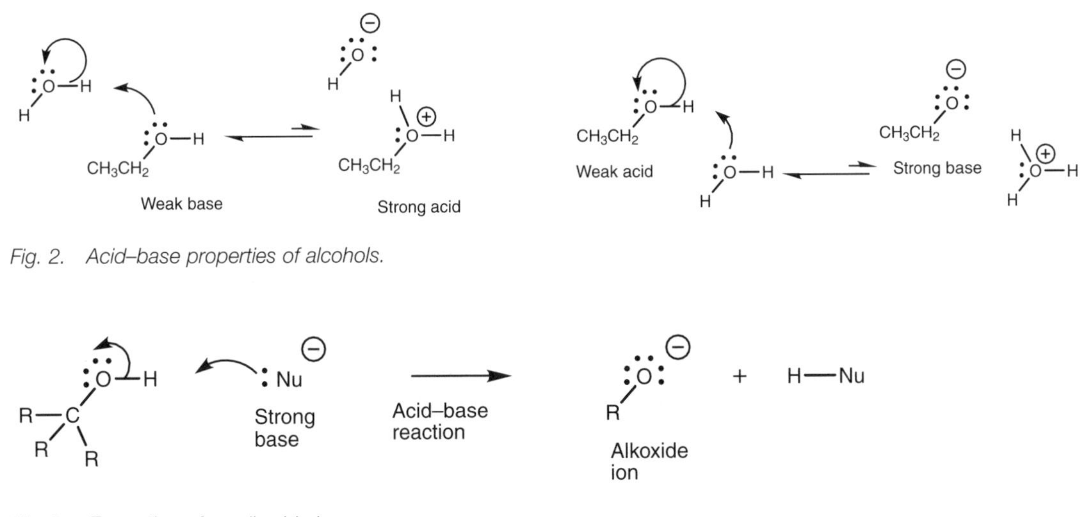

Fig. 2. Acid–base properties of alcohols.

Fig. 3. Formation of an alkoxide ion.

Nucleophiles which are also strong bases react with the electrophilic hydrogen of an alcohol rather than the electrophilic carbon. Nucleophilic attack at carbon would require the loss of a hydroxide ion in a nucleophilic substitution reaction (Topic L2). However, this is not favored since the hydroxide ion is a strong base and a poor leaving group (*Fig. 4*). Nevertheless, reactions which involve the cleavage of an alcohol's C–O bond are possible if the alcohol is first 'activated' such that the hydroxyl group is converted into a better leaving group (Topic M4; *Fig. 5*). One method is to react the alcohol under acidic conditions such that the hydroxyl group is protonated before the nucleophile makes its attack (*Fig. 5a*). Cleavage of

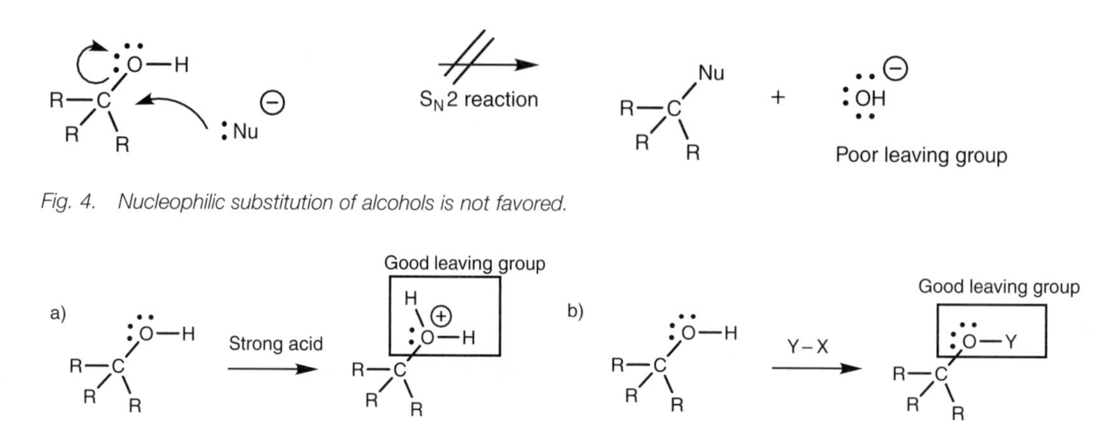

Fig. 4. Nucleophilic substitution of alcohols is not favored.

Fig. 5. Activation of an alcohol.

the C–O bond would then be more likely since the leaving group would be a neutral water molecule, which is a much better leaving group. Alternatively, the alcohol can be treated with an electrophilic reagent to convert the OH group into a different group (OY) which can then act as a better leaving group (*Fig. 5b*). In both cases, the alcohol must first act as a nucleophile, with the oxygen atom acting as the nucleophilic center. The intermediate formed can then react more readily as an electrophile at the carbon center.

Phenols

Phenols are compounds which have an OH group directly attached to an aromatic ring. Therefore, the oxygen is sp^3 hybridized and the aryl carbon is sp^2 hybridized. Although phenols share some characteristics with alcohols, they have distinct properties and reactions which set them apart from that functional group.

Phenols can take part in intermolecular hydrogen bonding, which means that they have a moderate water solubility and have higher boiling points than aromatic compounds lacking the phenolic group. Phenols are weakly acidic, and in aqueous solution an equilibrium exists between the phenol and the phenoxide ion (*Fig. 6a*). On treatment with a base, the phenol is fully converted to the phenoxide ion (*Fig. 6b*).

The phenoxide ion is stabilized by resonance and delocalization of the negative charge into the ring (Topic G2/G3), which means that phenoxide ions are weaker bases than alkoxide ions. This in turn means that phenols are more acidic than alcohols, but less acidic than carboxylic acids. The pK_a values of most phenols is in the order of 11, compared to 18 for alcohols and 4.74 for acetic acid. This means that phenols can be ionized with weaker bases than those required to ionize alcohols, but require stronger bases than those required to ionize carboxylic acids. For example, phenols are ionized by sodium hydroxide but not by the weaker base sodium hydrogen carbonate. Alcohols being less acidic are not ionized by

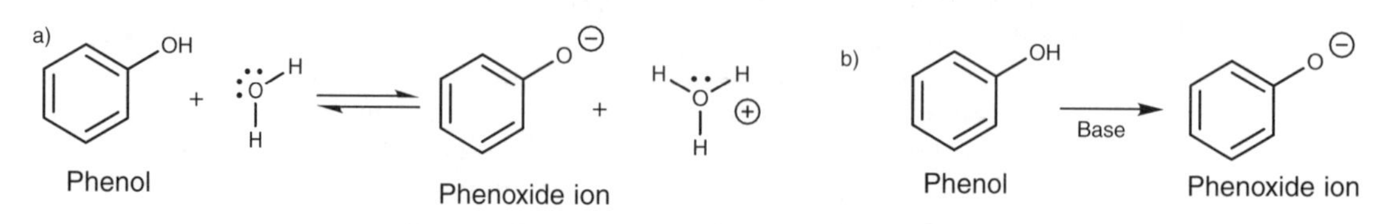

Fig. 6. Acidic reactions of phenol.

either base whereas carboxylic acids are ionized by both sodium hydroxide and sodium hydrogen carbonate solutions.

These acid–base reactions permit a simple way of distinguishing between most carboxylic acids, phenols, and alcohols. Since the salts formed from the acid–base reaction are water soluble, compounds containing these functional groups can be distinguished by testing their solubilities in sodium hydrogen carbonate and sodium hydroxide solutions. This solubility test is not valid for low molecular weight structures such as methanol or ethanol since these are water soluble and dissolve in basic solution because of their water solubility rather than their ability to form salts.

M4 REACTIONS OF ALCOHOLS

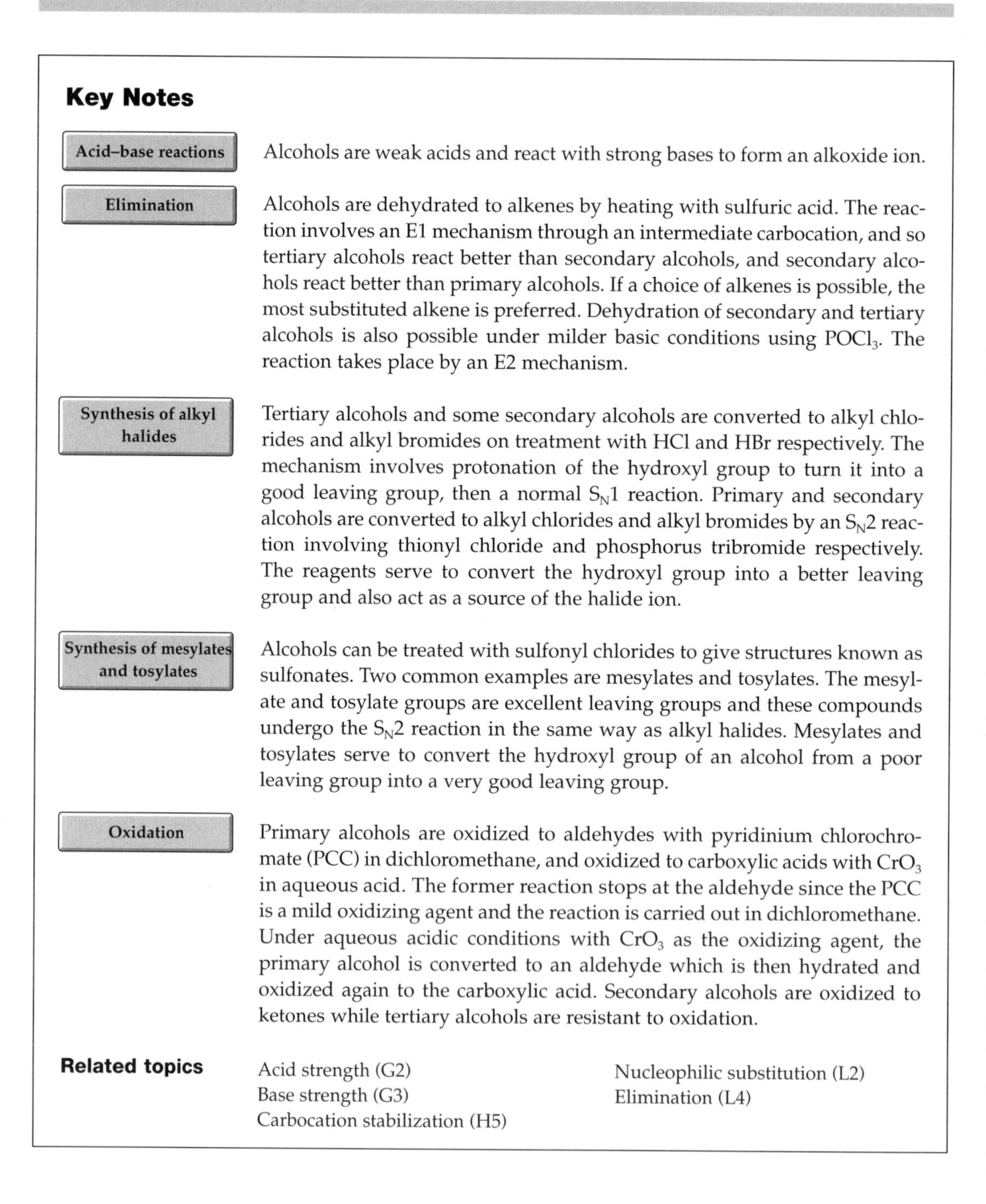

Key Notes

Acid–base reactions

Alcohols are weak acids and react with strong bases to form an alkoxide ion.

Elimination

Alcohols are dehydrated to alkenes by heating with sulfuric acid. The reaction involves an E1 mechanism through an intermediate carbocation, and so tertiary alcohols react better than secondary alcohols, and secondary alcohols react better than primary alcohols. If a choice of alkenes is possible, the most substituted alkene is preferred. Dehydration of secondary and tertiary alcohols is also possible under milder basic conditions using $POCl_3$. The reaction takes place by an E2 mechanism.

Synthesis of alkyl halides

Tertiary alcohols and some secondary alcohols are converted to alkyl chlorides and alkyl bromides on treatment with HCl and HBr respectively. The mechanism involves protonation of the hydroxyl group to turn it into a good leaving group, then a normal S_N1 reaction. Primary and secondary alcohols are converted to alkyl chlorides and alkyl bromides by an S_N2 reaction involving thionyl chloride and phosphorus tribromide respectively. The reagents serve to convert the hydroxyl group into a better leaving group and also act as a source of the halide ion.

Synthesis of mesylates and tosylates

Alcohols can be treated with sulfonyl chlorides to give structures known as sulfonates. Two common examples are mesylates and tosylates. The mesylate and tosylate groups are excellent leaving groups and these compounds undergo the S_N2 reaction in the same way as alkyl halides. Mesylates and tosylates serve to convert the hydroxyl group of an alcohol from a poor leaving group into a very good leaving group.

Oxidation

Primary alcohols are oxidized to aldehydes with pyridinium chlorochromate (PCC) in dichloromethane, and oxidized to carboxylic acids with CrO_3 in aqueous acid. The former reaction stops at the aldehyde since the PCC is a mild oxidizing agent and the reaction is carried out in dichloromethane. Under aqueous acidic conditions with CrO_3 as the oxidizing agent, the primary alcohol is converted to an aldehyde which is then hydrated and oxidized again to the carboxylic acid. Secondary alcohols are oxidized to ketones while tertiary alcohols are resistant to oxidation.

Related topics

Acid strength (G2)
Base strength (G3)
Carbocation stabilization (H5)
Nucleophilic substitution (L2)
Elimination (L4)

Acid–base reactions

Alcohols are slightly weaker acids than water which means that the conjugate base generated from an alcohol (the **alkoxide** ion) is a stronger base than the conjugate base of water (the hydroxide ion). As a result, it is not possible to generate an alkoxide ion using sodium hydroxide as base. Alcohols do not react with sodium bicarbonate or amines, and a stronger base such as sodium hydride or sodium amide is required to generate the alkoxide ion (*Fig. 1*). Alcohols can also be converted to alkoxide ions on treatment with potassium, sodium, or lithium metal. Some organic reagents can also act as strong bases, for example Grignard reagents and organolithium reagents.

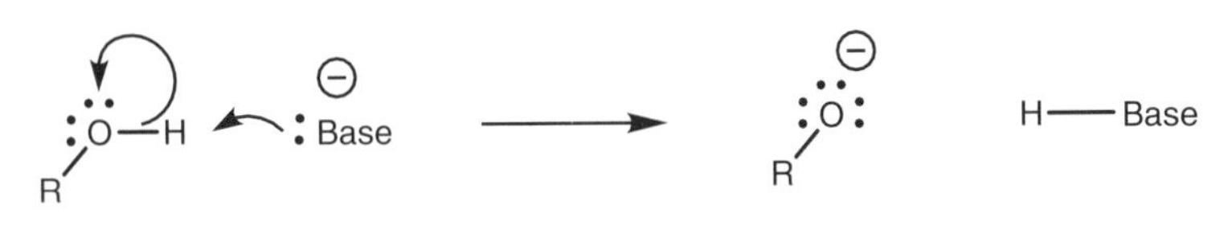

Fig. 1. Generation of an alkoxide ion.

Alkoxide ions are neutralized in water and so reactions involving these reagents should be carried out in the alcohol from which they were derived, that is reactions involving sodium ethoxide are best carried out in ethanol. Alcohols have a typical pK_a of 15.5–18.0 compared to pK_a values of 25 for ethyne, 38 for ammonia and 50 for ethane.

Elimination

Alcohols, like alkyl halides, can undergo elimination reactions to form alkenes (see Topic L4; *Fig. 2*). Since water is eliminated, the reaction is also known as a dehydration.

Like alkyl halides, the elimination reaction of an alcohol requires the presence of a susceptible proton at the β-position (*Fig. 3*).

Whereas the elimination of alkyl halides is carried out under basic conditions, the elimination of alcohols is carried out under acid conditions. Under basic conditions, an E2 elimination would require the loss of a hydroxide ion as a leaving group. Since the hydroxide ion is a strong base, it is not a good leaving group and so the elimination of alcohols under basic conditions is difficult to achieve.

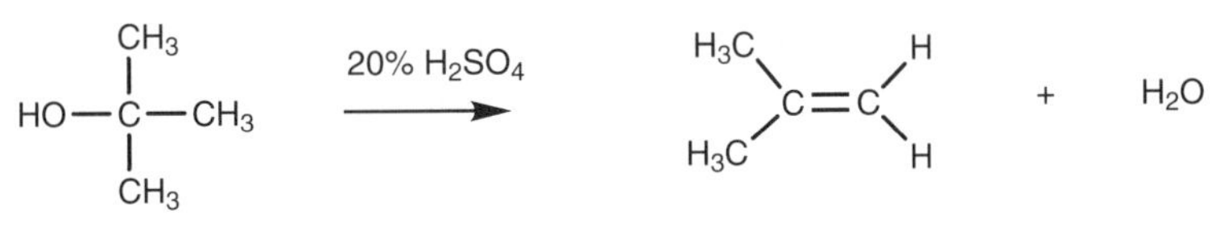

Fig. 2. Elimination of an alcohol.

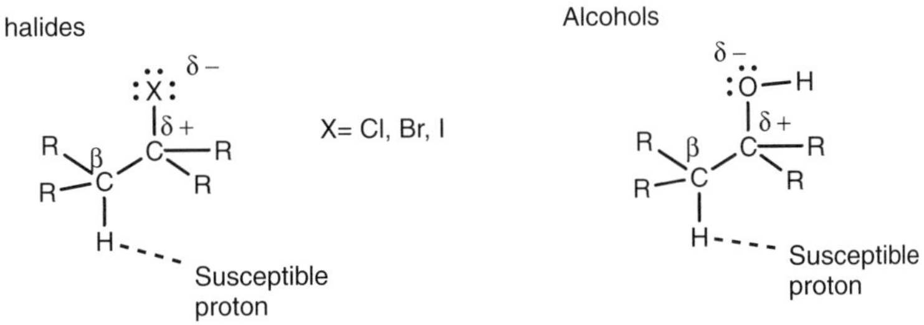

Fig. 3. Susceptible β-protons in an alkyl halide and an alcohol.

Elimination under acidic conditions is more successful since the hydroxyl group is first protonated and then departs the molecule as a neutral water molecule (**dehydration**) which is a much better leaving group. If different isomeric alkenes are possible, the most substituted alkene will be favored – another example of Zaitsev's rule (*Fig. 4*). The reaction works best with tertiary alcohols since the elimination proceeds by the E1 mechanism (Topic L4).

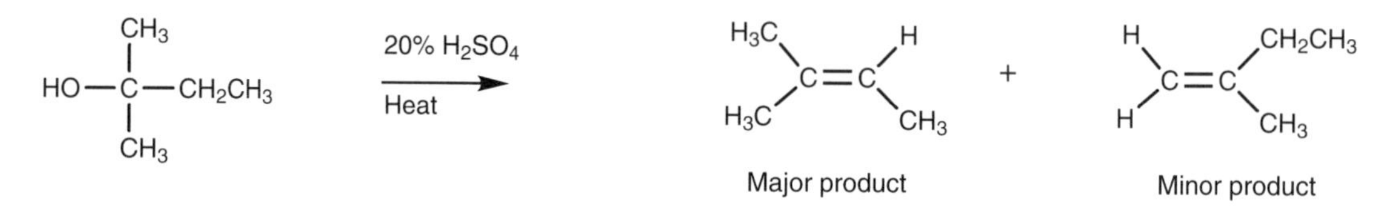

Fig. 4. Elimination of alcohols obeys Zaitsev's rule.

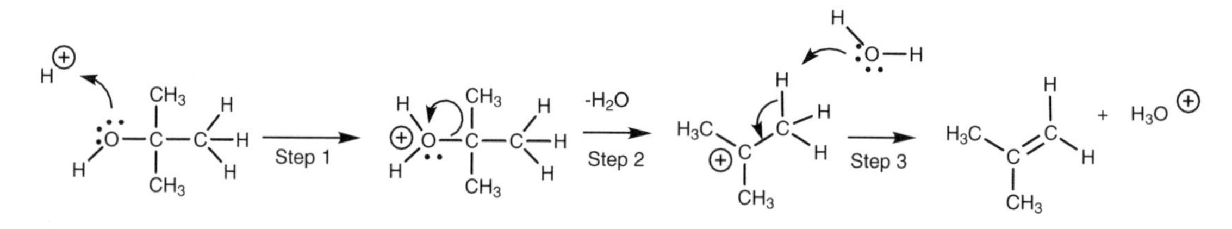

Fig. 5. E1 Elimination mechanism for alcohols.

The mechanism (*Fig. 5*) involves the nucleophilic oxygen of the alcohol using one of its lone pairs of electrons to form a bond to a proton to produce a charged intermediate (Step 1). Now that the oxygen is protonated, the molecule has a much better leaving group since water can be ejected as a neutral molecule. The E1 mechanism can now proceed as normal. Water is lost and a carbocation is formed (Step 2). Water then acts as a base in the second step, using one of its lone pairs of electrons to form a bond to the β-proton of the carbocation. The C–H bond is broken and both the electrons in that bond are used to form a π bond between the two carbons. Since this is an E1 reaction, tertiary alcohols react better than primary or secondary alcohols.

The E1 reaction is not ideal for the dehydration of primary or secondary alcohols since vigorous heating is required to force the reaction and this can result in

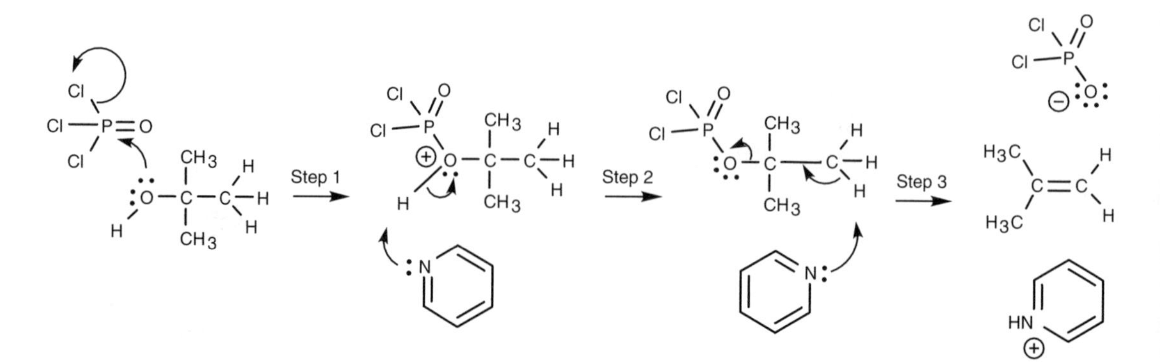

Fig. 6. Mechanism for the POCl3 dehydration of an alcohol.

rearrangement reactions. Therefore, alternative methods are useful. Reagents such as phosphorus oxychloride ($POCl_3$) dehydrate secondary and tertiary alcohols under mild basic conditions using pyridine as solvent (*Fig. 6*). The phosphorus oxychloride serves to activate the alcohol, converting the hydroxyl function into a better leaving group. The mechanism involves the alcohol acting as a nucleophile in the first step. Oxygen uses a lone pair of electrons to form a bond to the electrophilic phosphorus of $POCl_3$ and a chloride ion is lost (Step 1). Pyridine then removes a proton from the structure to form a dichlorophosphate intermediate (Step 2). The dichlorophosphate group is a much better leaving group than the hydroxide ion and so a normal E2 reaction can take place. Pyridine acts as a base to remove a β-proton and as this is happening, the electrons from the old C–H bond are used to form a π bond and eject the leaving group (Step 3).

Synthesis of alkyl halides

Tertiary alcohols can undergo the S_N1 reaction to produce tertiary alkyl halides (cf. Topic L2; *Fig. 7*). Since the reaction requires the loss of the hydroxide ion (a poor leaving group), a little bit of 'trickery' is required in order to convert the hydroxyl moiety into a better leaving group. This can be achieved under acidic conditions with the use of HCl or HBr. The acid serves to protonate the hydroxyl moiety as the first step and then a normal S_N1 mechanism takes place where water is lost from the molecule to form an intermediate carbocation. A halide ion then forms a bond to the carbocation center in the third step.

The first two steps of this mechanism are exactly the same as the elimination reaction described above. Both reactions are carried out under acidic conditions and one might ask why elimination does not occur. The difference here is that halide ions serve as good nucleophiles and are present in high concentration. The elimination reaction described earlier is carried out using concentrated sulfuric acid and only weak nucleophiles are present (i.e. water) in low concentration. Having said that, some elimination can occur and although the reaction of alcohols with HX produces mainly alkyl halide, some alkene by-product is usually present.

Since primary alcohols and some secondary alcohols do not undergo the S_N1 reaction, nucleophilic substitution of these compounds must involve an S_N2 mechanism. Once again, protonation of the OH group is required as a first step, then the reaction involves simultaneous attack of the halide ion and loss of water. The reaction proceeds with good nucleophiles such as the iodide or bromide ion, but fails with the weaker nucleophilic chloride ion. In this case, a Lewis acid (Topic G4) needs to be added to the reaction mixture. The Lewis acid forms a complex with the oxygen of the alcohol group, resulting in a much better leaving group for the subsequent S_N2 reaction.

Nevertheless, the reaction of primary and secondary alcohols with hydrogen halides can often be a problem since unwanted rearrangement reactions often take

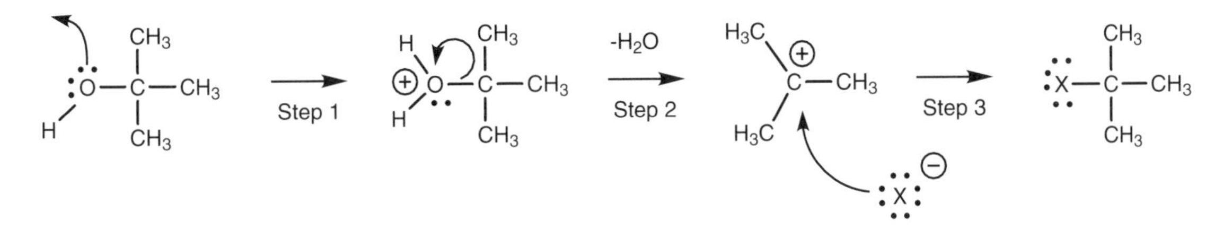

Fig. 7. Conversion of alcohols to alkyl halides.

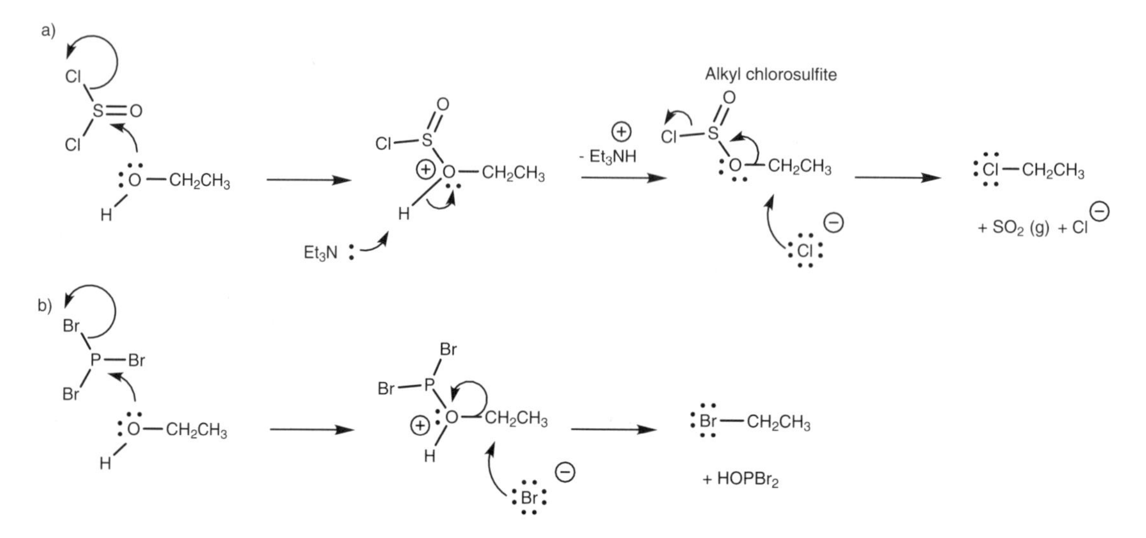

Fig. 8. Conversion of an alcohol to an alkyl halide using (a) thionyl chloride; (b) phosphorus tribromide.

place. To avoid this, alternative procedures carried out under milder basic conditions have been used with reagents such as thionyl chloride or phosphorus tribromide (*Fig. 8*). These reagents act as electrophiles and react with the alcoholic oxygen to form an intermediate where the OH moiety has been converted into a better leaving group. A halide ion is released from the reagent in this process, and this can act as the nucleophile in the subsequent S_N2 reaction.

In the reaction with thionyl chloride, triethylamine is present to mop up the HCl formed during the reaction. The reaction is also helped by the fact that one of the products (SO_2) is lost as a gas, thus driving the reaction to completion.

Phosphorus tribromide has three bromine atoms present and each PBr_3 molecule can react with three alcohol molecules to form three molecules of alkyl bromide.

Synthesis of mesylates and tosylates

It is sometimes convenient to synthesize an activated alcohol which can be used in nucleophilic substitution reactions like an alkyl halide. Mesylates and tosylates are examples of sulfonate compounds which serve this purpose. They are synthe-

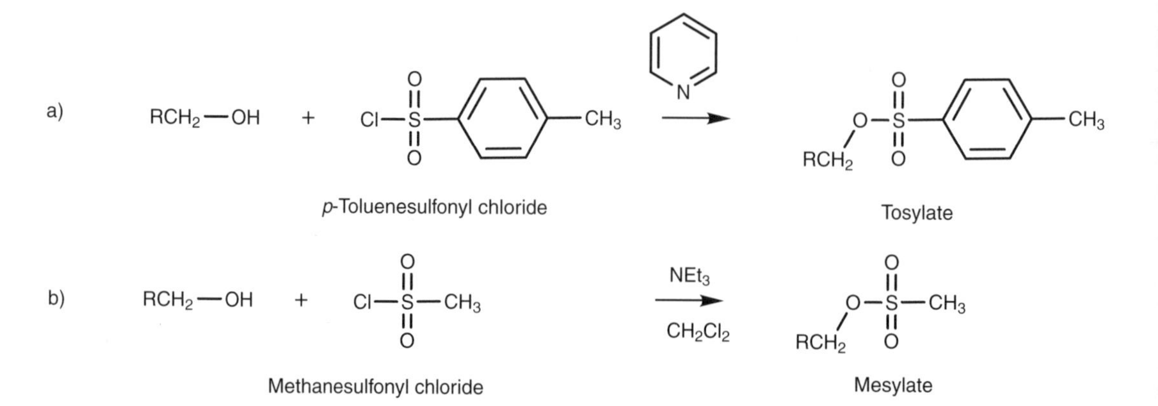

Fig. 9. Synthesis of (a) tosylate and (b) mesylate.

sized by treating alcohols with sulfonyl chlorides in the presence of a base such as pyridine or triethylamine (*Fig. 9*). The base serves to 'mop up' the HCl which is formed and prevents acid-catalyzed rearrangement reactions.

The reaction mechanism (*Fig. 10*) involves the alcohol oxygen acting as a nucleophilic center and substituting the chloride ion from the sulfonate. The base then removes a proton from the intermediate to give the sulfonate product. Neither of these steps affects the stereochemistry of the alcohol carbon and so the stereochemistry of chiral alcohols is retained.

The mesylate and tosylate groups are excellent leaving groups and can be viewed as the equivalent of a halide. Therefore mesylates and tosylates can undergo the S_N2 reaction in the same way as alkyl halides (*Fig. 11*).

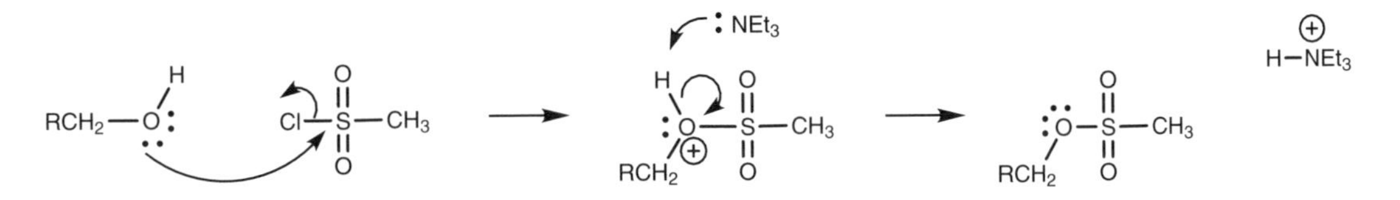

Fig. 10. Mechanism for the formation of a mesylate.

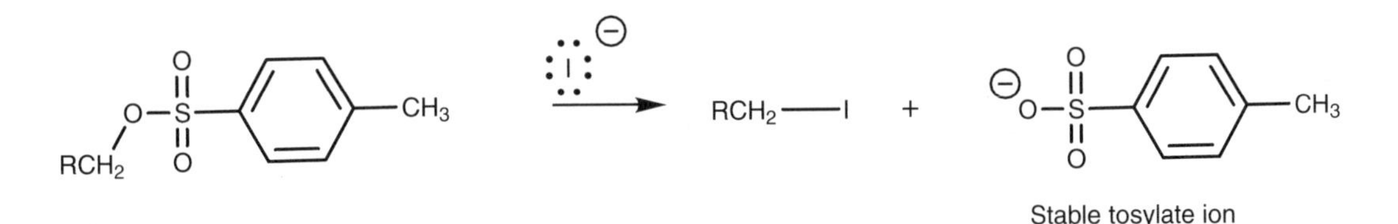

Fig. 11. Nucleophilic substitution of a tosylate.

Oxidation

The oxidation of alcohols is an extremely important reaction in organic synthesis. Primary alcohols can be oxidized to aldehydes, but the reaction is tricky since there is the danger of over-oxidation to carboxylic acids. With volatile aldehydes, the aldehydes can be distilled from the reaction solution as they are formed. However, this is not possible for less volatile aldehydes. This problem can be overcome by using a mild oxidizing agent called pyridinium chlorochromate (PCC; *Fig. 12a*). If a stronger oxidizing agent is used in aqueous conditions (e.g. CrO_3 in aqueous sulfuric acid), primary alcohols are oxidized to carboxylic acids (*Fig. 12b*), while secondary alcohols are oxidized to ketones (*Fig. 12c*).

The success of the PCC oxidation in stopping at the aldehyde stage is solvent dependent. The reaction is carried out in methylene chloride, whereas oxidation

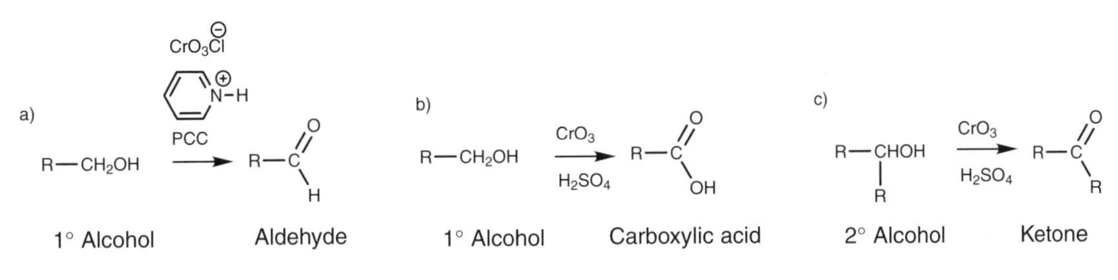

Fig. 12. Oxidations of alcohols.

with CrO_3 is carried out in aqueous acid. Under aqueous conditions, the aldehyde which is formed by oxidation of the alcohol is hydrated and this structure is more sensitive to oxidation than the aldehyde itself (*Fig. 13*). In methylene chloride, hydration cannot occur and the aldehyde is more resistant to oxidation.

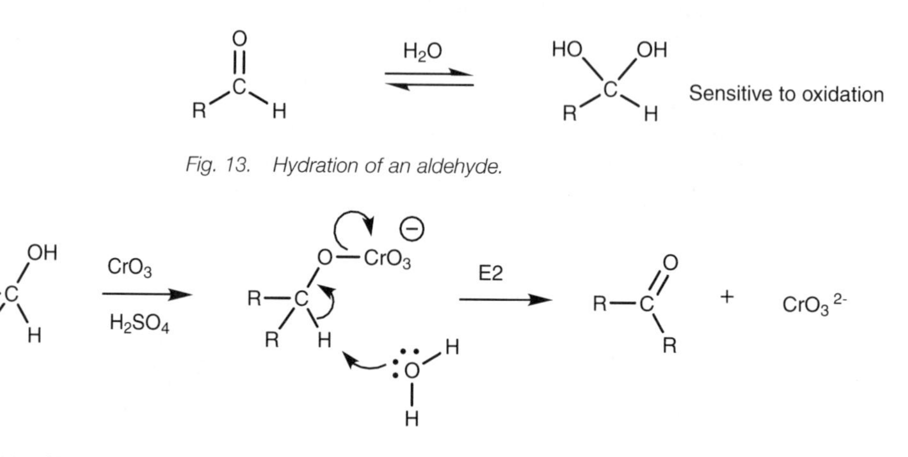

Fig. 13. Hydration of an aldehyde.

Fig. 14. Mechanism of oxidation of a secondary alcohol with CrO_3.

The mechanism of oxidation for a secondary alcohol with CrO_3 (*Fig. 14*) involves the nucleophilic oxygen reacting with the oxidizing agent to produce a charged chromium intermediate. Elimination then occurs where an α-proton is lost along with the chromium moiety to produce the carbonyl group. The mechanism can be viewed as an E2 mechanism, the difference being that different bonds are being created and broken. Since the mechanism requires an α-proton to be removed from the alcoholic carbon, tertiary alcohols cannot be oxidized since they do not contain such a proton. The mechanism also explains why an aldehyde product is resistant to further oxidation when methylene chloride is the solvent (i.e. no OH present to react with the chromium reagent). When aqueous conditions are used the aldehyde is hydrated and this generates two OH groups which are available to bond to the chromium reagent and result in further oxidation.

M5 REACTIONS OF PHENOLS

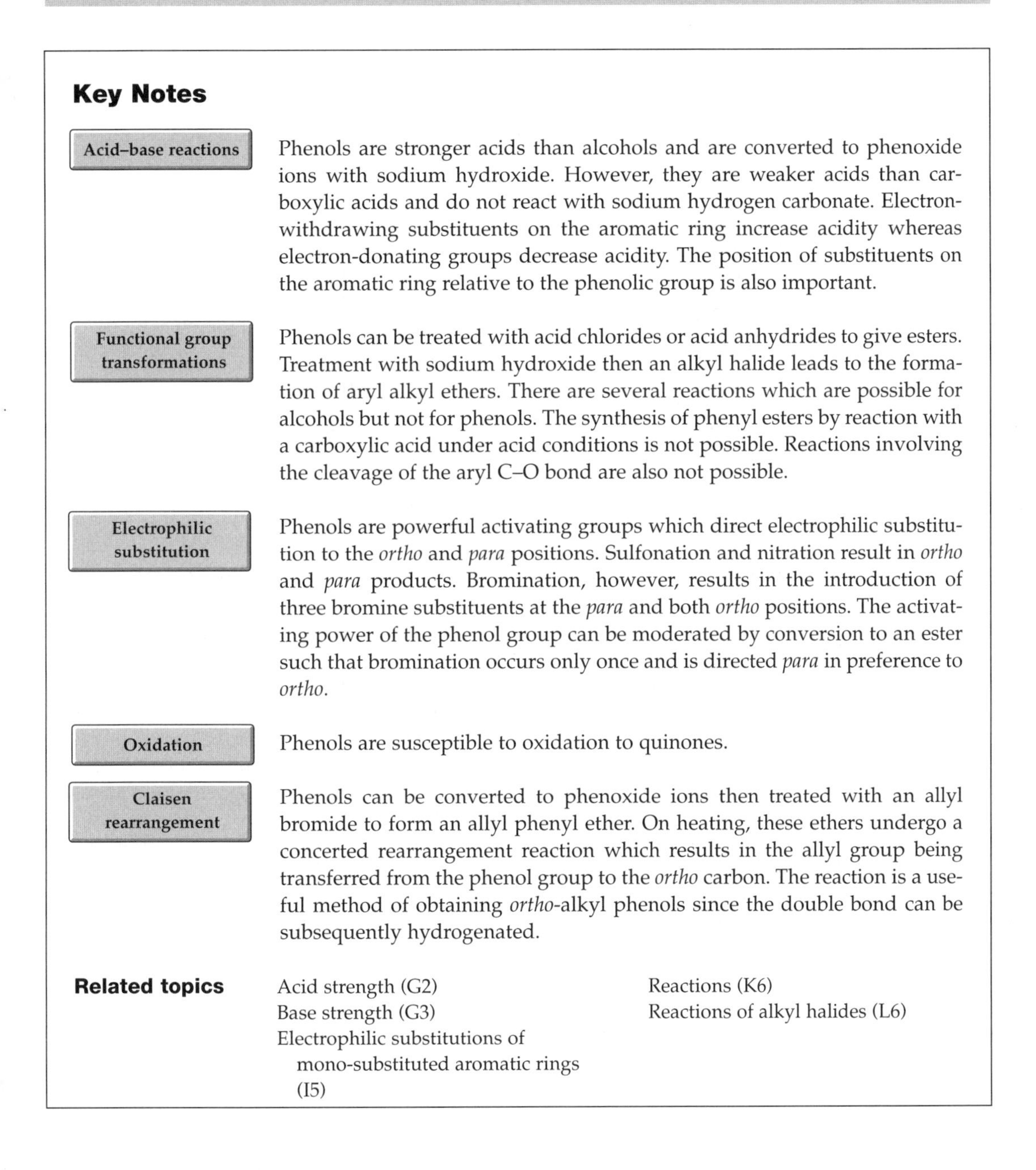

Key Notes

Acid–base reactions

Phenols are stronger acids than alcohols and are converted to phenoxide ions with sodium hydroxide. However, they are weaker acids than carboxylic acids and do not react with sodium hydrogen carbonate. Electron-withdrawing substituents on the aromatic ring increase acidity whereas electron-donating groups decrease acidity. The position of substituents on the aromatic ring relative to the phenolic group is also important.

Functional group transformations

Phenols can be treated with acid chlorides or acid anhydrides to give esters. Treatment with sodium hydroxide then an alkyl halide leads to the formation of aryl alkyl ethers. There are several reactions which are possible for alcohols but not for phenols. The synthesis of phenyl esters by reaction with a carboxylic acid under acid conditions is not possible. Reactions involving the cleavage of the aryl C–O bond are also not possible.

Electrophilic substitution

Phenols are powerful activating groups which direct electrophilic substitution to the *ortho* and *para* positions. Sulfonation and nitration result in *ortho* and *para* products. Bromination, however, results in the introduction of three bromine substituents at the *para* and both *ortho* positions. The activating power of the phenol group can be moderated by conversion to an ester such that bromination occurs only once and is directed *para* in preference to *ortho*.

Oxidation

Phenols are susceptible to oxidation to quinones.

Claisen rearrangement

Phenols can be converted to phenoxide ions then treated with an allyl bromide to form an allyl phenyl ether. On heating, these ethers undergo a concerted rearrangement reaction which results in the allyl group being transferred from the phenol group to the *ortho* carbon. The reaction is a useful method of obtaining *ortho*-alkyl phenols since the double bond can be subsequently hydrogenated.

Related topics

Acid strength (G2)
Base strength (G3)
Electrophilic substitutions of mono-substituted aromatic rings (I5)
Reactions (K6)
Reactions of alkyl halides (L6)

Acid–base reactions

Phenols are stronger acids than alcohols and react with bases such as sodium hydroxide to form phenoxide ions (Topic M3). However, they are weaker acids than carboxylic acids and do not react with sodium hydrogen carbonate.

Phenols are acidic since the oxygen's lone pair of electrons can participate in a resonance mechanism involving the adjacent aromatic ring (*Fig. 1*). Three

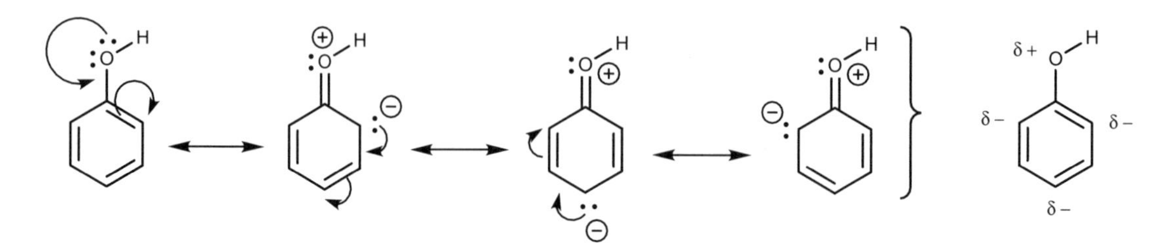

Fig. 1. Resonance structures for phenol.

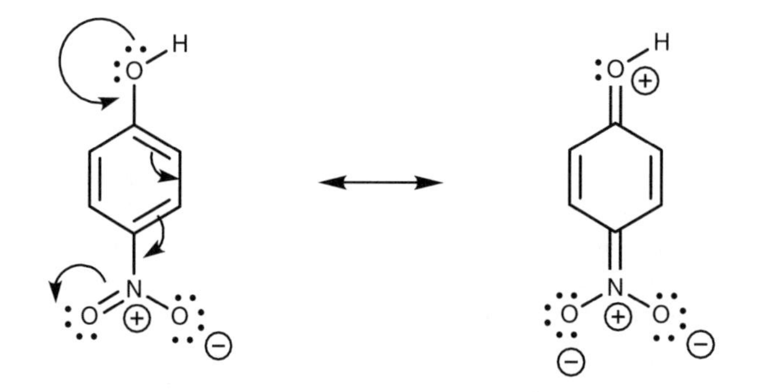

Fig. 2. Resonance effect of a para-*nitro group on a phenol.*

resonance structures are possible where the oxygen gains a positive charge and the ring gains a negative charge. The net result is a slightly positive charge on the oxygen which accounts for the acidity of its proton. There are also three aromatic carbons with slightly negative charges.

The type of substituents present on the aromatic ring can have a profound effect on the acidity of the phenol. This is because substituents can either stabilize or destabilize the partial negative charge on the ring. The better the partial charge is stabilized, the more effective the resonance will be and the more acidic the phenol will be. Electron-withdrawing groups such as a nitro substituent increase the acidity of the phenol since they stabilize the negative charge by an inductive effect. Nitro groups which are *ortho* or *para* to the phenolic group have an even greater effect. This is because a fourth resonance structure is possible which delocalizes the partial charge even further (*Fig. 2*).

Electron-donating substituents (e.g. alkyl groups) have the opposite effect and decrease the acidity of phenols.

Functional group transformations

Phenols can be converted into esters by reaction with acid chlorides or acid anhydrides (Topic K6), and into ethers by reaction with alkyl halides in the presence of base (Topic L6; *Fig. 3*). These reactions can be carried out under milder conditions than those used for alcohols due to the greater acidity of phenols. Thus

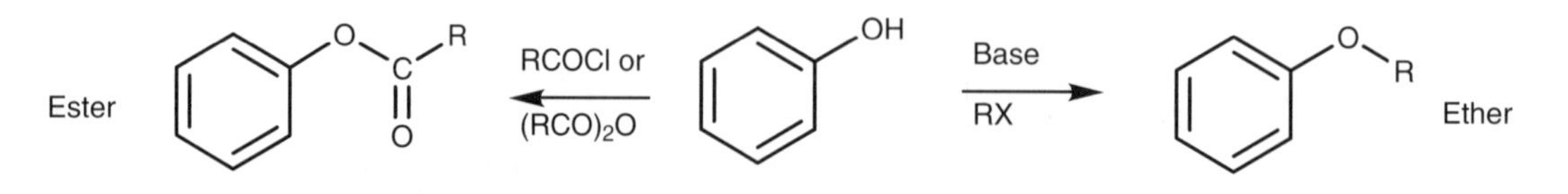

Fig. 3. Functional group transformations for a phenol.

phenols can be converted to phenoxide ions with sodium hydroxide rather than metallic sodium.

Although the above reactions are common to alcohols and phenols, there are several reactions which can be carried out on alcohols but not phenols, and *vice versa*. For example, unlike alcohols, phenols cannot be converted to esters by reaction with a carboxylic acid under acid catalysis. Reactions involving the cleavage of the C–O bond are also not possible for phenols. The aryl C–O bond is stronger than the alkyl C–O bond of an alcohol.

Electrophilic substitution

Electrophilic substitution is promoted by the phenol group which acts as an activating group and directs substitution to the *ortho* and *para* positions (Topic I5). Sulfonation and nitration of phenols are both possible to give *ortho* and *para* substitution products. On occasions, the phenolic groups may be too powerful an activating group and it is difficult to control the reaction to one substitution. For example, the bromination of phenol leads to 2,4,6-tribromophenol even in the absence of a Lewis acid (*Fig. 4*).

The activating power of the phenolic group can be decreased by converting the phenol to an ester which can be removed by hydrolysis once the electrophilic substitution reaction has been carried out (*Fig. 5*). Since the ester is a weaker activating group, substitution occurs only once. Furthermore, since the ester is a bulkier group than the phenol, *para* substitution is favored over *ortho* substitution.

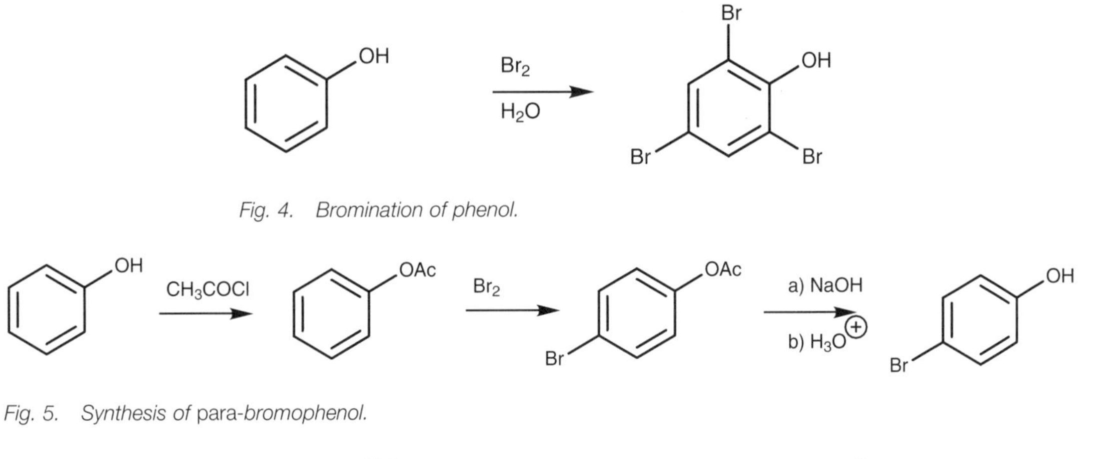

Fig. 4. Bromination of phenol.

Fig. 5. Synthesis of para*-bromophenol.*

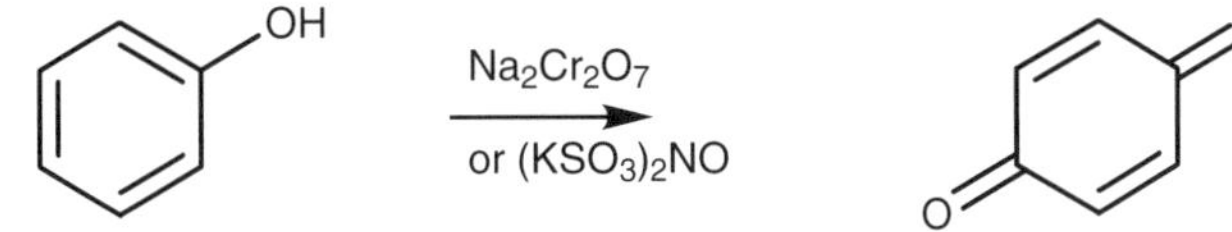

Fig. 6. Oxidation of phenol.

Oxidation

Phenols are susceptible to oxidation to quinones (*Fig. 6*).

Claisen rearrangement

A useful method of introducing an alkyl substituent to the *ortho* position of a phenol is by the Claisen rearrangement (*Fig. 7*). The phenol is converted to the phenoxide ion, then treated with 3-bromopropene (an allyl bromide) to form an ether. On heating, the allyl group ($-CH_2-CH=CH_2$) is transferred from the phenolic group to the *ortho* position of the aromatic ring. The mechanism involves a

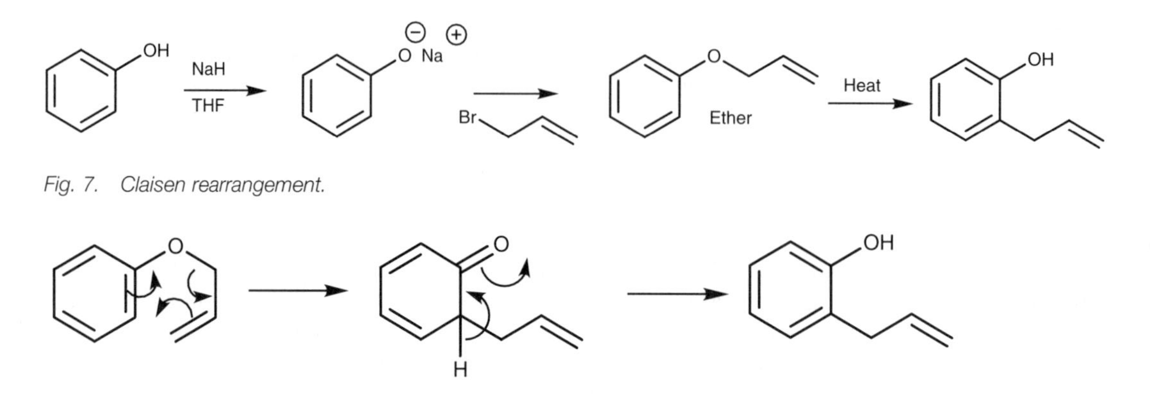

Fig. 7. Claisen rearrangement.

Fig. 8. Mechanism for the Claisen rearrangement.

concerted process of bond formation and bond breaking called a **pericyclic reaction** (*Fig. 8*). This results in a ketone structure which immediately tautomerizes to the final product. Different allylic reagents could be used in the reaction and the double bond in the final product could be reduced to form an alkane substituent without affecting the aromatic ring.

M6 Chemistry of thiols

Key Notes

Preparation

Thiols can be prepared by the reaction of an alkyl halide with KOH and an excess of hydrogen sulfide. A hydrogen sulfide anion is formed which undergoes an S_N2 reaction with the alkyl halide. Hydrogen sulfide has to be in excess in order to limit further reaction to a thioether. Alternatively, the alkyl halide can be treated with thiourea to form an *S*-alkylisothiouronium salt which is then hydrolyzed with aqueous base to give the thiol. Disulfides can be reduced to thiols with zinc and acid.

Properties

Hydrogen bonding is weak, resulting in boiling points which are lower than comparable alcohols and similar to comparable thioethers.

Reactivity

Thiols (RSH) contain a large polarizable sulfur atom. The S–H bond is weak compared to alcohols, making thiols prone to oxidation. Thiolate ions are extremely good nucleophiles whilst being weak bases. Thiols are stronger acids than alcohols.

Reactions

Thiols are oxidized by bromine or iodine to give disulfides. Treatment of a thiol with a base results in the formation of a thiolate ion.

Related topics

Intermolecular bonding (C3)
Nucleophilic substitution (L2)
Reactions of alkyl halides (L6)

Preparation

Thiols can be prepared by the treatment of alkyl halides with an excess of KOH and hydrogen sulfide (*Fig. 1a*). The preparation is an S_N2 reaction involving the

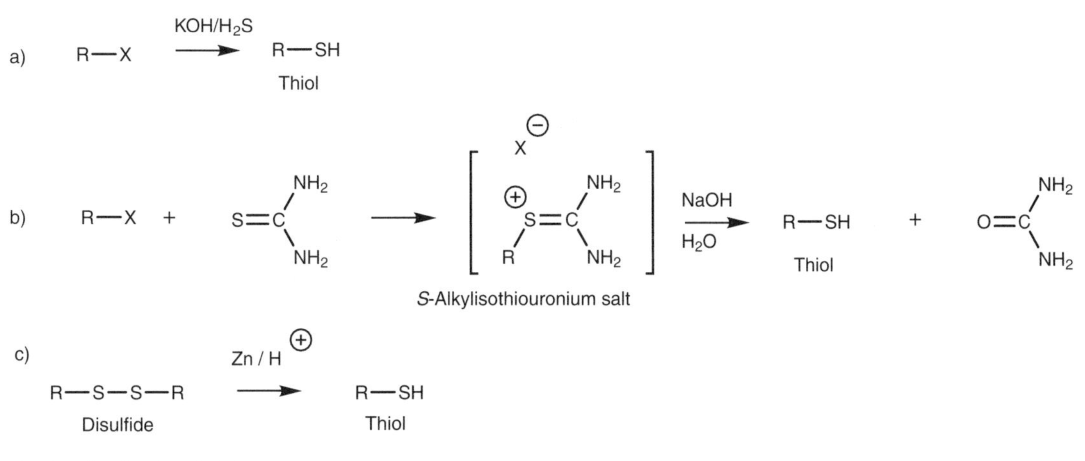

Fig. 1. Synthesis of thiols.

generation of a hydrogen sulfide anion (HS^-) as nucleophile. A problem with this reaction is the possibility of the product being ionized and reacting with a second molecule of alkyl halide to produce a thioether (RSR) as a byproduct. An excess of hydrogen sulfide is normally used to avoid this problem.

The problem of thioether formation can also be avoided by using an alternative procedure involving thiourea (*Fig. 1b*). The thiourea acts as the nucleophile in an S_N2 reaction to produce an *S*-alkylisothiouronium salt which is then hydrolyzed with aqueous base to give the thiol.

Thiols can also be formed by reducing disulfides with zinc in the presence of acid (*Fig. 1c*).

Properties

Thiols form extremely weak hydrogen bonds – much weaker than alcohols – and so thiols have boiling points which are similar to comparable thioethers and which are lower than comparable alcohols. For example, ethanethiol boils at 37°C whereas ethanol boils at 78°C.

Low molecular weight thiols are notorious for having disagreeable aromas.

Reactivity

Thiols are the sulfur equivalent of alcohols (RSH). The sulfur atom is larger and more polarizable than oxygen which means that sulfur compounds as a whole are more powerful nucleophiles than the corresponding oxygen compounds. Thiolate ions (e.g. $CH_3CH_2S^-$) are stronger nucleophiles and weaker bases than corresponding alkoxides ($CH_3CH_2O^-$). Conversely, thiols are stronger acids than corresponding alcohols.

The relative size difference between sulfur and oxygen also means that sulfur's bonding orbitals are more diffuse than oxygen's bonding orbitals. This results in a poorer bonding interaction between sulfur and hydrogen, than between oxygen and hydrogen. As a result, the S–H bond of thiols is weaker than the O–H bond of alcohols (80 kcal mol^{-1} vs. 100 kcal mol^{-1}). This in turn means that the S–H bond of thiols is more prone to oxidation than the O–H bond of alcohols.

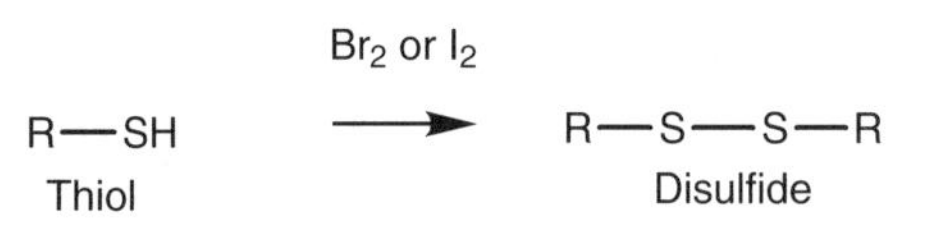

Fig. 2. Oxidation of thiols.

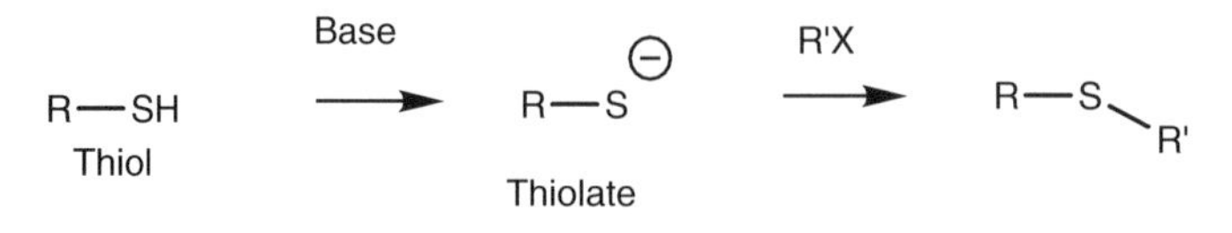

Fig. 3. Formation of thiolate ions.

Reactions

Thiols are easily oxidized by mild oxidizing agents such as bromine or iodine to give disulfides (*Fig. 2*).

Thiols react with base to form thiolate ions which can act as powerful nucleophiles (Topic L6; *Fig. 3*).

N1 PREPARATION OF ETHERS, EPOXIDES, AND THIOETHERS

Key Notes

Ethers

Ethers can be prepared by the S_N2 reaction of an alkyl halide with an alkoxide ion. The reaction works best for primary alkyl halides. Alcohols and alkyl halides can be reacted in the presence of silver oxide to give an ether. Alkenes can be treated with alcohols in the presence of mercuric trifluoroacetate to form ethers by electrophilic addition.

Epoxides

Epoxides can be synthesized from alkenes and *meta*-chloroperbenzoic acid, or by converting the alkene to a halohydrin and treating the product with base to induce an intramolecular S_N2 reaction which displaces the halogen atom. Aldehydes and ketones can be converted to epoxides on treatment with a sulfur ylide.

Thioethers

Thioethers are prepared by the S_N2 reaction between an alkyl halide and a thiolate ion. Symmetrical thioethers can be prepared by treating the alkyl halide with KOH and hydrogen sulfide where the latter is not in excess.

Related topics

Electrophilic addition to symmetrical alkenes (H3)
Reduction and oxidation of alkenes (H6)
Nucleophilic substitution (L2)
Reactions of alkyl halides (L6)
Reactions of alcohols (M4)
Chemistry of thiols (M6)
Reactions of ethers, epoxides, and thioethers (N3)

Ethers

The Williamson ether synthesis is the best method of preparing ethers (*Fig. 1a*). The procedure involves the S_N2 reaction between a metal alkoxide and a primary alkyl halide or tosylate (Topic L2). The alkoxide required for the reaction is prepared by treating an alcohol with a strong base such as sodium hydride (Topic M4). An alternative procedure is to treat the alcohol directly with the alkyl halide in the presence of silver oxide, thus avoiding the need to prepare the alkoxide beforehand (*Fig. 1b*).

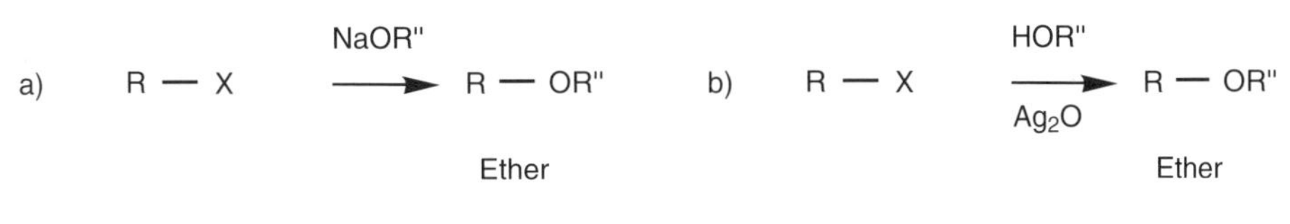

Fig. 1. Synthesis of ethers.

If an unsymmetrical ether is being synthesized, the most hindered alkoxide should be reacted with the simplest alkyl halide, rather than the other way round

(*Fig. 2*). Since this is an S_N2 reaction, primary alkyl halides react better then secondary or tertiary alkyl halides.

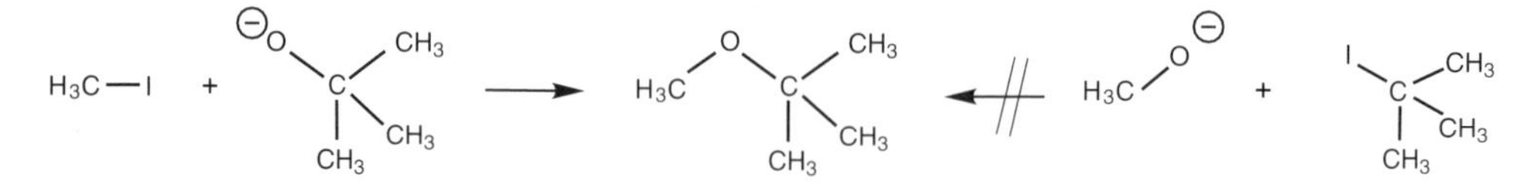

Fig. 2. Choice of synthetic routes to an unsymmetrical ether.

Alkenes can be converted to ethers by the electrophilic addition of mercuric trifluoroacetate, followed by addition of an alcohol. An organomercuric intermediate is obtained which can be reduced with sodium borohydride to give the ether (Topic H3; *Fig. 3*).

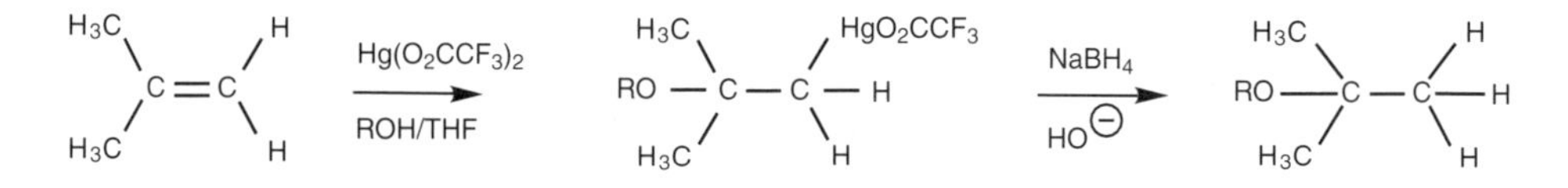

Fig. 3. Synthesis of an ether from an alkene and an alcohol.

Epoxides

Epoxides can be synthesized by treating aldehydes or ketones with sulfur ylides (Topic N3). They can also be prepared from alkenes by reaction with *m*-chloroperoxybenzoic acid (Topic H6). Alternatively they can be obtained from alkenes in a two-step process (*Fig. 4*). The first step involves electrophilic addition of a halogen in aqueous solution to form a halohydrin (Topic H3). Treatment of the halohydrin with base then ionizes the alcohol group, which can then act as a nucleophile (*Fig. 5*). The oxygen uses a lone pair of electrons to form a bond to the neighboring electrophilic carbon, thus displacing the halogen by an intramolecular S_N2 reaction.

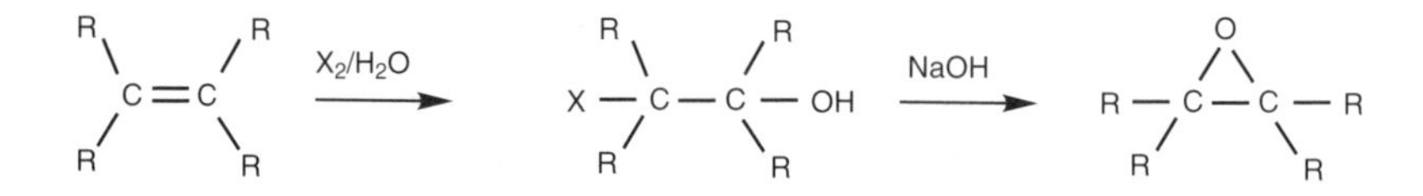

Fig. 4. Synthesis of an epoxide via a halohydrin.

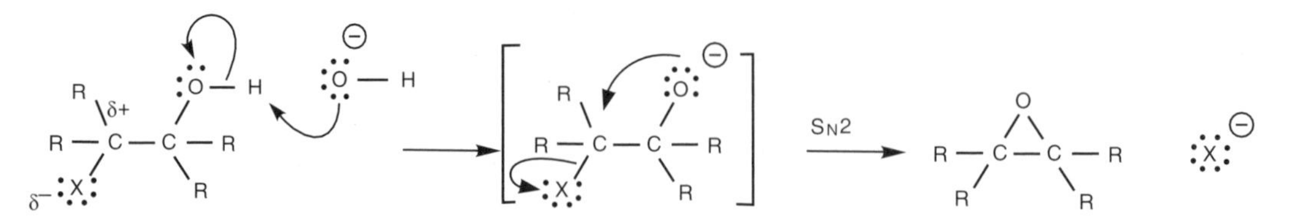

Fig. 5. Mechanism of epoxide formation from a halohydrin.

Thioethers

Thioethers (or sulfides) are prepared by the S_N2 reaction of primary or secondary alkyl halides with a thiolate anion (RS^-), (Topic L6; *Fig. 6*). The reaction is similar to the Williamson ether synthesis.

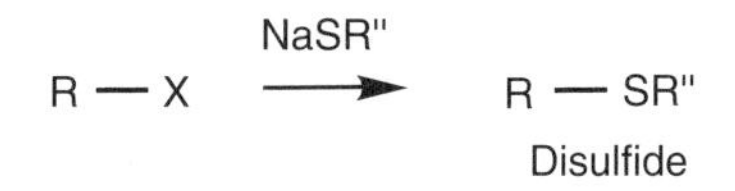

Fig. 6. Synthesis of a disulfide from an alkyl halide.

Symmetrical thioethers can be prepared by treating an alkyl halide with KOH and an equivalent of hydrogen sulfide. The reaction produces a thiol which is ionized again by KOH and reacts with another molecule of alkyl halide (Topic M6).

N2 PROPERTIES OF ETHERS, EPOXIDES AND THIOETHERS

Key Notes

Ethers

Ethers consist of an sp^3 hybridized oxygen linked to two carbon atoms by a single σ bond. Alkyl ethers are ethers where two alkyl groups are linked to the oxygen. Aryl ethers are ethers where one or two aromatic rings are attached to the oxygen. Since ethers cannot form hydrogen bonds, they have lower boiling points than comparable alcohols, and similar boiling points to comparable alkanes. However, hydrogen bonding is possible with protic solvents which means that ethers have water solubilities similar to alcohols of equivalent molecular weight. Ethers are relatively unreactive since they have weak nucleophilic and electrophilic centers.

Epoxides

Epoxides are three-membered cyclic ethers which are more reactive than other cyclic or acyclic ethers due to the ring strain inherent in three-membered rings. They will react with nucleophiles by an S_N2 reaction at the electrophilic carbons.

Thioethers

Thioethers are the sulfur equivalents of ethers. The polarizable sulfur can stabilize a negative charge on an adjacent carbon making protons attached to that carbon acidic.

Related topics

sp^3 Hybridization (A3)
sp^2 Hybridization (A4)
Intermolecular bonding (C3)
Properties and reactions (C4)
Organic structures (E4)
Reactions of ethers, epoxides, and thioethers (N3)

Ethers

Ethers consist of an oxygen linked to two carbon atoms by σ bonds. In aliphatic ethers (ROR), the three atoms involved are sp^3 hybridized and have a bond angle of 112°. Aryl ethers are ethers where the oxygen is linked to one or two aromatic rings (ArOR or ArOAr) in which case the attached carbon(s) is sp^2 hybridized.

The C–O bonds are polarized such that the oxygen is slightly negative and the carbons are slightly positive. Due to the slightly polar C–O bonds, ethers have a small dipole moment. However, ethers have no X–H groups (X=heteroatom) and cannot interact by hydrogen bonding. Therefore, they have lower boiling points than comparable alcohols and similar boiling points to comparable alkanes. However, hydrogen bonding is possible to protic solvents resulting in solubilities similar to alcohols of comparable molecular weight.

The oxygen of an ether is a nucleophilic center and the neighboring carbons are electrophilic centers, but in both cases the nucleophilicity or electrophilicity is weak (*Fig. 1*). Therefore, ethers are relatively unreactive.

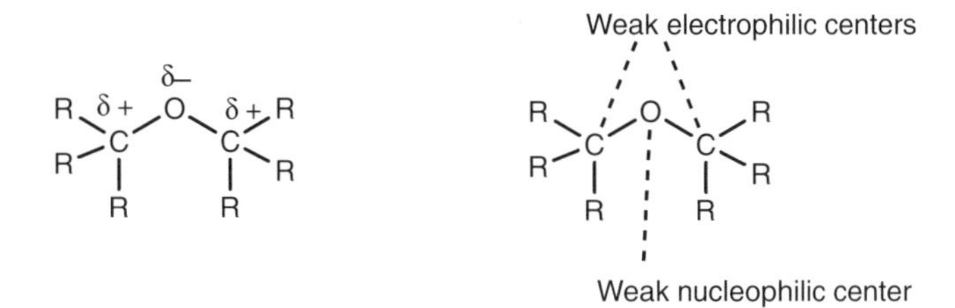

Fig. 1. Properties of ethers.

Epoxides

Epoxides (or oxiranes) are three-membered cyclic ethers and differ from other cyclic and acyclic ethers in that they are reactive to various reagents. The reason for this reactivity is the strained three-membered ring. Reactions with nucleophiles can result in ring opening and relief of strain. Nucleophiles will attack either of the electrophilic carbons present in an epoxide by an S_N2 reaction (*Fig. 2*).

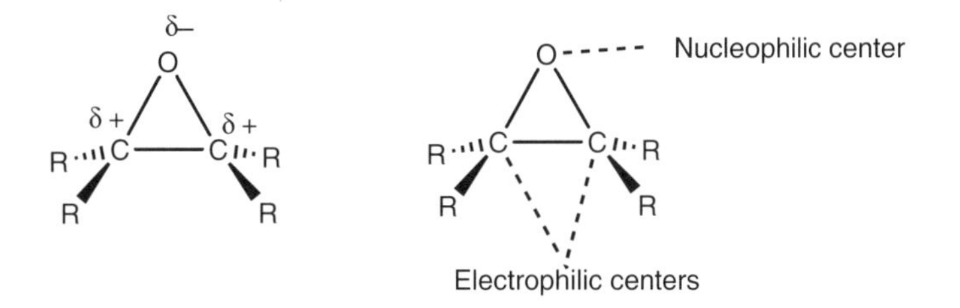

Fig. 2. Properties of an epoxide.

Thioethers

Thioethers (or sulfides; RSR) are the sulfur equivalents of ethers (ROR). Since the sulfur atoms are polarizable, they can stabilize a negative charge on an adjacent carbon atom. This means that hydrogens on this carbon are more acidic than those on comparable ethers.

N3 REACTIONS OF ETHERS, EPOXIDES, AND THIOETHERS

Key Notes

Ethers

Ethers are unreactive functional groups, but can be cleaved by strong acids such as HI or HBr. Primary and secondary ethers react by the S_N2 mechanism to produce the least substituted alkyl halide and an alcohol. If the alcohol is primary, further reaction may occur to convert this to an alkyl halide as well. Tertiary ethers are cleaved by the S_N1 reaction under milder conditions. However, an elimination reaction (E1) may occur in preference to the S_N1 reaction resulting in formation of an alcohol and an alkene. Trifluoroacetic acid can be used in such situations in place of HX. Most ethers react slowly with atmospheric oxygen to produce peroxides and hydroperoxides which can prove to be explosive.

Epoxides

Epoxides are more reactive to nucleophiles than ethers since an S_N2 reaction relieves ring strain by opening up the ring. Hydrolysis under acidic or basic conditions converts epoxides to 1,2-diols which are *trans* to each other in cyclic systems. Treatment with hydrogen halides produces 1,2-halohydrins and treatment with Grignard reagents allows the formation of C–C bonds with simultaneous formation of an alcohol. Nucleophiles will attack unsymmetrical epoxides at the least substituted carbon when basic reaction conditions are employed. Under acidic reaction conditions, nucleophiles will prefer to attack the most substituted carbon atom.

Thioethers

Thioethers are nucleophilic. The sulfur atom can act as a nucleophilic center and take part in an S_N2 reaction with alkyl halides to form a trialkylsulfonium salt ($R_2SR'^+$). This in turn can be treated with base to form a sulfur ylide (R_2S^+-CR_2^-) where the sulfur can stabilize the neighboring negative charge. Sulfur ylides can be used to synthesize epoxides from aldehydes or ketones. Thioethers can be oxidized to sulfoxides and sulfones, and can be reduced to alkanes.

Related topics

Nucleophilic substitution (L2)
Elimination (L4)
Reactions of alcohols (M4)
Properties of ethers, epoxides, and thioethers (N2)

Ethers

Ethers are generally unreactive functional groups and the only useful reaction which they undergo is cleavage by strong acids such as HI and HBr to produce an alkyl halide and an alcohol (*Fig. 1*). The ether is first protonated by the acid, then nucleophilic substitution takes place where the halide ion acts as the nucleophile. Primary and secondary ethers react by the S_N2 mechanism (Topic L2) and the halide reacts at the least substituted carbon atom to produce an alkyl halide and

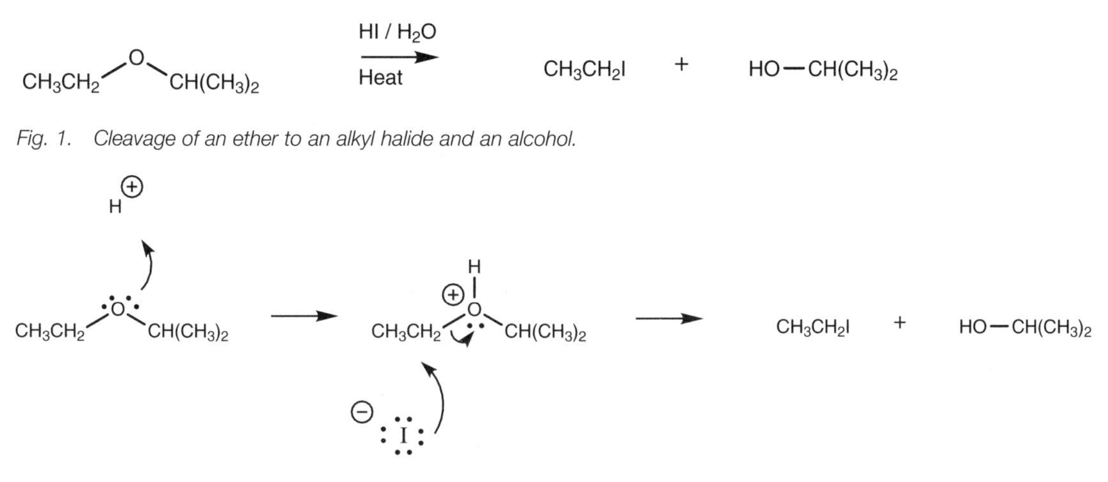

Fig. 1. Cleavage of an ether to an alkyl halide and an alcohol.

Fig. 2. Mechanism for the cleavage of an ether.

an alcohol (*Fig. 2*). The initial protonation is essential since it converts a poor leaving group (an alkoxide ion) into a good leaving group (the alcohol).

Primary alcohols formed from this reaction may be converted further to an alkyl halide (Topic M4). Tertiary ethers react by the S_N1 mechanism to produce the alcohol. However, an alkene may also be formed due to E1 elimination (Topic L4) and this may be the major product (*Fig. 3*).

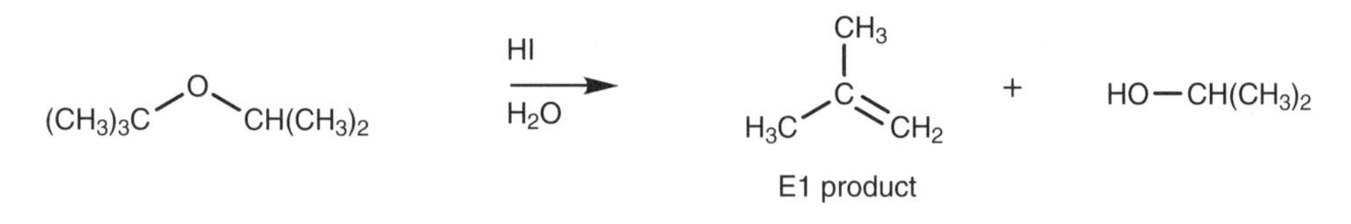

Fig. 3. Cleavage of a tertiary ether.

Trifluoroacetic acid can be used instead of HX, resulting in an E1 reaction and production of the alcohol and the alkene.

A problem with most ethers is their slow reaction with atmospheric oxygen by a radical process to form **hydroperoxides** (ROOH) and **peroxides** (ROOR). These products can prove to be explosive if old solvents are concentrated to dryness.

Epoxides

Epoxides are cyclic ethers, but they are more reactive than normal ethers because of the ring strain involved in a three-membered ring. Therefore, ring opening through an S_N2 nucleophilic substitution is a common reaction of epoxides. For example, epoxides can be ring-opened under acidic or basic conditions to give a 1,2-diol (*Fig. 4*). In both cases, the reaction involves an S_N2 mechanism (Topic L2) with the incoming nucleophile attacking the epoxide from the opposite direction of the epoxide ring. This results in a *trans* arrangement of the diol system when the reaction is carried out on cycloalkane epoxides. Under acidic conditions (*Fig. 5*), the epoxide oxygen is first protonated, turning it into a better leaving group (Step 1). Water then acts as the nucleophile and attacks one of the electrophilic carbon atoms of the epoxide. Water uses a lone pair of electrons to form a new bond to carbon and as it does so, the C–O bond of the epoxide cleaves with both electrons moving onto the epoxide oxygen to neutralize the positive charge (Step 2).

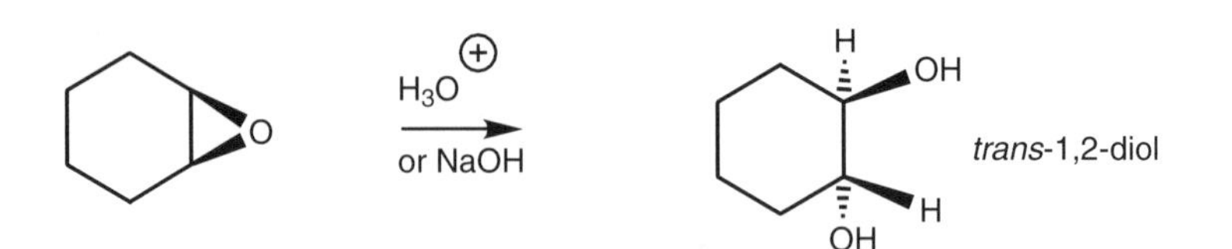

Fig. 4. Ring opening of an epoxide under acidic or basic conditions.

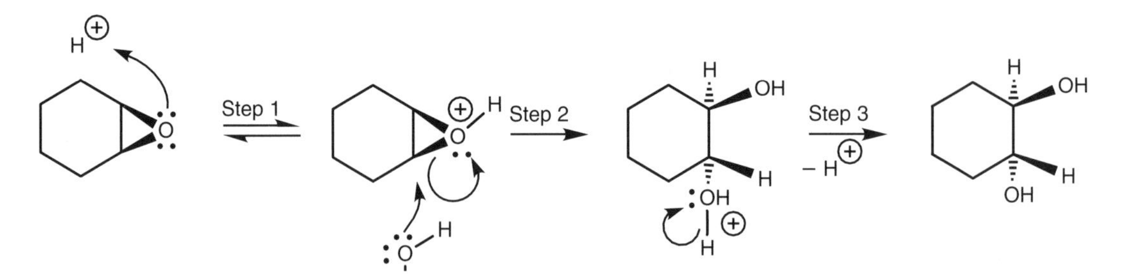

Fig. 5. Mechanism for the ring opening of an epoxide under acidic conditions.

Although water is a poor nucleophile, the reaction is favored due to the neutralization of the positive charge on oxygen and the relief of ring strain once the epoxide is opened up. Unlike other S_N2 reactions of course, the leaving group is still tethered to the molecule.

Ring opening under basic conditions is also possible with heating, but requires the loss of a negatively charged oxygen (*Fig. 6*). This is a poor leaving group and would not occur with normal ethers. It is only possible here because the reaction opens up the epoxide ring and relieves ring strain.

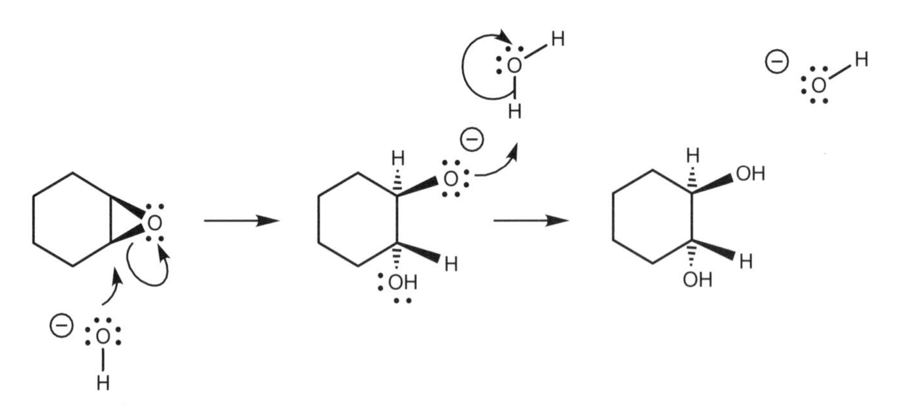

Fig. 6. Mechanism for the ring opening of an epoxide under basic conditions.

Ring opening by the S_N2 reaction is also possible using nucleophiles other than water. With unsymmetrical epoxides, the S_N2 reaction will occur at the least substituted position if it is carried out under basic conditions (*Fig. 7*). However, under acidic conditions, the nucleophile will usually attack the most substituted position (*Fig. 8*). This is because the positive charge in the protonated intermediate is shared between the oxygen and the most substituted carbon. This makes the more substituted carbon more reactive to nucleophiles.

The reaction of epoxides with hydrogen halides (*Fig. 9*) is analogous to the reaction of normal ethers with HX. Protonation of the epoxide with acid is followed

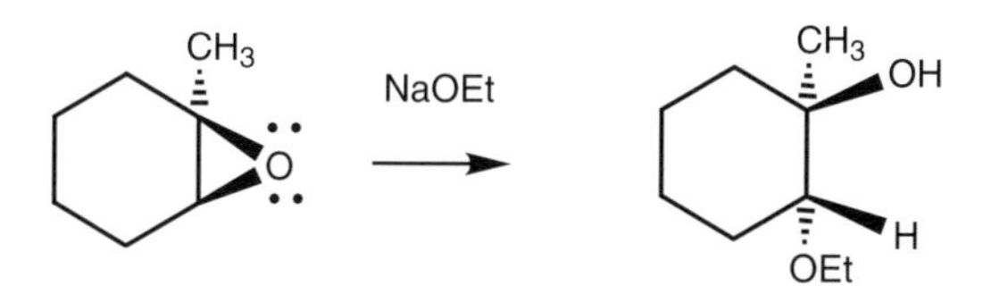

Fig. 7. Ring opening of an epoxide with the ethoxide ion.

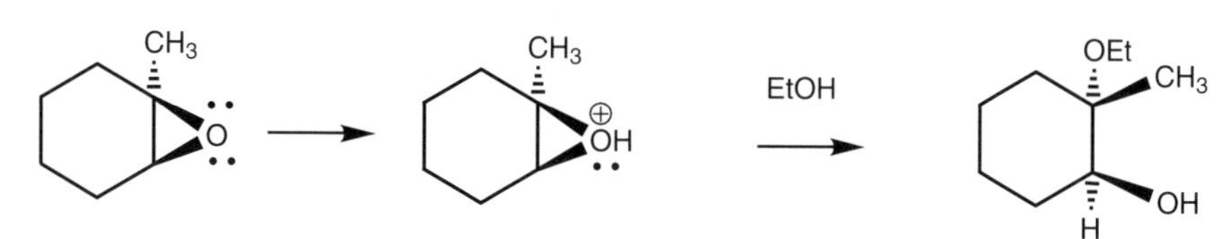

Fig. 8. Ring opening of an epoxide with ethanol under acidic conditions.

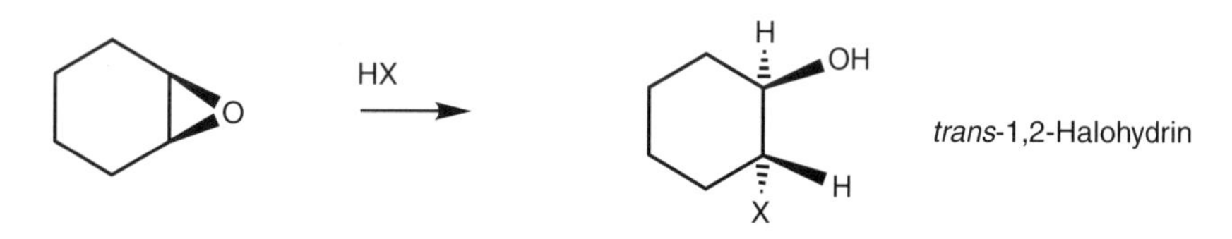

Fig. 9. Reaction of an epoxide with HX.

by nucleophilic attack by a halide ion resulting in 1,2-halohydrins. The halogen and alcohol groups will also be in a *trans* arrangement if the reaction is done on an epoxide linked to a cyclic system, for example.

Unlike ethers, epoxides undergo the S_N2 reaction with a Grignard reagent (*Fig. 10*).

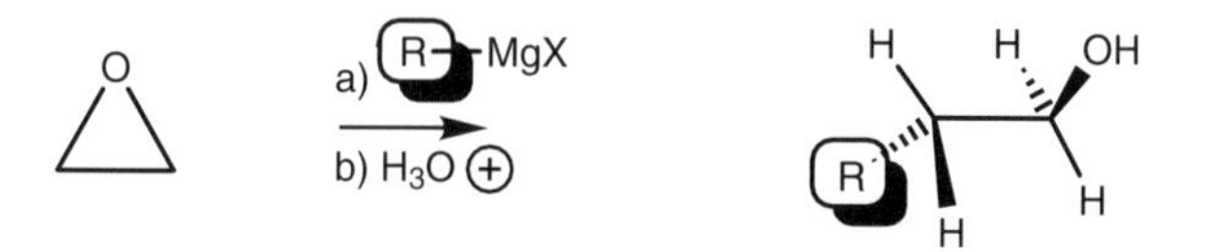

Fig. 10. Grignard reaction with an epoxide.

Thioethers

Unlike ethers, thioethers make good nucleophiles due to the sulfur atom. This is because the sulfur atom has its valence electrons further away from the nucleus. As a result, these electrons experience less attraction from the nucleus, making them more polarizable and more nucleophilic. Since they are good nucleophiles, thioethers can react with alkyl halides to form trialkylsulfonium salts (R_3S^+; *Fig. 11*) – a reaction which is impossible for normal ethers. Sulfur is also able to stabilize a negative charge on a neighboring carbon atom, especially when the sulfur itself is positively charged. This makes the protons on neighboring carbons acidic, allowing them to be removed with base to form sulfur ylides.

Sulfur ylides are useful in the synthesis of epoxides from aldehydes or ketones (*Fig. 12*). They undergo a typical nucleophilic addition with the carbonyl group to form the expected tetrahedral intermediate (Topic J3). This intermediate now has a very good thioether leaving group which also creates an electrophilic carbon atom at the neighboring position. Therefore, the molecule is set up for further reaction which involves the nucleophilic oxygen anion displacing the thioether and forming an epoxide.

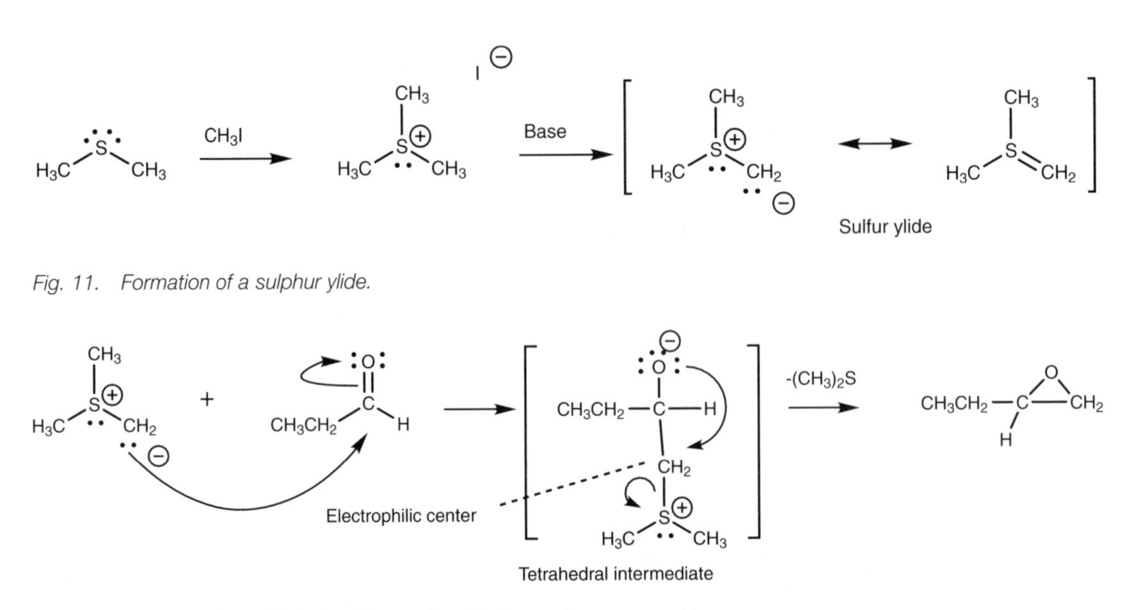

Fig. 11. Formation of a sulphur ylide.

Fig. 12. Reaction of an aldehyde with a sulfur ylide to produce an epoxide.

Thioethers can also be oxidized with hydrogen peroxide to give a sulfoxide (R_2SO) which, on oxidation with a peroxyacid, gives a sulfone (R_2SO_2; *Fig. 13*).

Thioethers can be reduced using Raney nickel – a catalyst which has hydrogen gas adsorbed onto the nickel surface (*Fig. 14*). This reaction is particularly useful for reducing thioacetals or thioketals since this provides a means of converting aldehydes or ketones to alkanes (Topic J7). The mechanism of the reduction reaction is radical based and is not fully understood.

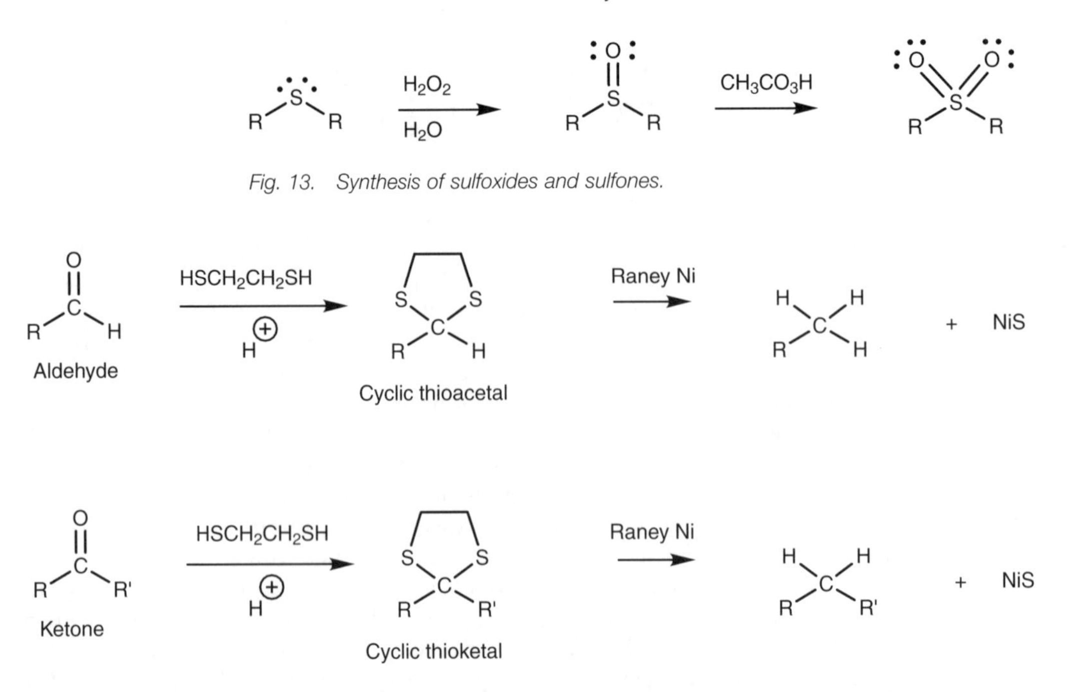

Fig. 13. Synthesis of sulfoxides and sulfones.

Fig. 14. Reduction of aldehydes and ketals via cyclic thioacetals and cyclic thioketals respectively.

O1 PREPARATION OF AMINES

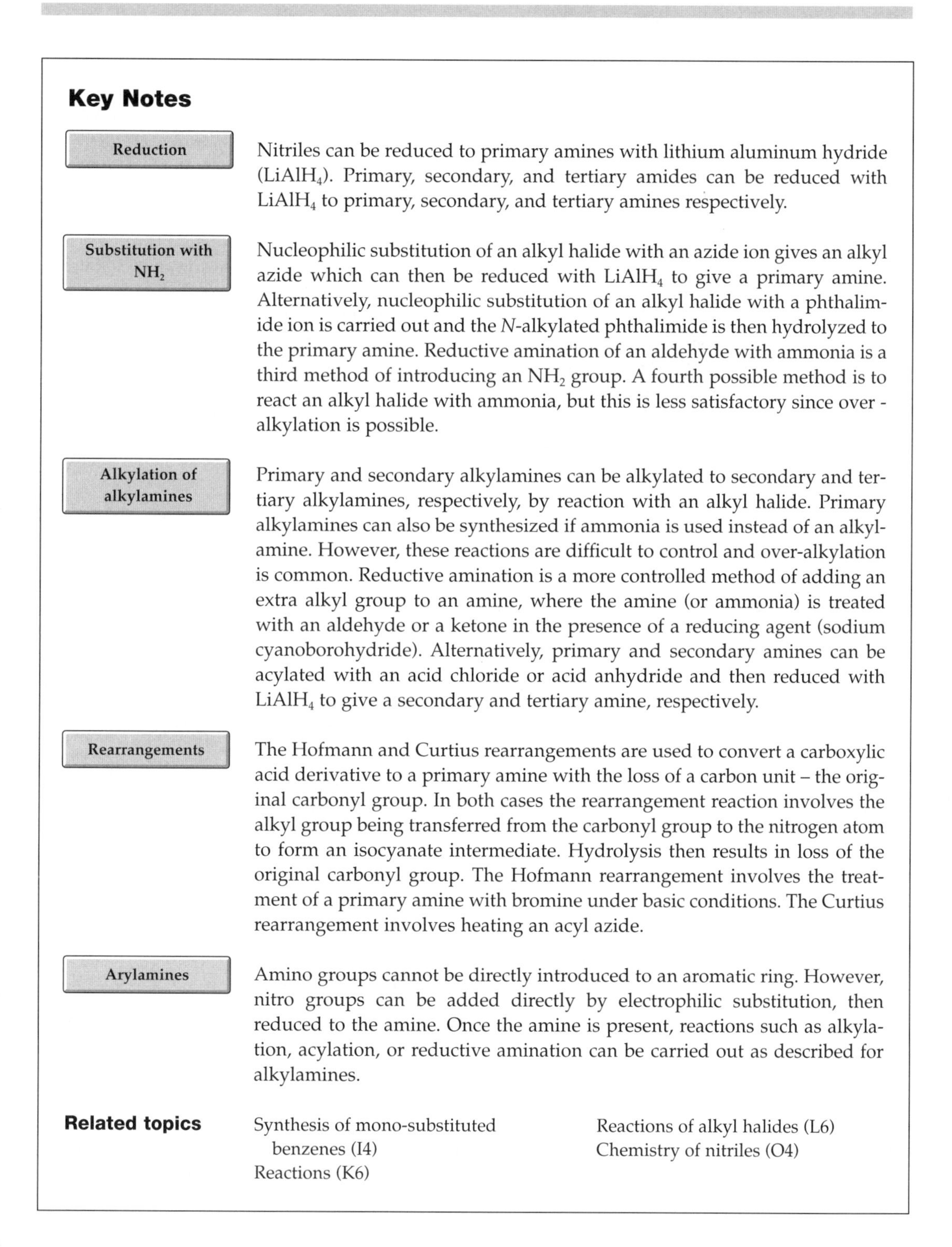

Key Notes

Reduction

Nitriles can be reduced to primary amines with lithium aluminum hydride ($LiAlH_4$). Primary, secondary, and tertiary amides can be reduced with $LiAlH_4$ to primary, secondary, and tertiary amines respectively.

Substitution with NH_2

Nucleophilic substitution of an alkyl halide with an azide ion gives an alkyl azide which can then be reduced with $LiAlH_4$ to give a primary amine. Alternatively, nucleophilic substitution of an alkyl halide with a phthalimide ion is carried out and the *N*-alkylated phthalimide is then hydrolyzed to the primary amine. Reductive amination of an aldehyde with ammonia is a third method of introducing an NH_2 group. A fourth possible method is to react an alkyl halide with ammonia, but this is less satisfactory since over - alkylation is possible.

Alkylation of alkylamines

Primary and secondary alkylamines can be alkylated to secondary and tertiary alkylamines, respectively, by reaction with an alkyl halide. Primary alkylamines can also be synthesized if ammonia is used instead of an alkylamine. However, these reactions are difficult to control and over-alkylation is common. Reductive amination is a more controlled method of adding an extra alkyl group to an amine, where the amine (or ammonia) is treated with an aldehyde or a ketone in the presence of a reducing agent (sodium cyanoborohydride). Alternatively, primary and secondary amines can be acylated with an acid chloride or acid anhydride and then reduced with $LiAlH_4$ to give a secondary and tertiary amine, respectively.

Rearrangements

The Hofmann and Curtius rearrangements are used to convert a carboxylic acid derivative to a primary amine with the loss of a carbon unit – the original carbonyl group. In both cases the rearrangement reaction involves the alkyl group being transferred from the carbonyl group to the nitrogen atom to form an isocyanate intermediate. Hydrolysis then results in loss of the original carbonyl group. The Hofmann rearrangement involves the treatment of a primary amine with bromine under basic conditions. The Curtius rearrangement involves heating an acyl azide.

Arylamines

Amino groups cannot be directly introduced to an aromatic ring. However, nitro groups can be added directly by electrophilic substitution, then reduced to the amine. Once the amine is present, reactions such as alkylation, acylation, or reductive amination can be carried out as described for alkylamines.

Related topics

Synthesis of mono-substituted benzenes (I4)
Reactions (K6)
Reactions of alkyl halides (L6)
Chemistry of nitriles (O4)

Reduction

Nitriles and amides can be reduced to alkylamines using lithium aluminum hydride ($LiAlH_4$; Topics O4 and K6). In the case of a nitrile, a primary amine is the only possible product. Primary, secondary, and tertiary amines can be prepared from primary, secondary, and tertiary amides, respectively.

Substitution with NH_2

Primary alkyl halides and some secondary alkyl halides can undergo S_N2 nucleophilic substitution with an azide ion (N_3^-) to give an alkyl azide (Topic L6). The azide can then be reduced with $LiAlH_4$ to give a primary amine (*Fig. 1*).

The overall reaction is equivalent to replacing the halogen atom of the alkyl halide with an NH_2 unit. Another method of achieving the same result is the **Gabriel synthesis** of amines. This involves treating phthalimide with KOH to abstract the N–H proton (*Fig. 2*). The N–H proton of phthalimide is more acidic (pK_a 9) than the N–H proton of an amide since the anion formed can be stabilized by resonance with both neighboring carbonyl groups. The phthalimide ion can then be alkylated by treating it with an alkyl halide in a nucleophilic substitution (Topic L2). Subsequent hydrolysis (Topic K6) releases a primary amine (*Fig. 3*).

A third possible method is to react an alkyl halide with ammonia, but this is less satisfactory since over-alkylation is possible (see below). The reaction of an aldehyde with ammonia by reductive amination is a fourth method of obtaining primary amines (see below).

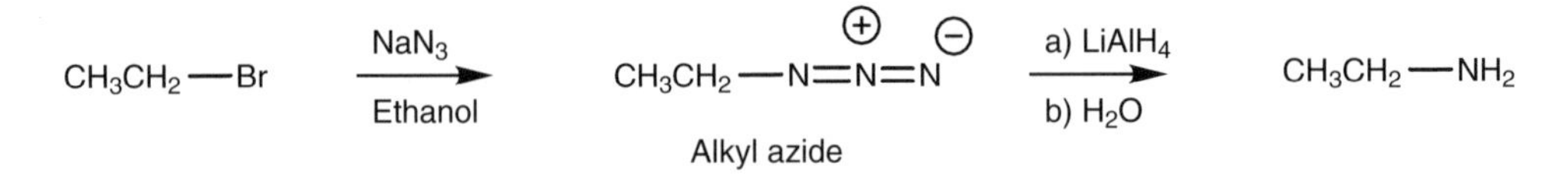

Fig. 1. Synthesis of a primary amine from an alkyl halide via an alkyl azide.

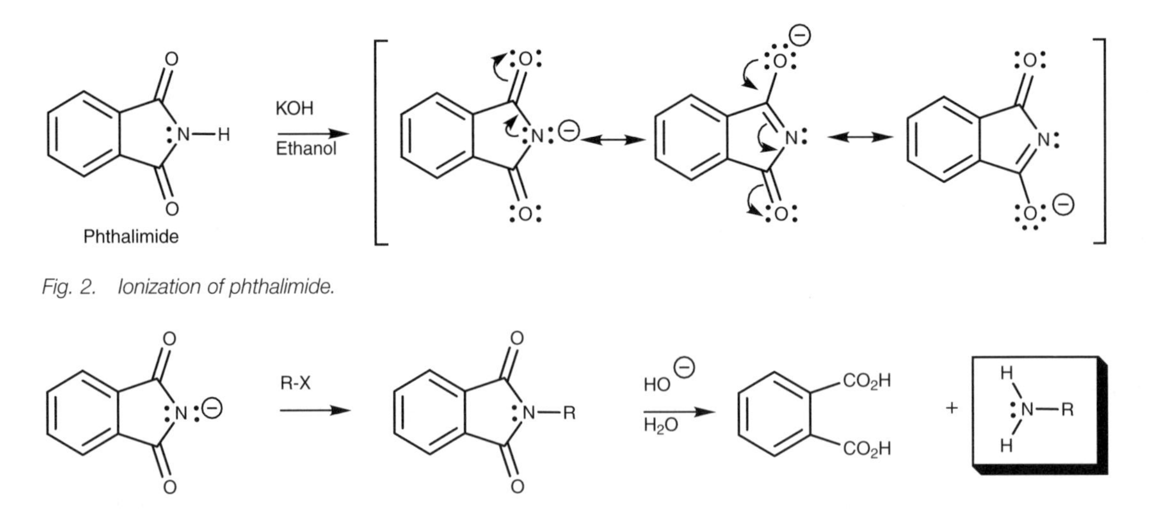

Fig. 2. Ionization of phthalimide.

Fig. 3. Gabriel synthesis of primary amines.

Alkylation of alkylamines

It is possible to convert primary and secondary amines to secondary and tertiary amines respectively, by alkylation with alkyl halides by the S_N2 reaction (Topic L6). However, over-alkylation can be a problem and better methods of amine synthesis are available.

Reductive amination is a more controlled method of adding an extra alkyl group to an alkylamine (*Fig. 4*). Primary and secondary alkylamines can be treated with a ketone or an aldehyde in the presence of a reducing agent called sodium cyanoborohydride. The alkylamine reacts with the carbonyl compound by nucleophilic addition followed by elimination to give an imine or an iminium ion (Topic J6) which is immediately reduced by sodium cyanoborohydride to give the final amine. Overall, this is the equivalent of adding one extra alkyl group to the amine. Therefore, primary amines are converted to secondary amines and secondary amines are converted to tertiary amine. The reaction is also suitable for the synthesis of primary amines if ammonia is used instead of an alkylamine. The reaction goes through an imine intermediate if ammonia or a primary amine is used (*Fig. 4a*). When a secondary amine is used, an iminium ion intermediate is involved (*Fig. 4b*).

An alternative way of alkylating an amine is to acylate the amine to give an amide (Topic K5), and then carry out a reduction with $LiAlH_4$ (Topic K6; *Fig. 5*). Although two steps are involved, there is no risk of over-alkylation since acylation can only occur once.

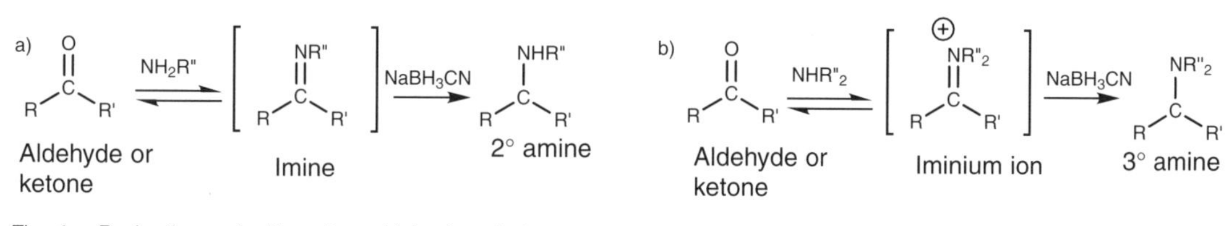

Fig. 4. Reductive amination of an aldehyde or ketone.

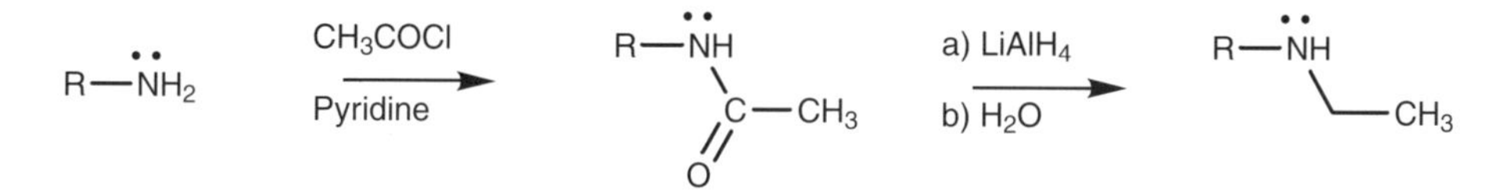

Fig. 5. Alkylation of an amide via an amine.

Rearrangements

There are two rearrangement reactions which can be used to convert carboxylic acid derivatives into primary amines where the carbon chain in the product has been shortened by one carbon unit (*Fig. 6*). These are known as the Hofmann and the Curtius rearrangements. The Hofmann rearrangement involves the treatment of a primary amide with bromine under basic conditions, while the Curtius rearrangement involves heating an acyl azide. The end result is the same – a primary amine with loss of the original carbonyl group.

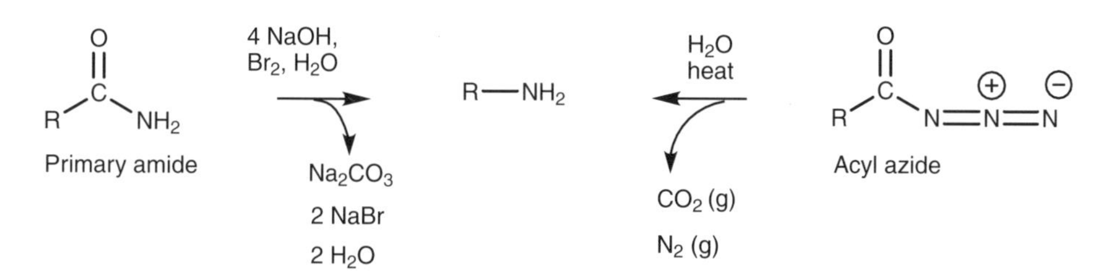

Fig. 6. Hofmann rearrangement (left) and Curtius rearrangement (right).

In both reactions, the alkyl group (R) is transferred from the carbonyl group to the nitrogen to form an intermediate isocyanate (O=C=N–R). This is then hydrolyzed by water to form carbon dioxide and the primary amine. The Curtius rearrangement has the added advantage that nitrogen is lost as a gas which helps to drive the reaction to completion.

Arylamines

The direct introduction of an amino group to an aromatic ring is not possible. However, nitro groups can be added directly by electrophilic substitution and then reduced to the amine (Topic I4; *Fig. 7*). The reduction is carried out under acidic conditions resulting in an arylaminium ion as product. The free base can be isolated by basifying the solution with sodium hydroxide to precipitate the arylamine.

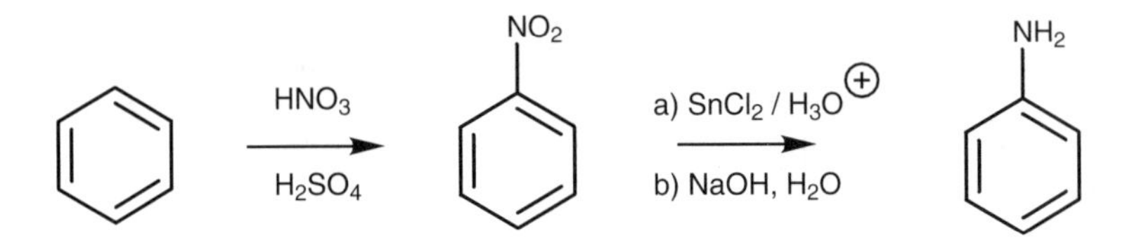

Fig. 7. Introduction of an amine to an aromatic ring.

Once an amino group has been introduced to an aromatic ring, it can be alkylated with an alkyl halide, acylated with an acid chloride or converted to a higher amine by reductive animation as described for an alkylamine.

O2 Properties of amines

Key Notes

Structure

Amines consist of an sp^3 hybridized nitrogen linked to three substituents by σ bonds. The functional group is pyramidal in shape with bond angles of approximately 109°. If the substituents are alkyl groups, the amine is aliphatic or an alkylamine. If one or more of the substituents is aromatic, the amine is aromatic or an arylamine. If the amine has only one alkyl or aryl substituent, it is defined as primary. If there are two such substituents, the amine is secondary, and if there are three such groups, the amine is tertiary.

Pyramidal inversion

Amines can be chiral if they have three different substituents. However, it is not possible to separate enantiomers since they can easily interconvert by pyramidal inversion. The process involves a planar intermediate where the nitrogen has changed from sp^3 hybridization to sp^2 hybridization and the lone pair of electrons are in a *p* orbital. Pyramidal inversion is not possible for chiral quaternary ammonium salts and enantiomers of these structures can be separated.

Physical properties

Amines are polar compounds with higher boiling points than comparable alkanes. They have similar water solubilities to alcohols due to hydrogen bonding, and low molecular weight amines are completely miscible with water. Low molecular weight amines have an offensive fishy smell.

Basicity

Amines are weak bases which are in equilibrium with their ammonium ion in aqueous solution. The basic strength of an amine is indicated by its pK_b value. There are two main effects on basic strength. Alkyl groups have an inductive effect which stabilizes the ammonium ion and results in increased basicity. Solvation of the ammonium ion by water stabilizes the ion and increases basicity. The more hydrogen bonds which are possible between the ammonium ion and water, the greater the stability and the greater the basicity. The alkyl inductive effect is greatest for ammonium ions formed from tertiary amines, whereas the solvation effect is greatest for ammonium ions formed from primary amines. In general, primary and secondary amines are stronger bases than tertiary amines. Aromatic amines are weaker bases than aliphatic amines since nitrogen's lone pair of electrons interacts with the π system of the aromatic ring, and is less likely to form a bond to a proton. Aromatic substituents affect basicity. Activating substituents increase electron density in the aromatic ring which helps to stabilize the ammonium ion and increase basic strength. Deactivating groups have the opposite effect. Substituents capable of interacting with the aromatic ring by resonance have a greater effect on basicity if they are at the *ortho* or *para* positions.

Reactivity

Amines react as nucleophiles or bases since they have a readily available lone pair of electrons which can participate in bonding. Primary and secondary amines can act as weak electrophiles or acids with a strong base, by losing an N–H proton to form an amide anion (R_2N^-).

Related topics

sp^3 Hybridization (A3)
sp^2 Hybridization (A4)
Intermolecular bonding (C3)
Properties and reactions (C4)
Organic structures (E4)
Base strength (G3)
Reactions of amines (O3)

Structure

Amines consist of an sp^3 hybridized nitrogen linked to three substituents by three σ bonds. The substituents can be hydrogen, alkyl, or aryl groups, but at least one of the substituents has to be an alkyl or aryl group. If only one such group is present, the amine is defined as primary. If two groups are present, the amine is secondary. If three groups are present, the amine is tertiary (Topic C6). If the substituents are all alkyl groups, the amine is defined as being an alkylamine. If there is at least one aryl group directly attached to the nitrogen, then the amine is defined as an arylamine.

The nitrogen atom has four sp^3 hybridized orbitals pointing to the corners of a tetrahedron in the same way as an sp^3 hybridized carbon atom. However, one of the sp^3 orbitals is occupied by the nitrogen's lone pair of electrons. This means that the atoms in an amine functional group are pyramidal in shape. The C–N–C bond angles are approximately 109° which is consistent with a tetrahedral nitrogen. However, the bond angle is slightly less than 109° since the lone pair of electrons demands a slightly greater amount of space than a σ bond.

Pyramidal inversion

Since amines are tetrahedral, they are chiral if they have three different substituents. However, it is not possible to separate the enantiomers of a chiral amine since amines can easily undergo pyramidal inversion – a process which interconverts the enantiomers (*Fig. 1*). The inversion involves a change of hybridization where the nitrogen becomes sp^2 hybridized (Topic A4) rather than sp^3 hybridized. As a result, the molecule becomes planar and the lone pair of electrons occupy a *p* orbital. Once the hybridization reverts back to sp^3, the molecule can either revert back to its original shape or invert.

Although the enantiomers of chiral amines cannot be separated, such amines can be alkylated to form quaternary ammonium salts where the enantiomers **can** be separated. Once the lone pair of electrons is locked up in a σ bond, pyramidal inversion becomes impossible and the enantiomers can no longer interconvert.

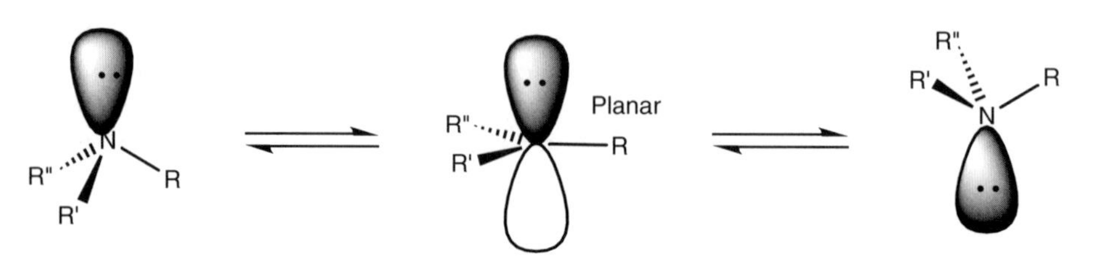

Fig. 1. Pyramidal inversion.

Physical properties

Amines are polar compounds and intermolecular hydrogen bonding is possible for primary and secondary amines. Therefore, primary and secondary amines have higher boiling points than alkanes of similar molecular weight. Tertiary amines also have higher boiling points than comparable alkanes, but have slightly lower boiling points than comparable primary or secondary amines since they cannot take part in intermolecular hydrogen bonding.

However, all amines can participate in hydrogen bonding with protic solvents, which means that amines have similar water solubilities to comparable alcohols. Low molecular weight amines are freely miscible with water. Low molecular weight amines have an offensive fish-like odor.

Basicity

Amines are weak bases but they are more basic than alcohols, ethers, or water. As a result, amines act as bases when they are dissolved in water and an equilibrium is set up between the ionized form (the **ammonium** ion) and the unionized form (the free base; *Fig. 2*).

The basic strength of an amine can be measured by its pK_b value (typically 3–4; Topic G3). The lower the value of pK_b, the stronger the base. The pK_b for ammonia is 4.74, which compares with pK_b values for methylamine, ethylamine, and propylamine of 3.36, 3.25 and 3.33, respectively. This demonstrates that larger alkyl groups increase base strength. This is an inductive effect whereby the ion is stabilized by dispersing some of the positive charge over the alkyl group (*Fig. 3*). This shifts the equilibrium of the acid base reaction towards the ion, which means that the amine is more basic. The larger the alkyl group, the more significant this effect.

Fig. 2. Acid–base reaction between an amine and water.

Fig. 3. Inductive effect of an alkyl group on an alkylammonium ion.

Further alkyl substituents should have an even greater inductive effect and one might expect secondary and tertiary amines to be stronger bases than primary amines. This is not necessarily the case and there is no direct relationship between basicity and the number of alkyl groups attached to nitrogen. The inductive effect of more alkyl groups is counterbalanced by a **solvation effect**.

Once the ammonium ion is formed, it is solvated by water molecules – a stabilizing factor which involves hydrogen bonding between the oxygen atom of water and any N–H group present in the ammonium ion (*Fig. 4*). The more hydrogen bonds which are possible, the greater the stabilization. As a result, solvation and solvent stabilization is stronger for alkylaminium ions formed from primary amines than for those formed from tertiary amines. The solvent effect tends to be more important than the inductive effect as far as tertiary amines are concerned and so tertiary amines are generally weaker bases than primary or secondary amines.

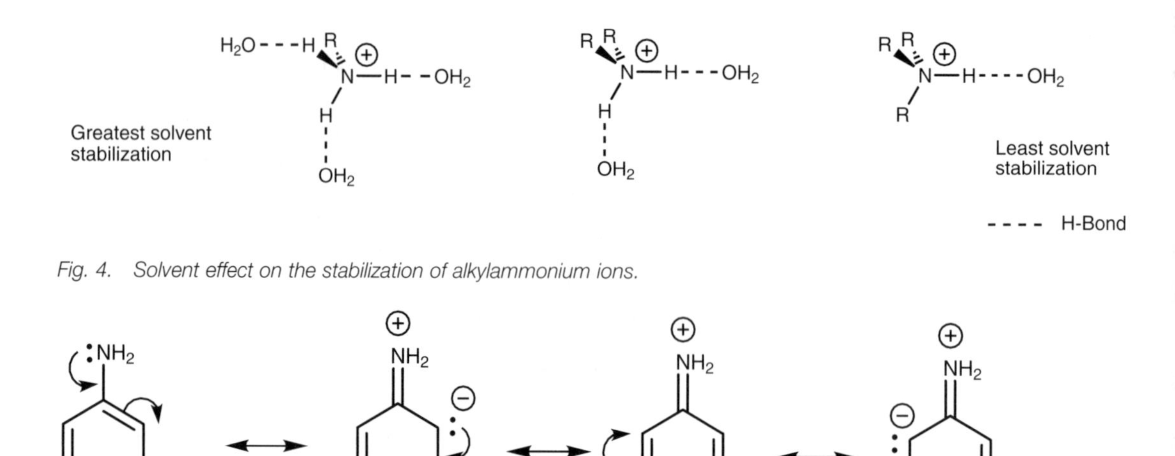

Fig. 4. Solvent effect on the stabilization of alkylammonium ions.

Fig. 5. Resonance interaction between nitrogen's lone pair and the aromatic ring.

Aromatic amines (arylamines) are weaker bases than alkylamines since the orbital containing nitrogen's lone pair of electrons overlaps with the π system of the aromatic ring. In terms of resonance, the lone pair of electrons can be used to form a double bond to the aromatic ring, resulting in the possibility of three **zwitterionic** resonance structures (*Fig. 5*). (A zwitterion is a molecule containing a positive and a negative charge.) Since nitrogen's lone pair of electrons is involved in this interaction, it is less available to form a bond to a proton and so the amine is less basic.

The nature of aromatic substituent also affects the basicity of aromatic amines. Substituents which deactivate aromatic rings (e.g. NO_2, Cl, or CN) lower electron density in the ring, which means that the ring will have an electron-withdrawing effect on the neighboring ammonium ion. This means that the charge will be destabilized and the amine will be a weaker base. Substituents which activate the aromatic ring enhance electron density in the ring which means that the ring will have an electron-donating effect on the neighboring charge. This has a stabilizing effect and so the amine will be a stronger base. The relative position of aromatic substituents can be important if resonance is possible between the aromatic ring and the substituent. In such cases, the substituent will have a greater effect if it is at the *ortho* or *para* position. For example, *para*-nitroaniline is a weaker base than *meta*-nitroaniline. This is because one of the possible resonance structures for the *para* isomer is highly disfavored since it places a positive charge immediately next to the ammonium ion (*Fig. 6*). Therefore, the number of feasible resonance structures for the *para* isomer is limited to three, compared to four for the *meta* isomer. This means that the *para* isomer experiences less stabilization and so the amine will be less basic.

If an activating substituent is present, capable of interacting with the ring by resonance, the opposite holds true and the *para* isomer will be a stronger base than the *meta* isomer. This is because the crucial resonance structure mentioned above would have a negative charge immediately next to the ammonium ion and this would have a stabilizing effect.

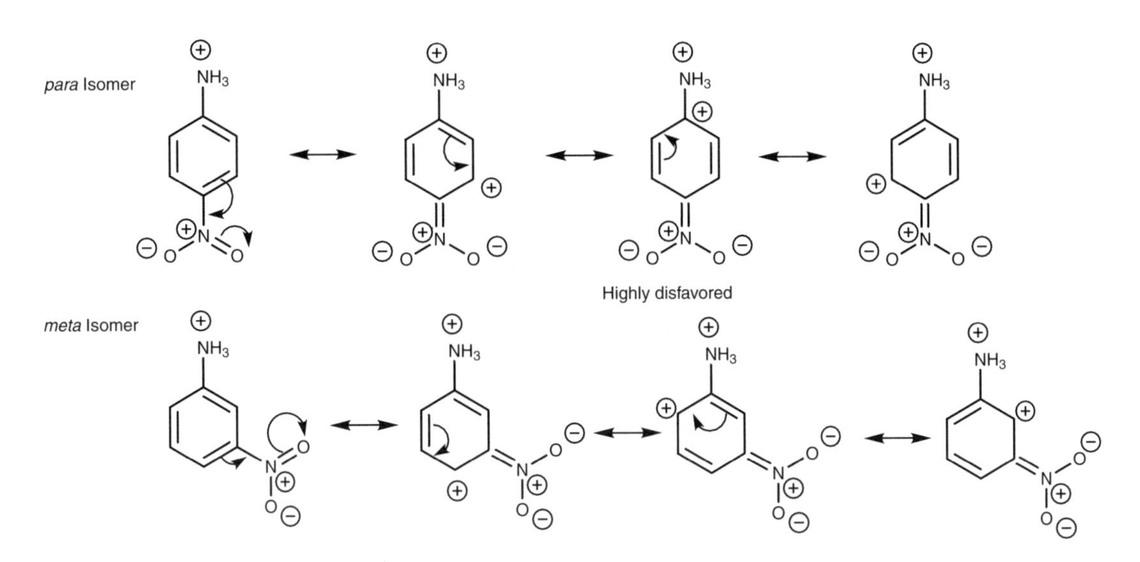

Fig. 6. Resonance structures for para-nitroaniline and meta-nitroaniline.

Reactivity

Amines react as nucleophiles or bases, since the nitrogen atom has a readily available lone pair of electrons which can participate in bonding (*Fig. 7*). As a result, amines react with acids to form water soluble salts. This allows the easy separation of amines from other compounds. A crude reaction mixture can be extracted with dilute hydrochloric acid such that any amines present are protonated and dissolve into the aqueous phase as water-soluble salts. The free amine can be recovered by adding sodium hydroxide to the aqueous solution such that the free amine precipitates out as a solid or as an oil.

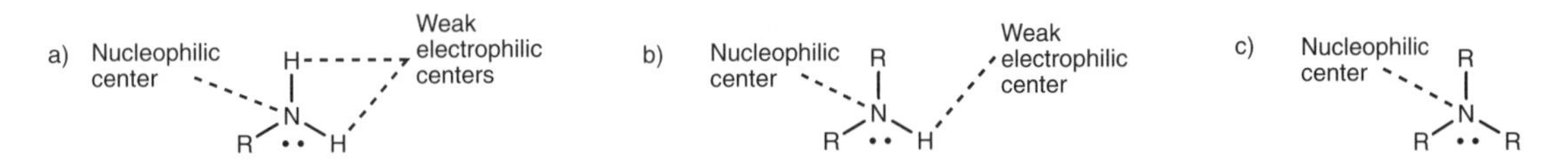

Fig. 7. Nucleophilic and electrophilic centers in (a) primary, (b) secondary, and (c) tertiary amines.

Amines will also react as nucleophiles with a wide range of electrophiles including alkyl halides, aldehydes, ketones, and acid chlorides.

The N–H protons of primary and secondary amines are weakly electrophilic or acidic and will react with a strong base to form amide anions. For example, diisopropylamine (pK_a ~40) reacts with butyllithium to give lithium diisopropylamide (LDA) and butane.

O3 REACTIONS OF AMINES

Key Notes

Alkylation

Ammonia, primary amines, and secondary amines can be alkylated with alkyl halides to give primary, secondary, and tertiary amines, respectively. However, over-alkylation usually occurs and mixtures are obtained. The method is best used for converting tertiary amines to quaternary ammonium salts. A better method of alkylating a primary or secondary amine is to treat the amine with an aldehyde or ketone in the presence of a reducing agent. Reaction of the amine with the carbonyl compound produces an intermediate imine which is reduced to the amine. No over-alkylation takes place.

Acylation

Primary and secondary amines can be acylated with an acid chloride or acid anhydride to give secondary and tertiary amides, respectively.

Sulfonylation

Primary and secondary amines can be sulfonylated with a sulfonyl chloride to give a sulfonamide.

Elimination

Primary amines can be converted to alkenes if the amine is first methylated to a quaternary ammonium salt, then treated with silver oxide. Elimination of triethylamine takes place to give the least substituted alkene. The reaction is known as the Hofmann elimination. The reaction can also be carried out on secondary and tertiary amines although a mixture of alkenes may be formed depending on the substituents present. Aromatic amines will also react if they contain a suitable *N*-alkyl substituent.

Electrophilic aromatic substitution

Aromatic amines undergo electrophilic aromatic substitutions. The amino group is strongly activating and directs substitution to the *ortho* and *para* positions. Nitration, sulfonation, and bromination are all possible, but bromination may occur more than once. Friedel–Crafts alkylation and acylation are not possible since the amino group complexes the Lewis acid involved in the reaction. The problems of excess bromination and Lewis acid complexation can be overcome by converting the amine to an amide before carrying out the substitution reaction. The amide can be hydrolyzed back to the amine once the substitution reaction has been carried out.

Diazonium salts

Aromatic primary amines can be converted to diazonium salts on treatment with nitrous acid. These salts are extremely important in aromatic chemistry since they can be converted to a variety of other substituents. Diazonium salts also react with phenols or aromatic amines in a process called diazonium coupling to produce a highly conjugated system which is usually colored. Such products are often used as dyes.

Related topics	Synthesis of di- and tri-substituted benzenes (I6)	Reactions (K6)
	Preparations of carboxylic acid derivatives (K5)	Reactions of alkyl halides (L6)
		Preparation of amines (O1)

Alkylation

Ammonia, primary amines, and secondary amines (both aromatic and aliphatic) can undergo the S_N2 reaction with alkyl halides to produce a range of primary, secondary, and tertiary amines (Topic L6). Primary, secondary, and tertiary amines are produced as ammonium salts which are converted to the free amine by treatment with sodium hydroxide (*Fig. 1a*).

In theory, it should be possible to synthesize primary amines from ammonia, secondary amines from primary amines, and tertiary amines from secondary amines. In practice, over-alkylation is common. For example, reaction of ammonia with methyl iodide leads to a mixture of primary, secondary, and tertiary amines along with a small quantity of the quaternary ammonium salt (*Fig. 2*).

Alkylation of tertiary amines by this method is a good way of obtaining quaternary ammonium salts (*Fig. 1b*) since no other products are possible. However, alkylation of lower order amines is not so satisfactory.

A better method of alkylating a primary or secondary amine is to treat the amine with a ketone or an aldehyde in the presence of a reducing agent – sodium cyanoborohydride. This reaction is known as a **reductive amination** and is described in Topic O1. Over-alkylation cannot occur by this method.

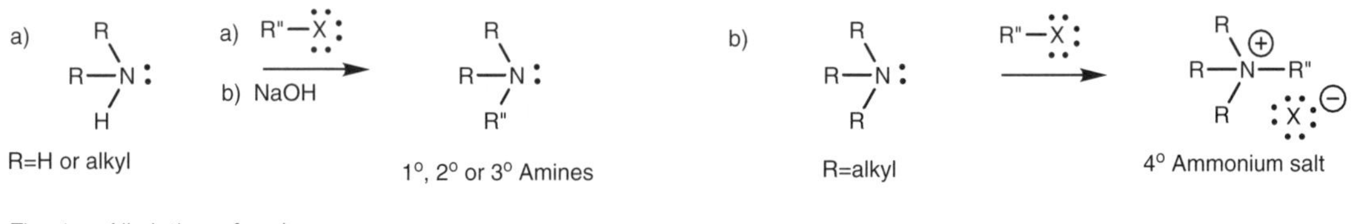

Fig. 1. Alkylation of amines.

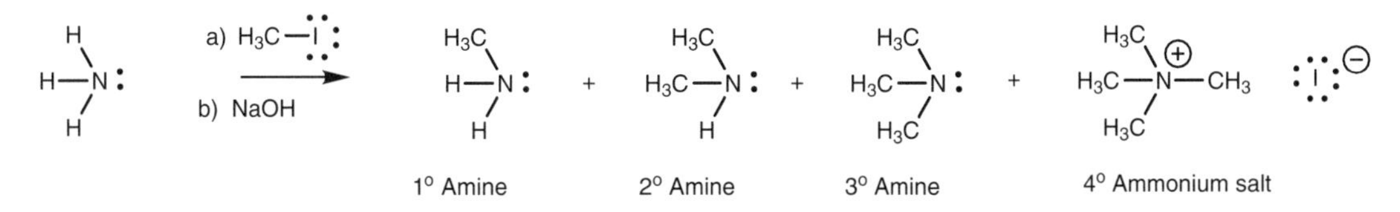

Fig. 2. Alkylation of ammonia with methyl iodide.

Acylation

Primary and secondary amines (both aromatic and aliphatic) can be acylated with an acid chloride or acid anhydride to form secondary and tertiary amides, respectively (Topic K5). This reaction can be viewed as the acylation of an amine or as the nucleophilic substitution of a carboxylic acid derivative.

Sulfonylation

In a similar reaction to acylation, primary and secondary amines (both aromatic and aliphatic) can be treated with a sulfonyl chloride to give a sulfonamide (*Fig. 3*). Tertiary amines do not give a stable product and are recovered unchanged.

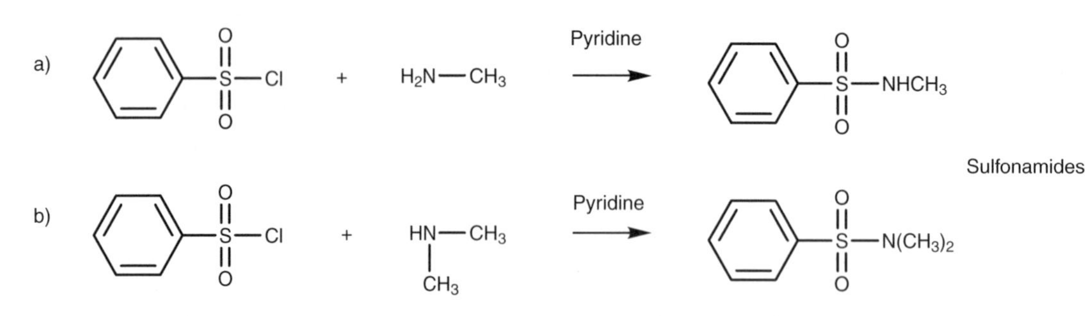

Fig. 3. Reaction of benzenesulfonyl chloride with (a) primary amine; (b) secondary amine.

Elimination

Primary amines could be converted to alkenes if it was possible to eliminate ammonia from the molecule. However, the direct elimination of ammonia is not possible. A less direct method of achieving the same result is to exhaustively methylate the amine by the S_N2 reaction (Topic L2) to give a quaternary ammonium salt. Once this is formed, it is possible to eliminate triethylamine in the presence of silver oxide and to form the desired alkene. The reaction is called the **Hofmann elimination** (*Fig. 4*). The silver oxide provides a hydroxyl ion which acts as the base for an E2 elimination (Topic L4; *Fig. 5*). However, unlike most E2 eliminations, the less substituted alkene is preferred if a choice is available (*Fig. 6*). The reason for this preference is not fully understood, but may have something to do with the large bulk of the triethylamine leaving group hindering the approach of the hydroxide ion such that it approaches the least hindered β-carbon.

Secondary and tertiary amines can also be exhaustively methylated then treated

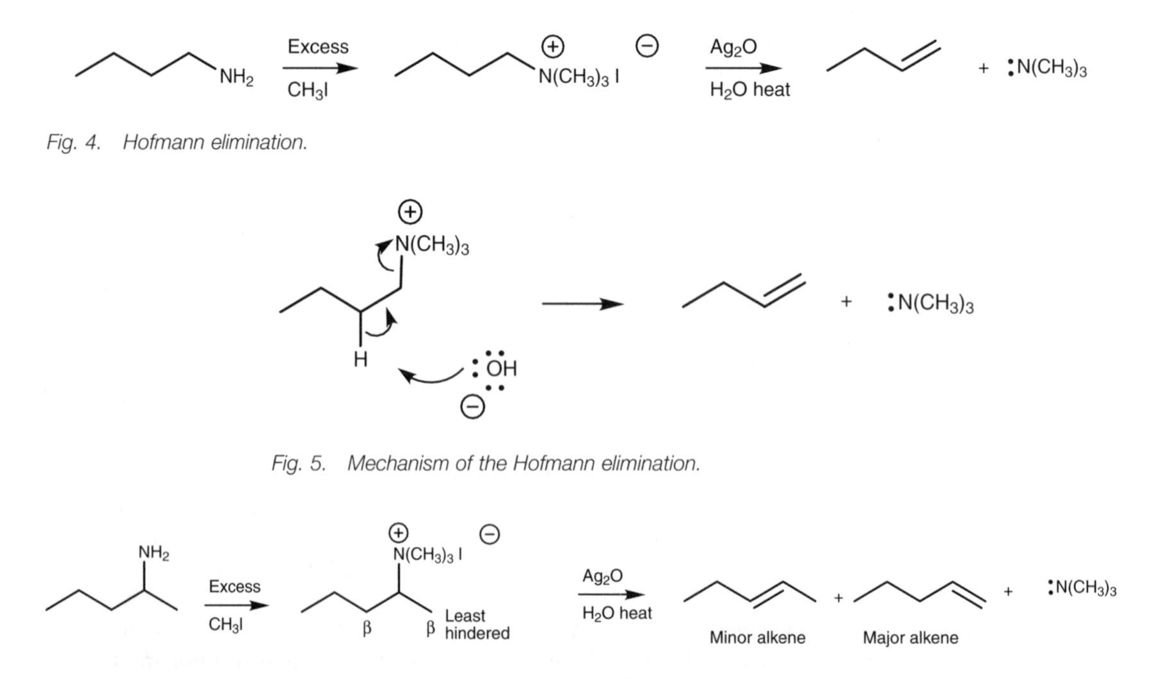

Fig. 4. Hofmann elimination.

Fig. 5. Mechanism of the Hofmann elimination.

Fig. 6. The less substituted alkene is preferred in the Hofmann elimination.

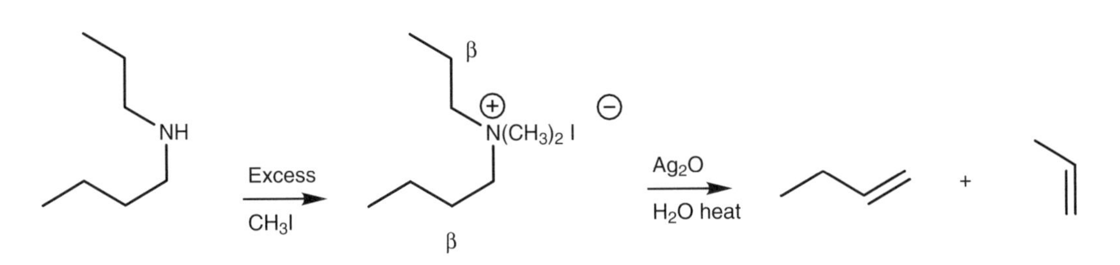

Fig. 7. Hofmann elimination of a secondary amine.

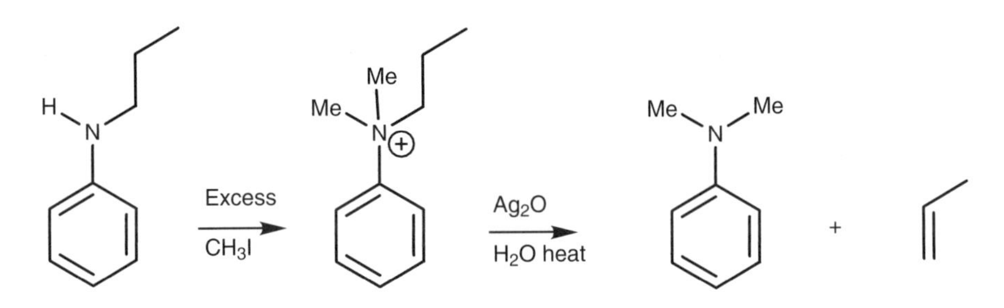

Fig. 8. Hofmann elimination of an aromatic amine.

with silver oxide. However, mixtures of different alkenes may be obtained if the *N*-substituents are different alkyl groups (*Fig. 7*).

The Hofmann elimination is not possible with primary arylamines, but secondary and tertiary arylamines will react if one of the substituents is a suitable alkyl group. Elimination of the aromatic amine can then occur such that the alkyl substituent is converted to the alkene (*Fig. 8*).

Electrophilic aromatic substitution

Aromatic amines such as aniline undergo electrophilic substitution reactions where the amino group acts as a strongly activating group, directing substitution to the *ortho* and *para* positions (Topic I6). Like phenols, the amino group is such a strong activating group that more than one substitution may take place. For example, reaction of aniline with bromine results in a tribrominated structure as the only product. This problem can be overcome by converting the amine to a less activating group. Typically, this involves acylating the group to produce an amide (Topic K5; *Fig. 9*). This group is a weaker activating group and so mono-substitution takes place. Furthermore, since the amide group is bulkier than the original amino group, there is more of a preference for *para* substitution over *ortho* substitution. Once the reaction has been carried out, the amide can be hydrolyzed back to the amino group (Topic K6).

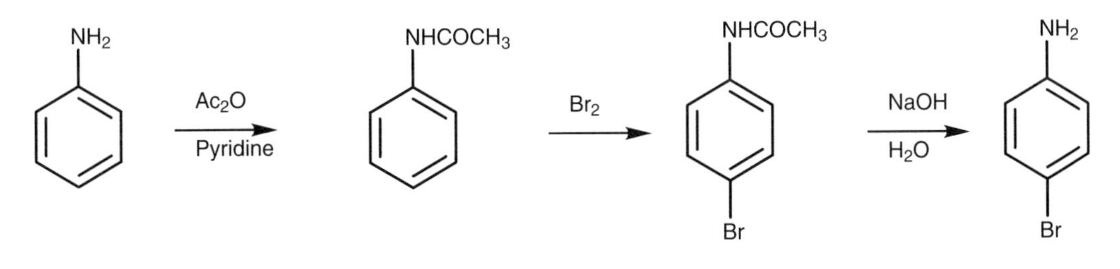

Fig. 9. Synthesis of para-bromoaniline.

Anilines can be sulfonated and nitrated, but the Friedel–Crafts alkylation and acylation are not possible since the amino group forms an acid base complex with the Lewis acid required for this reaction. One way round this is to convert the aniline to the amide as above before carrying out the reaction.

Diazonium salts

Primary arylamines or anilines can be converted to diazonium salts, which in turn can be converted to a large variety of substituents (*Fig. 10*).

Reaction of an aniline with nitrous acid results in the formation of the stable diazonium salt in a process called **diazotization** (*Fig. 11*). In the strong acid conditions used, the nitrous acid dissociates to form an ^{+}NO ion which can then act as an electrophile. The aromatic amine uses its lone pair of electrons to form a bond to this ^{+}NO ion. Loss of a proton from the intermediate formed, followed by a proton shift leads to the formation of a diazohydroxide. The hydroxide group is now protonated turning it into a good leaving group, and a lone pair from the aryl nitrogen forms a second π bond between the two nitrogen atoms and expels water.

Once the diazonium salt has been formed, it can be treated with various nucleophiles such as Br^-, Cl^-, I^-, ^-CN and ^-OH (*Fig. 12*). The nucleophile displaces nitrogen from the aromatic ring and the nitrogen which is formed is lost from the

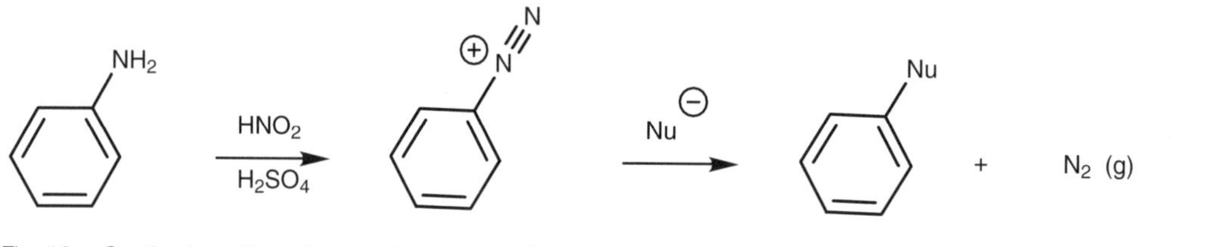

Fig. 10. Synthesis and reactions of diazonium salts.

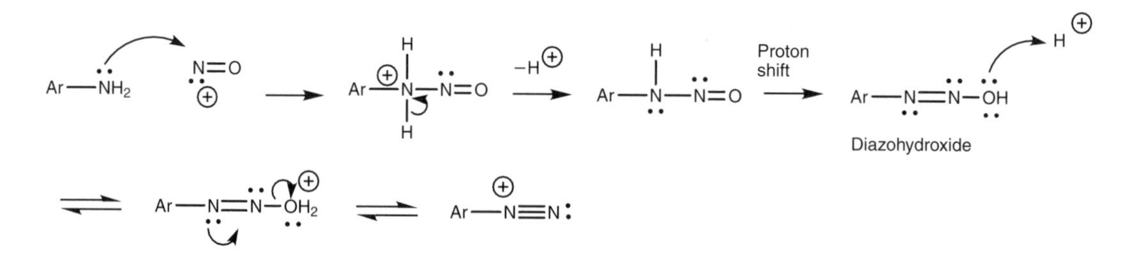

Fig. 11. Mechanism of diazotization.

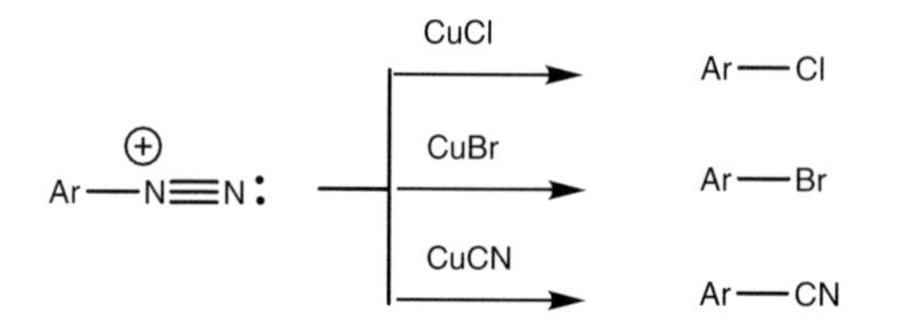

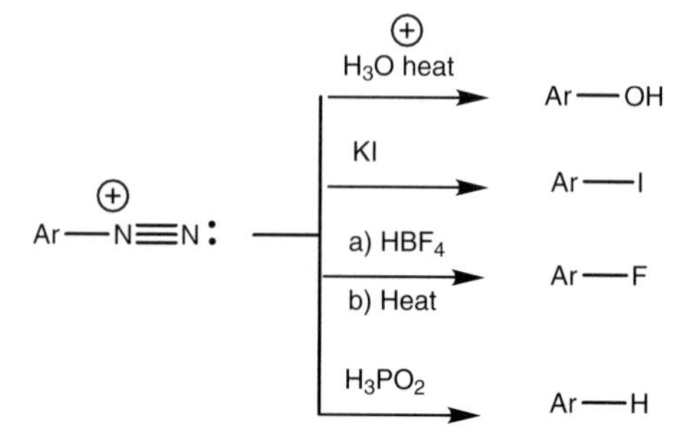

Fig. 12. Reactions of diazonium salts.

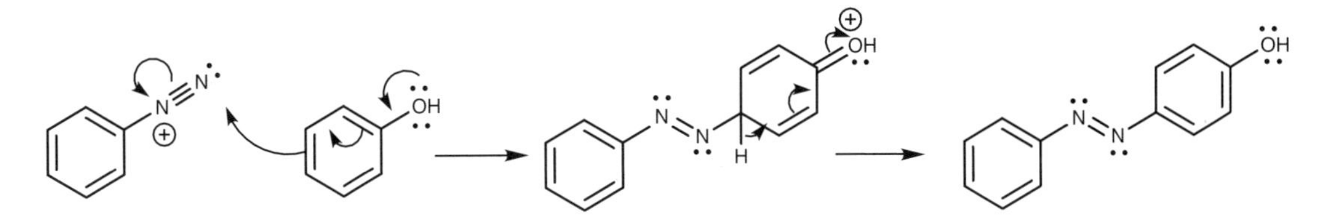

Fig. 13. *Diazonium coupling.*

reaction mixture as a gas, thus helping to drive the reaction to completion. Those reactions involving Cu(I) are also known as the **Sandmeyer reaction**.

Diazonium salts are also used in a reaction called **diazonium coupling** where the diazonium salt is coupled to the *para* position of a phenol or an arylamine (*Fig. 13*). The azo products obtained have an extended conjugated system which includes both aromatic rings and the N=N link. As a result, these compounds are often colored and are used as dyes.

The above coupling is more efficient if the reaction is carried out under slightly alkaline conditions (NaOH) such that the phenol is ionized to a phenoxide ion (ArO^-). Phenoxide ions are more reactive to electrophilic addition than phenols themselves. Strong alkaline conditions cannot be used since the hydroxide ion adds to the diazonium salt and prevents coupling. If the *para* position of the phenol is already occupied, diazo coupling can take place at the *ortho* position instead.

Aliphatic amines, as well as secondary and tertiary aromatic amines, react with nitrous acid, but these reactions are less useful in organic synthesis.

O4 Chemistry of nitriles

Key Notes

Preparation

Primary alkyl halides can be treated with the cyanide ion and converted to nitriles of general formula RCH_2CN. Primary amides can be dehydrated with thionyl chloride or phosphorus pentoxide to give a nitrile.

Properties

The nitrile group is linear in shape with both the carbon and nitrogen being *sp* hybridized. The triple bond is made up of one σ bond and two π bonds. The nitrile group is polarized such that the nitrogen is a nucleophilic center and the carbon is an electrophilic center. Nucleophiles react with nitriles at the electrophilic carbon center to form an imine intermediate which reacts further depending on the reaction conditions.

Reactions

Nitriles can be hydrolyzed to carboxylic acids on treatment with aqueous acid or base. The reaction involves the formation of an intermediate primary amide. Nitriles can be converted to primary amines by reduction with lithium aluminum hydride. The reaction involves addition of two hydride ions. With a milder reducing agent such as DIBAH, only one hydride ion is added and an aldehyde is obtained. Nitriles react with Grignard reagents to produce ketones. The Grignard reagent provides the equivalent of a carbanion which reacts with the electrophilic carbon center of the nitrile group.

Related topics

sp Hybridization (A5)
Organic structures (E4)
Preparations of carboxylic acids (K4)
Nucleophilic substitution (L2)
Reactions of alkyl halides (L6)
Preparation of amines (O1)

Preparation

Nitriles are commonly prepared by the S_N2 reaction of a cyanide ion with a primary alkyl halide (Topic L6). However, this limits the nitriles which can be synthesized to those having the following general formula RCH_2CN. A more general synthesis of nitriles involves the dehydration of primary amides with reagents such as thionyl chloride or phosphorus pentoxide (*Fig. 1*).

Properties

The nitrile group (CN) is linear in shape with both the carbon and the nitrogen atoms being *sp* hybridized. The triple bond linking the two atoms consists of one σ bond and two π bonds. Nitriles are strongly polarized. The nitrogen is a

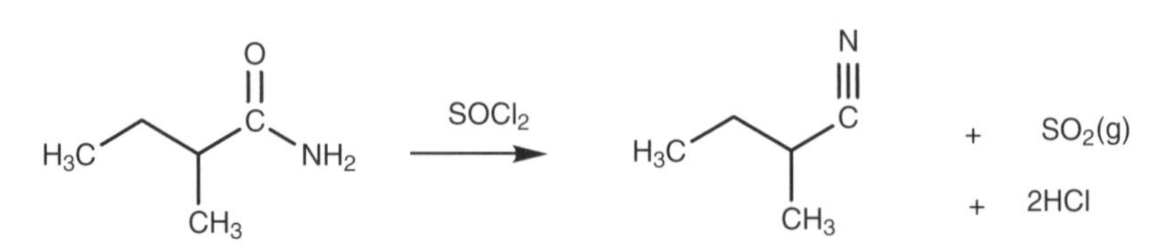

Fig. 1. Dehydration of primary amides with thionyl chloride.

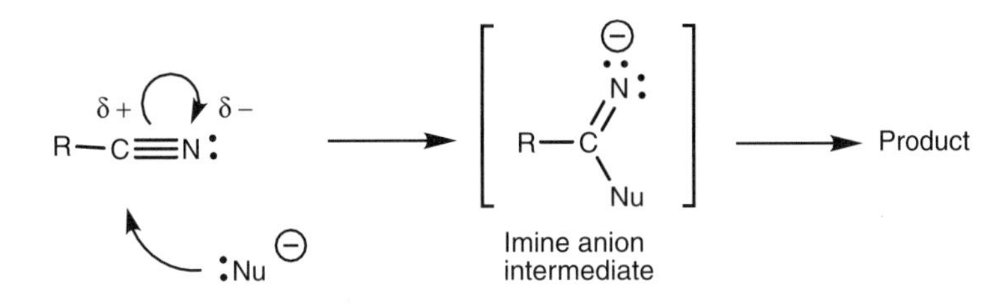

Fig. 2. Reaction between nucleophile and nitriles.

nucleophilic center and the carbon is an electrophilic center. Nucleophiles react with nitriles at the electrophilic carbon (*Fig. 2*). Typically, the nucleophile will form a bond to the electrophilic carbon resulting in simultaneous breaking of one of the π bonds. The π electrons end up on the nitrogen to form an sp^2 hybridized imine anion which then reacts further to give different products depending on the reaction conditions used.

Reactions

Nitriles (RCN) are hydrolyzed to carboxylic acids (RCO_2H) in acidic or basic aqueous solutions. The mechanism of the acid-catalyzed hydrolysis (*Fig. 3*) involves initial protonation of the nitrile's nitrogen atom. This activates the nitrile group towards nucleophilic attack by water at the electrophilic carbon. One of the nitrile π bonds breaks simultaneously and both the π electrons move onto the nitrogen resulting in a hydroxy imine. This rapidly isomerizes to a primary amide which is hydrolyzed under the reaction conditions (Topic K6) to give the carboxylic acid and ammonia.

Nitriles (RCN) can be reduced to primary amines (RCH_2NH_2) with lithium aluminum hydride which provides the equivalent of a nucleophilic hydride ion. The reaction can be explained by the nucleophilic attack of two hydride ions (*Fig. 4*).

With a milder reducing agent such as DIBAH (diisobutylaluminum hydride), the reaction stops after the addition of one hydride ion, and an aldehyde is obtained instead (RCHO).

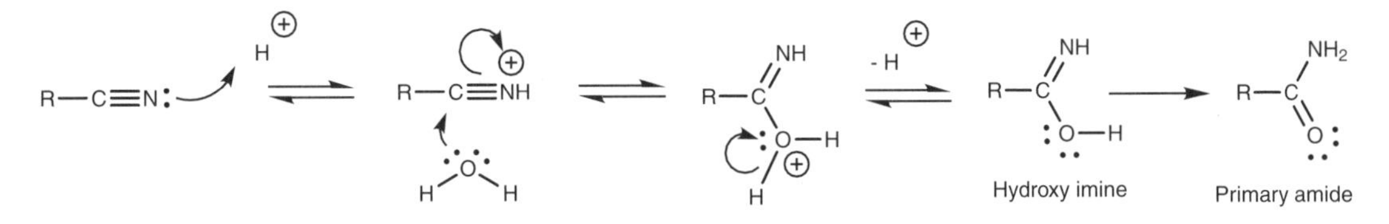

Fig. 3. Acid-catalyzed hydrolysis of nitrile to carboxylic acid.

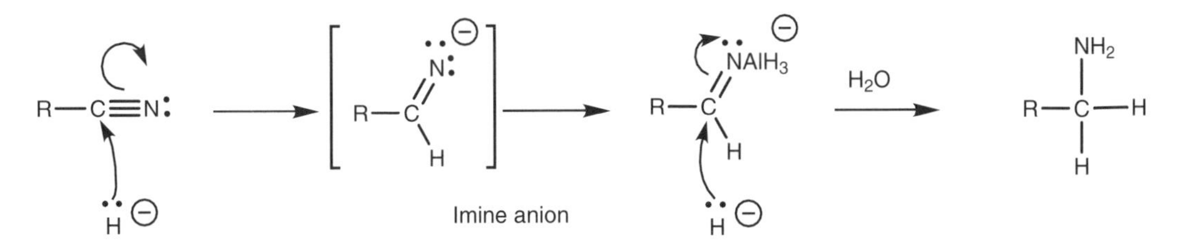

Fig. 4. Reduction of nitriles to form primary amines.

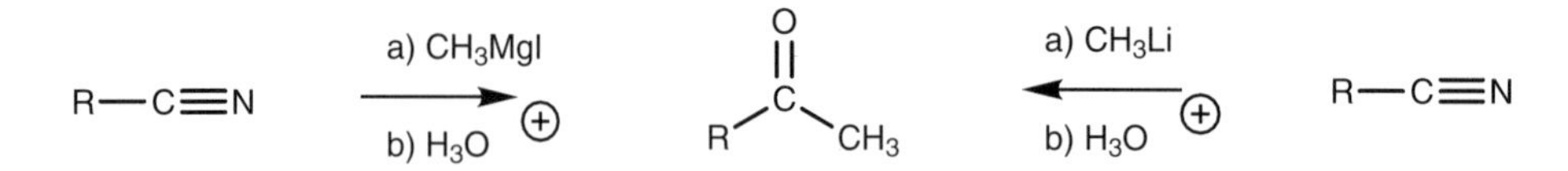

Fig. 5. Nitriles react with Grignard reagent or organolithium reagents to produce ketones.

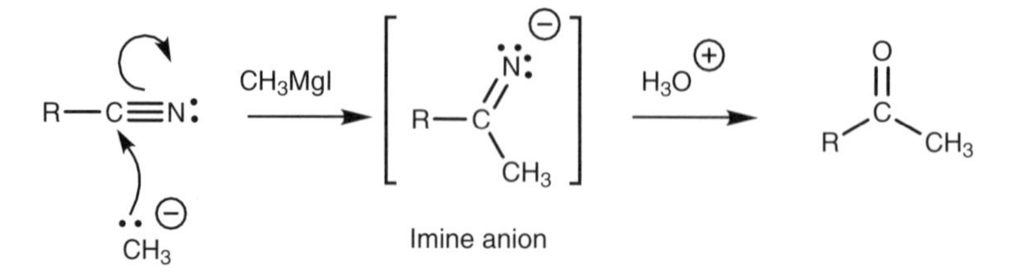

Fig. 6. Mechanism of the Grignard reaction on a nitrile group.

Grignard reaction

Nitriles can be treated with Grignard reagents or organolithium reagents (Topic L7) to give ketones (*Fig. 5*).

Grignard reagents provide the equivalent of a nucleophilic carbanion which can attack the electrophilic carbon of a nitrile group (*Fig. 6*). One of the π bonds is broken simultaneously resulting in an intermediate imine anion which is converted to a ketone when treated with aqueous acid.

Further Reading

General reading

McMurray, J. (2000) *Organic Chemistry*, 5th edn. Brooks/Cole Publishing Co., Pacific Grove, CA.

Morrison, R.T. and Boyd, R.N. (2000) *Organic Chemistry*, 7th edn. Prentice Hall International, Inc., New York (in press).

Solomons, T.W.G. (2000) *Organic Chemistry*, 7th edn. John Wiley & Sons, Inc., New York.

Self learning texts

Patrick, G.L. (1997) *Beginning Organic Chemistry 1*. Oxford University Press, Oxford.

Patrick, G.L. (1997) *Beginning Organic Chemistry 2*. Oxford University Press, Oxford.

INDEX